Pacific Northwest Cenozoic Biostratigraphy

Edited by

JOHN M. ARMENTROUT

Mobil Exploration and Producing Services, Inc.
P.O. Box 900
Dallas, Texas 75221

SPECIAL PAPER
184

THE GEOLOGICAL SOCIETY OF AMERICA
BOULDER, COLORADO

Library of Congress Catalog Card Number 80-82937
ISBN 0-8137-2184-9

Published by
THE GEOLOGICAL SOCIETY OF AMERICA, INC.
P.O Box 9140, 3300 Penrose Place
Boulder, Colorado 80301

Printed in the United States of America

KATHERINE VAN WINKLE PALMER

Katherine Van Winkle Palmer is one of the pioneers of Pacific Northwest Cenozoic biostratigraphy. Born in 1895 in Oakville, Washington, she began her career in paleontology working on the Eocene molluscan fauna of the Cowlitz Formation which crops out along the rivers near her family's home. The research "Fauna from the Eocene of Washington" (1922) was carried out and published jointly with Professor C. E. Weaver at the University of Washington.

Katherine Van Winkle continued her education at Cornell where she received her doctorate. She had intended to return to Washington to pursue the work she had begun as an undergraduate, but her marriage to Cornell Professor E. Laurence Palmer and the rearing of their two sons kept her in Ithaca, New York.

At Cornell, Dr. Katherine Palmer worked as an assistant to Professor A. D. Harris and helped him to establish the Paleontological Research Institution (PRI) in 1932. She subsequently served as director of PRI from 1952 to 1978, overseeing the publication of more than 150 *Bulletins of American Paleontology* and more than 20 issues of *Palaeontographica Americana*. Dr. Palmer was always supportive of Pacific Northwest paleontological research, and two of the most recent *Bulletins of American Paleontology* published under her supervision were Louie Marincovich's *Cenozoic Naticidae (Mollusca; Gastropoda) of the Northeastern Pacific* (1977, Vol. 70) and Carole Hickman's *Paleogene Marine Gastropods of the Keasey Formation in Oregon* (1980, Vol. 78).

Dr. Palmer's own research has resulted in over 150 publications. Among these are *The Veneridae of Eastern America, Cenozoic and Recent* (1972), *Gastropoda of the Claibornian Mid-Eocene of the Southern United States* (1937), *The Mollusca of the Jackson Eocene of the Mississippi Embayment (Sabine River to Alabama River)* (1946), and *Catalogue of the Paleocene and Eocene Mollusca of the Southern and Eastern United States* (1965). As with all of her work, each of these publications demonstrated Dr. Palmer's attention to detail and documentation and made the task of subsequent workers in the area far easier.

Dr. Palmer received the highest recognition of the Paleontological Society of America in 1972 and of the Western Society of Malacologists in 1974. Tulane University awarded her an honorary Doctor of Science degree in 1978.

This volume is dedicated to Dr. Katherine Van Winkle Palmer in recognition of her past contributions and her continuing support of Pacific Northwest paleontology and biostratigraphy.

Contents

Acknowledgments

The editorial assistance of the authors who read each other's papers is gratefully acknowledged, as is the assistance of the following colleagues: C. J. S. Hickman, University of California-Berkeley; J. T. Whetten, University of Washington; W. N. Orr, University of Oregon; L. H. Burckle, Lamont-Doherty Geophysical Observatory; E. J. Moore, R. Z. Poore, and K. McDougall, U.S. Geological Survey; S. McCoy, Jr. and D. Englehart, Amoco Production Company; W. C. Elsik and R. L. Fleisher, Esso Production Research; and J. F. Stone, R. L. Pierce, K. Verschoor, J. R. Rider, R. W. Morin, S. F. Percival, and R. J. Echols of Mobil Oil Corporation.

Geological Society of America
Special Paper 184
1981

Introduction

JOHN M. ARMENTROUT
Mobil Exploration and Producing Services, Inc., P.O. Box 900, Dallas, Texas 75221

In 1977, at the Geological Society of America Annual Meeting (held that year in Seattle, Washington), a symposium titled "Pacific Northwest Cenozoic Biostratigraphy" was convened by John M. Armentrout of Mobil Oil Corporation and Kristin McDougall of the U.S. Geological Survey.

The purpose of the symposium was to focus on new data concerning the age relationships of the Cenozoic strata of Oregon and Washington and the correlation of the local strata within a worldwide time framework. Not since the efforts of the C. E. Weaver committee (Weaver and others, 1944) had such an effort been undertaken. Sixteen papers were presented. Topics ranged from biostratigraphy to paleoecology based on studies of fossil mollusks, foraminifers, echinoderms, palynomorphs, and megafloras. Eight of the papers presented comprise the bulk of this volume. Two papers have been added: a historical overview of molluscan biostratigraphic research by W. O. Addicott and a synthesis by J. M. Armentrout of the biostratigraphic and geochronologic data that were presented at the symposium.

The symposium was limited to Oregon and Washington, as this area comprises a Cenozoic biogeographic province (Addicott, 1976). The provinciality of the faunas and floras restricts application of biostratigraphic time scales based on these organisms to areas within the biogeographic province. Much work has been done since Weaver and others (1944) presented their ideas about the correlations and age relationships of West Coast Cenozoic strata. In 1944, Weaver (in Weaver and others, 1944, p. 570) commented, "Two schools of thought exist on the Pacific Coast regarding the classification of Cenozoic formations." One was that of paleontologists working with megafossils, the second that of paleontologists working with foraminifera. These two schools of thought existed because the rock record of the Pacific Coast was either rich in microfossils and lacking in megafossils or vice versa. The microfossil and megafossil schemes were integrated by the Weaver committee, and the resulting correlations of West Coast formations were depicted on a chart. Subsequent work revealed that several of the "stage" and "formational" units were coeval biofacies.

To overcome the biofacies problems, workers are now using refined time-stratigraphic biostratigraphy, radiometric dating, planktonic biostratigraphy, and magnetostratigraphy. These geochronologic tools, combined in interdisciplinary biostratigraphic studies, provide for a recalibration of the local time-stratigraphy and have resulted in major changes in the interpreted age of Pacific Northwest Cenozoic strata (Fig. 1). The papers in this volume reflect the nature of current research efforts.

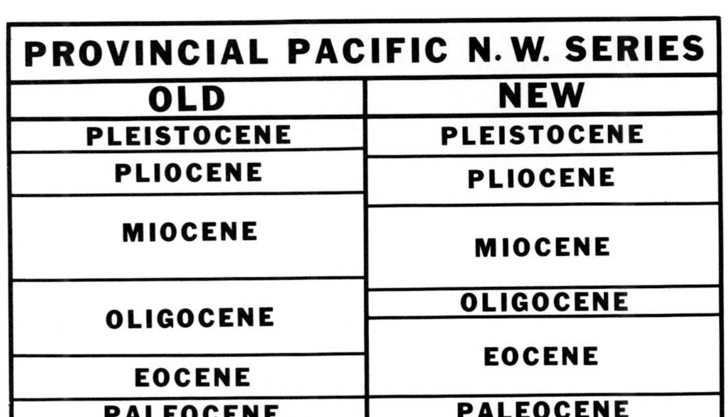

Figure 1. Relationship of European series as used in Oregon and Washington: old from Weaver and others (1944); new from Armentrout (this volume).

REFERENCES CITED

Addicott, W. O., 1976, Neogene molluscan stages of Oregon and Washington, *in* Fritsche, A. E., and others, eds., Neogene symposium: Society of Economic Paleontologists and Mineralogists, Pacific Section, San Francisco, California, April 1976, p. 95–115.

Armentrout, John, 1981, Correlation and ages of Cenozoic chronostratigraphic units in Oregon and Washington: Geological Society of America Special Paper 184 (this volume).

Weaver, C. E., and others, 1944, Correlation of the marine Cenozoic formations of western North America: Geological Society of America Bulletin, v. 55, p. 569-598.

MANUSCRIPT RECEIVED BY THE SOCIETY MARCH 14, 1980
MANUSCRIPT ACCEPTED MAY 7, 1980

Printed in U.S.A.

Geological Society of America
Special Paper 184
1981

Brief history of Cenozoic marine biostratigraphy of the Pacific Northwest

WARREN O. ADDICOTT
U.S. Geological Survey, 345 Middlefield Road, MS 52, Menlo Park, California 94025

ABSTRACT

Biostratigraphic study of marine Cenozoic rocks of the Pacific Northwest began some 75 years ago, in part as a consequence of the discovery of petroleum in neighboring California during the late 1800s. It was preceded, however, by an almost equally long period of reconnaissance and early description that began during the middle 1800s. Paleontological study of the marine Cenozoic was initiated by T. A. Conrad's descriptions of Oligocene and Miocene fossils from Oregon. The first biochronologies were developed by C. E. Weaver and, to a lesser extent, by Ralph Arnold and Harold Hannibal during the early 1900s. Their molluscan-based horizons and formations formed the principal means of age determination and correlation until the mid-1900s when the California-based benthic foraminiferal zonations were adapted to the Pacific Northwest Province by Weldon Rau. Recent development of worldwide planktonic microfossil zonations has created a renewed interest in biostratigraphic study of the Pacific Northwest in addition to a means of calibrating the benthic chronologies.

INTRODUCTION

Marine Cenozoic deposits of the Pacific Coast of the United States and Canada were discovered near the mouth of the Columbia River almost 150 years ago. Thus, Cenozoic biostratigraphy, as practiced on the West Coast, can trace its origins to the Pacific Northwest. The hub of activity in this field has been in California, where the discovery of petroleum during the late 1800s gave tremendous impetus to research and the training of stratigraphic paleontologists. Nevertheless, the refinement of biostratigraphic zonations of the Coast Ranges of Oregon and Washington has played an important and continuing role in deciphering the Cenozoic history of the continental margin of western North America.

The growth and development of Cenozoic biostratigraphy in this region is highlighted by five significant phases: (1) initial descriptions and age determinations of mollusks, 1848 to 1857; (2) horizon and formation biochronologies, 1912 to 1916; (3) Paleogene stage and zone biochronology, 1942 to 1944; (4) benthic foraminifer zones, 1958 to 1967; and (5) planktonic calibration of the benthic chronologies, 1972 to 1979. The intervening years, particularly the period following 1916, were marked by several important contributions. These five phases in the Pacific Northwest correspond, in a general way, to periods of development in California (Addicott, 1979), but in almost every case, investigations in the Pacific Northwest follow those begun in California by a decade or so. This lag reflects the causal relation between petroleum exploration and biostratigraphic research and the fact that the petroleum-rich Cenozoic basins of the California Coast Ranges have always been the focal point of this exchange. This overview isolates some of the significant advances in knowledge, principally work through the mid-1940s, and focuses attention on biostratigraphers and geologists who have played decisive roles in building the marine Cenozoic zonations of this region. Often the personal side of the great achievers enhances the significance of their contributions in the light of the human characteristics that had to be dealt with in their careers. Inevitably there is a sensitive, less than perfect person behind the mask of scientific enlightenment, precision, and diligent work. I have chosen to touch on some of these human aspects to provide added insight into the accomplishments of some of the early leaders.

INITIAL DESCRIPTIONS

TIMOTHY A. CONRAD (1818-1889)

Marine fossils were first collected from the Pacific Northwest coast by J. K. Townsend in the mid-1830s at or near Astoria, Oregon (Townsend, 1839), from strata now included in the Astoria Formation. The collection of fossil

Figure 1. Timothy A. Conrad.

mollusks was sent to Timothy A. Conrad (Fig. 1), invertebrate paleontologist associated with the Philadelphia Academy of Sciences. Conrad (1848, 1849) correctly assigned the first Cenozoic collections from the Pacific Coast States to the Miocene, naming many new species and comparing several of them with mollusks from the Miocene of Maryland.

Conrad was the foremost authority on Tertiary paleontology in the United States during the mid-1800s. His description of West Coast Cenozoic faunas was incidental to his primary research on the Atlantic and Gulf Coastal Plains. in fact, Conrad never set foot in the Pacific Coast States. His pioneering work on Pacific Coast fossils collected during the Pacific Railroad Expeditions to California (Conrad, 1855a, 1855b, 1857a–1857d) brought compensation ranging from $10 to $50 for each of several published reports (Wheeler, 1935). Conrad's modesty was such that he included this notation with a bill of $50 for an 1854 report: "If you think 50 dollars too much make what deductions you think proper" (Wheeler, 1935, p. 72).

Conrad was a brilliant, self-trained naturalist given to periods of melancholy, severe self-doubt, and poor health that had a crippling effect on his ability to work and even on his will to live. "This moodiness was recurrent, an infection which victimized his happiness and discolored his views of life. He often worked by the side of unappreciative men. His best friends lost patience with his opinions or his peculiarities" (Wheeler, 1935, p. 112). He was highly regarded by the eminent British geologist Sir Charles Lyell, who sought out Conrad on each of his two visits to the United States. Lyell once wrote to Dr. James Hall, of the New York Geological Survey, "I have been so much pleased with Conrad, and am sure that were he not so isolated and could have frequent intercourse with congenial souls, he would no longer see difficulties or dwell on his constitutional maladies" (Clarke, 1921, p. 115).

Most of Conrad's early life was marked by poverty; he worked for several years without salary. His early travels to the field were financed by borrowing money from Academy members, whom he repaid with specimens, and he often relied upon the hospitality of friends for lodging (Moore, 1971). One of several stories about this overly modest naturalist concerns his portrait (Fig. 1). When a close friend once requested his picture, Conrad replied, "If you ever ask me for my picture again, I will never speak to you as long as I live" (Wheeler, 1935, p. 113). The portrait of Conrad (Fig. 1), the only one known to have been made, was drawn by an artist in the family from a photograph taken of him laid out after death.

Conrad wrote to his fellow paleontologist and friend, Fielding B. Meek, in 1863, "I go Monday to help H. [Dr. James Hall, Director of the New York Geological Survey] ferret out my skulking species of paleozoic shells. May the recording angel help me. God and I knew them once, and the Almighty may know them still. A man's memory is no part of his soul" (Merrill, 1924, p. 158).

A few additional mollusks of Eocene and Miocene age, from various localities in Oregon, were described, but not illustrated, by Shumard (1858). Thereafter, some 30 to 40 years passed during which little research on the marine Cenozoic faunas of the Pacific Northwest was completed and published. Yet this interval was not without progress; Conrad had brought the Cenozoic faunas of the Pacific Northwest to the attention of the geologic profession. A prominent controversy in the late 1800s was over the age of the Astoria Formation. It may have been touched off by Meek (1864), who held that the strata at Astoria could be of Mesozoic age. This may have led Conrad (1865a, 1865b), who valued Meek as a personal friend, to reassess his age determination; in any event, he reidentified an *Aturia* as the Atlantic Coastal species *A. ziczac* and revised the age to Eocene. The age of the Astoria Formation was debated for the next 70 years until Howe's (1926) definitive Miocene age assignment (Moore, 1963, Fig. 1).

Following the Conrad period, most of the early stratigraphic studies of the Cenozoic rocks of western Washington were oriented toward coal resources. Although primarily concerned with nonmarine Tertiary formations, coal research included incidental treatment of marine Paleogene mollusks from the Duwamish River near Seattle (White, 1889) and fossiliferous Miocene strata of the northwest Olympic Peninsula (Arnold, 1905, 1906a).

Thomas Condon (1822-1907)

During the 1870s and 1880s, Thomas Condon (Fig. 2), the famed pioneer geologist of Oregon, carried out reconnaissance studies of the marine Cenozoic rocks of the Coast Ranges while professor of natural science at the University of Oregon.

An Irish immigrant to New York City, Condon came west with his wife in 1852, settling in northwest Oregon, where he held a series of pastorates as a Congregational minister. Fascinated by trilobites and other fossils since his childhood in New York State, he took to carrying a

Figure 2. Thomas Condon.

geologic pick with him when, for lack of a study or library at The Dalles, Oregon, he prepared his sermons while on long walks in the solitude of the nearby hill country. His collections of fossils and his enthusiasm for them soon inspired others, and before long "he had many of the soldiers at a nearby fort, teamsters, trappers, and packers collecting for him" (McCornack, 1928, p. xii). He later exchanged fossil specimens with the Smithsonian Institution in return for publications and gave lectures on geology throughout the state. His enthusiasm and self-taught expertise may explain the high esteem in which he was held by everyone interested in geology who came in contact with him—scientist or lay person.

A widely respected expert on the geology of the entire state, particularly Tertiary vertebrates and plants of central Oregon, Condon was sought out by leading paleontologists and field geologists of the period for assistance and advice. Though author of very few scientific papers, his influence on geologic research of that period was profound. Condon corresponded with leading paleontologists throughout the United States and provided many of them with specimens for their research. He played an instrumental role, for example, in studies of Tertiary formations of western Oregon by W. H. Dall and J. S. Diller. Dall acknowledged Professor Condon's early work at Astoria, Oregon, that led to the naming of the Astoria Formation [as Astoria shales (Condon in Cope 1880)] as indispensable in providing material from localities that were inaccessible at the time of Dall's visit to the state in 1890 (Dall, 1909, p. 9). Condon's daughter wrote, "This pleasant visit from Dr. Dall was a real help to Professor Condon, and it was probably at this time that they strolled together to the banks of the Willamette River a mile or more east of the University [of Oregon], where Dr. Dall found a large and beautiful fossil Scalaria shell which he later named Scalaria Condoni" (McCornack, 1928, p. 255-256).

Josiah S. Diller (1850-1924)

Reconnaissance of the Oregon coast by J. S. Diller in the late 1800s rekindled interest in Cenozoic deposits of this region. Diller was assisted with paleontologic determinations by C. A. White and W. H. Dall. His investigations mark the beginning of the U.S. Geological Survey's studies of the Coast Ranges of the Pacific Northwest. Although Diller's main interest lay in the Mesozoic terrane of the Klamath Mountains, he contributed substantially to the description of Pleistocene terraces of the Oregon coast (Diller, 1896) and to the Tertiary geology of the Coos Bay-Cape Blanco part of the southwest Oregon coast (Diller, 1901, 1902, 1903) and northwest Oregon (Diller, 1896).

William H. Dall (1845-1927)

For a time William H. Dall (Fig. 3), a legendary figure to Cenozoic mollusk specialists by virtue of his broad expertise on mollusks and unequaled volume of scientific publications, included the Pacific Northwest in his studies. Although his principal work on Pacific Coast Tertiary faunas was in Alaska, he visited the Oregon coast several times during the period 1890 to 1910. His report on the Miocene faunas of Coos Bay and Astoria (Dall, 1909) stands as the first major monograph on a Pacific North-

Figure 3. William Healy Dall.

west invertebrate fauna and is still the definitive work on the late Miocene of this region. Dall contributed paleontologic support to U.S. Geological Survey mapping projects in Oregon and Washington (Diller, 1896, 1898, 1899, 1902; Washburne, 1914) and described several species from this region in his reports.

Insight into Dall's enormous publication record—more than 1,600 titles—and his influence on malacology is evident in this brief assessment of his personal qualities by his biographers: "Dall was endowed with indefatigable energy and unusual powers of organization. He was a balanced inheritant of the able, strict, scrupulously honest, Puritanic, determined, and positive factors contributed by the mother, softened by those of the dreamy, idealistic, poetic, gentle father" (Bartsch and others, 1946). An extremely orderly person, Dall "never was hurried or harrassed.... He was almost always writing and smoking a long-stemmed German pipe" (Woodring, 1958, p. 105). His devotion to science crossed over into his personal life. His honeymoon involved the first part of his trip from the East Coast to his summer work in Arctic Alaska; his bride accompanied him as far as Sitka, Alaska, at which point she returned home to Washington and he continued on to survey the Alaska Coast (Bartsch and others, 1946). It is said that Dall kept a small swivel table by his place at the dining room table. When his wife excused herself to prepare the next course during dinner, Dall would swing the table top in place and work on mollusk specimens until the next course was brought to him.

Origins of Initial Zonations in California

The initial zonations of the Cenozoic rocks of the Pacific Northwest, which allowed detailed geologic mapping of the often monotonous sequences of shale and siltstone, can trace their origins to California. The discovery of oil in California in 1875 led to a gradual increase of interest in the marine Tertiary rocks along the entire Pacific Coast; the discovery of minor coal deposits in marine Tertiary formations of the Oregon and Washington coast (Arnold, 1905; Diller, 1905) played a lesser role. During the 1890s, the two major universities in California began training paleontologists. James P. Smith arrived at Stanford University in 1892, fresh from receiving his doctorate at Gottingen, Germany (Anderson, 1932). J. C. Merriam enrolled at the nearby Berkeley campus of the University of California in 1894, completing several studies on Pacific Coast Tertiary invertebrates before going to Germany for advanced training. He returned to Berkeley in 1902 after receiving his doctorate at Munich under the guidance of Professor Carl von Zittel. Though primarily researchers in other fields, Smith and Merriam were instrumental in launching biostratigraphy on the Pacific Coast and in training students who made the bench-mark zonations of the Cenozoic during the period from 1906 to 1916. Both were inspiring teachers who played formative roles in the first biostratigraphic zonations published by their students: Arnold (1906b), Arnold and Hannibal (1913), and Weaver (1912, 1916a, 1916c).

First University Classes in Stratigraphic Paleontology

The densely forested Coast Ranges of Oregon and Washington seem to have held a fascination for California geologists over the years. Arnold and Hannibal's (1913) zonation of the Oregon and Washington Tertiary was the first of many California-based geologic studies of Pacific Northwest stratigraphy and faunas. Other early workers in the Pacific Northwest included Anderson and Martin (1914), who described several species of Miocene invertebrates from Oregon. Many native Northerners enrolled at Stanford and Berkeley for advanced training in paleontology in the Northwest, Charles E. Weaver at Seattle and Earl L. Packard at Eugene, Oregon, who sent their students to Stanford or Berkeley for advanced training. As the University of Washington faculty was small, Weaver felt that excellence in undergraduate teaching was his first concern and encouraged his students interested in postgraduate study in paleontology to go to Berkeley (Barksdale, 1974).

HORIZON AND FORMATION BIOCHRONOLOGIES—THE WEAVER PERIOD

Charles E. Weaver (1880-1958)

Possibly the single most important development in Cenozoic studies of the Pacific Northwest during the early 1900s was the appointment of Charles E. Weaver (Fig. 4) to the faculty of geology at the University of Washington. Weaver was a 1907 graduate of the University of California, where he studied the Paleocene and Miocene

Figure 4. Charles Edwin Weaver.

sections near Mount Diablo under the guidance of Professor J. C. Merriam. Moving to Seattle later that year, he began mapping the Tertiary formations of western Washington. Within a few years he had completed extensive biostratigraphic studies of those Tertiary rocks and published the first geologic map of western Washington at a scale of 1 in. to 30 mi with his preliminary report (Weaver, 1912). Also included were definitions of seven marine formations and enumeration of their faunas. This report was followed, four years later, by a report including maps of western Washington at scales of 7.5 and 2 in. to 1 mi and faunal and stratigraphic documentation (Weaver, 1916b). From these data, Weaver (1916a, 1916b) developed a series of faunal horizons with designated type sections and characteristic species (Weaver, 1916a, 1916c). Weaver's statement "The term horizon is used in the sense of a deposit formed at a particular time and identified by distinctive fossils" (Weaver, 1916a, p. 4) implies an early appreciation of time-stratigraphic classification long before its popularization by Schenck and Muller (1936, 1941). Weaver's work permitted, for the first time, definition of biostratigraphic units necessary to map the poorly exposed, predominantly siltstone sequences of the Coast Ranges.

Weaver was a shy and reserved person, deeply interested in books and widely read in the field of geology. He was very strong physically, leading summer field-mapping classes to the parched, inhospitable California Coast Ranges following his return from Argentina and Germany in the late 1920s until his retirement in 1950. Even after his retirement, he had the reputation of being able to walk a much younger man into the ground. As a student at Berkeley, Weaver spent weekends and vacations mapping near Martinez and on the north side of San Pablo Bay. He would depart the campus on foot, reaching his destination by walking to Martinez, crossing the Carquinez Straits by ferry, and continuing on foot to whatever quadrangle he was then mapping, a distance of 35 to 50 km. When night fell, he would simply stop and sleep on the ground. His phenomenal capacity for work and physical stamina characterized his early studies in western Washington. Weaver preferred to take the train or bus to his field area rather than drive. Once there, he would get around on foot, catching rides on logging railroads when possible. Being a very frugal person, except where his geologic interests or scientific library were concerned, he would further economize by taking meals in the least expensive local cafes (Barksdale, 1974).

Weaver's shyness may explain why he prepared his lectures so well that he did not use notes. It certainly was the reason for his lecturing, in his early years at least, from the rear of the classroom so as to avoid facing his students. An interesting aspect of his teaching career concerns his color blindness. Never a very confident driver, Weaver ceased driving with the advent of traffic lights in the 1920s owing to this condition. Curiously, he had taught optical crystallography and petrography for many years at Seattle despite this handicap.

Although his field shifted to South America in 1919, when he was hired as a consultant, and, much later to California, Weaver continued to refine his Tertiary subdivisions and geologic history of the Pacific Northwest over the following 25 years (Weaver, 1937, 1942, 1945; Weaver and others, 1944). Clearly, he is the founder of Tertiary biostratigraphy of the Pacific Northwest. He interested and trained many students who have contributed significantly to the advancement of knowledge in stratigraphic paleontology, among them: R. Stanley Beck, J. P. Buwalda, J. Wyatt Durham, Donald L. Frizzell, Charles V. Fullmer, Earl L. Packard, Katherine V. W. Palmer, and Ruth Todd.

Ralph Arnold (1875-1961)

A few years before Weaver reached Seattle, Ralph Arnold (Fig. 5), a 1904 graduate of Stanford University, completed his biostratigraphic zonation of the California Tertiary (Arnold, 1906b). Most of Arnold's scientific work was done during the 10 years that he worked for the U.S. Geological Survey, from 1900 to 1909. Arnold was to California marine Cenozoic stratigraphy what Weaver was to Pacific Northwest stratigraphy. In addition to developing his enormously significant biozonation, Arnold utilized biostratigraphy in mapping several California Tertiary basins, constructed the first paleogeographic maps of the Pacific Coast and drew the first paleoclimate inferences from the succession on molluscan faunas (Addicott, 1972). Although most of Arnold's assignments were in California, he worked briefly on coal resources and associated marine stratigraphy of the northwest coast of the Olympic Peninsula (Arnold, 1905, 1906a; Arnold and Hannibal, 1913; Clark and Arnold, 1923).

Figure 5. Ralph Arnold.

Harold Hannibal (1889-1965)

Arnold, after leaving the U.S. Geological Survey in 1909 to go into consulting geology, took an interest in Harold Hannibal (Fig. 6), an unusually gifted student of his former professor at Stanford University, J. P. Smith. Hannibal had a reputation as an exceptionally able fossil collector, an accomplished field stratigrapher, and a brilliant but controversial taxonomist (Taylor and Smith, 1971). These qualities interested Arnold, who took a liking to Hannibal despite the reclusive tendencies that alienated most of his peers. Around 1910, at Arnold's urging and sponsorship, Hannibal took a year off from his studies at Stanford to study under Weaver (E. L. Packard, March 1979, oral commun.) at the University of Washington. While in Washington, Hannibal obtained much of the material upon which his controversial paper on the fresh-water Cenozoic mollusks of the West Coast was based (Hannibal, 1912), including specimens borrowed from the State Geological Survey (A. M. Keen, March 1979, oral commun.). He also obtained knowledge of locality and stratigraphic data on the marine Cenozoic sections that Weaver had acquired up to that time.

During the summer of 1911, Hannibal returned to the Pacific Northwest to make stratigraphic studies and collections of fossils along the coast. In the spring of the following year, he made his headquarters in Seattle, once again under Arnold's sponsorship, and took field trips to western Washington and Vancouver Island. Later that year, he moved south to the Oregon coast before returning to Stanford in late October or early November. Whenever possible, Hannibal conducted his field work by using a bicycle, which he checked in the baggage car while traveling from one point to the next on the railroad.

Hannibal's field work was truly remarkable, as it resulted in a definitive resume of Cenozoic stratigraphy of the Pacific Northwest (Arnold and Hannibal, 1913). It was a more ambitious, if less well documented, undertaking than Weaver's (1912, 1916a, 1916b, 1916c) reports, because it took in northernmost California, Oregon, and British Columbia, in addition to western Washington. Arnold and Hannibal's 1913 report set the stage for what must have been severely strained relations with Weaver over priority of zonations. Although none of Weaver's published comments on their work bears this out, the timing of publications on the Tertiary stratigraphic divisions of the Pacific Northwest and the status of Hannibal as an interloper in Weaver's collections and stratigraphic records clearly point to such a conflict. Weaver's (1912) preliminary report on the stratigraphy of western Washington was followed within a year by Arnold and Hannibal's (1913) massive report on the Tertiary stratigraphy of the Pacific Northwest. One of the most remarkable aspects of their report is that Hannibal could have amassed so much stratigraphic data and have reached so many critical fossil localities in the few months of his reconnaissance in this vast region. It is possible that he made extensive use of Weaver's data, which he had access to during his stay at Seattle.

Figure 6. Harold Hannibal.

Hannibal's life was marked by tragedy. Beginning in 1914, he was afflicted with a series of severe mental disorders, perhaps brought on by a fall from a motorcycle (Taylor and Smith, 1971). However they may have started, these episodes soon led to the institutionalization that cut short a most promising career. At the onset of his mental difficulties, Hannibal was completing a major descriptive report on the marine Tertiary faunas of Oregon and Washington that would have documented his 1913 report with Arnold. The penciled original manuscripts and specimens designated as new species are still with the Stanford collections. A few of his new species have been described by other workers (Hertlein, 1925).

A measure of Hannibal's competence is evident in the Clark and Arnold (1923) monograph on the mollusks from the Sooke Formation of southwest Vancouver Island, British Columbia. According to Ralph Arnold, this report was largely Hannibal's work; after completing the manuscript, Hannibal gave it to Arnold, who, in turn, sent it to Clark at the University of California for polishing (A. M. Keen, March 1979, oral commun.).

One of the sad stories about this tragic figure concerns an escape from the mental institution in the early 1920s that led to an unscheduled visit to the Stanford geology department. Frederick P. Vickery, a student who was working on the mollusks of the upper Miocene Briones Sandstone of the Mont Diablo area, California, had his specimens laid out, unnumbered, on tables on the third floor of the Geology Corner of the Stanford quadrangle. When Hannibal saw these, he dumped them on the floor. Faced with chaos, Vickery reportedly abandoned his paleontologic thesis and completed his doctorate in structural geology. Not long before his death in 1965, Hannibal, still institutionalized, started corresponding with G Dallas Hanna of the California Academy of Sciences. Apparently his interest in geology never dwindled, for he wrote Hanna that he wanted to publish a report on the geologic hazards of a proposed bridge crossing over the southern part of San Francisco Bay.

Earl L. Packard (1885-)

As the initial work on Cenozoic zonations of western Oregon and Washington was being completed, Earl L. Packard (Fig. 7) began a long and distinguished career with the State University system of Oregon, arriving in Eugene in 1916, less than 10 years after Condon's death. Packard became one of the most influential teachers of stratigraphy and paleontology on the West Coast. An early student of C. E. Weaver at Seattle, where he completed his master's degree in 1912, Packard undertook his doctorate at the University of California, studying under J. C. Merriam (Fig. 7). Though a Mesozoic-invertebrate specialist, and not directly involved in Cenozoic marine stratigraphy following his graduate work, Packard worked from time to time on Cenozoic marine vertebrates (see Packard and Kellogg, 1934). He was instrumental in developing the early careers of many stratigraphers, including O. L. Bandy, L. G. Hertlein, E. J. Moore, S. W. Muller, H. G. Schenck, and H. E. Wheeler.

Hubert G. Schenck (1897-1960)

Packard's most promising early student was Hubert G. Schenck, to whom he assigned a study of exposures of the Eugene Formation near Springfield, Oregon, as this was about the only well-recorded Tertiary invertebrate locality of T. L. Condon. This work, initiated in 1917, formed the cornerstone for Schenck's later studies on the stratigraphy of the marine Oligocene exposures of western Oregon (Schenck, 1928). Schenck went on to become an expert on stratigraphic classification and probably the most influential teacher of biostratigraphy on the Pacific Coast before World War II. He was almost singlehandedly responsible for the training of foraminiferal micropaleontologists on the Pacific Coast during that period.

MOLLUSCAN PALEONTOLOGY IN THE 1920s AND 1930s

Developments in larger invertebrate biostratigraphy during the 1920s and 1930s slowed considerably with Weaver's shift of field to South America and California. Nonetheless, the two decades that preceded World War II were marked by a series of faunal monographs that strengthened Paleogene correlations in the Pacific Northwest. Among these classic works are Clark and Arnold's (1923) definitive report on the upper Oligocene Sooke Formation of southwestern Vancouver Island, British Columbia, Tegland's (1931) monograph on the fauna of the Blakeley Formation of Weaver (1912) near Seattle, Turner's (1938) study of Eocene mollusks of southwestern Oregon, and Effinger's (1938) report on the Gries Ranch fauna of western Washington. Many taxonomic studies completed during this period were helpful in developing the stage and zone chronology that followed in the early 1940s (Clark, 1925; Durham, 1937; Merriam, 1941; Schenck, 1926, 1931, 1936; Tegland, 1929). Almost all of these works emanated from the University of California, Berkeley, where Bruce L. Clark (Fig. 7) and his students were refining Paleogene molluscan zonations of the Pacific Coast Tertiary. The California Paleogene chronology (Clark, 1926, 1929; Clark and Vokes, 1936) was helpful as a standard for age determination and correlation in the Pacific Northwest.

Bruce L. Clark (1880-1945)

Clark was a warm-hearted and generous person, quick with enthusiasm for new geologic interpretations, and particularly effective with students. He somehow embued them with a measure of his intense excitement with sedimentary geology and his enthusiasm for new ideas. He backed up his students with generous financial assistance during the dark days of the Depression. At one point in the early 1930s, he and his wife were on the verge of cancelling a long-anticipated sabbatical year in Europe that a graduate student of his might not have to withdraw from school for lack of funds. At the last minute, the student received other funding. Clark's paleontologic and geologic reports were, and still are, subjects of debate in many quarters. He is said to have had so very active a mind that his thoughts raced well ahead of his ability to write them down. His office was cluttered, and he had the habit of using any available items as book marks—especially rock and fossil specimens and even old pipes. Charges of carelessness levied at Clark, whether deserved or not, are, I believe, far outweighed by his substantial contributions to Cenozoic zonations, which were backed up by an impressive series of monographs on critical faunas from California to Alaska. Campus politics operated against this warm, creative scientist in his later years. Even as he lay dying in the mid-1940s, a concerted effort by his colleague Charles Camp to get him promoted from associate to full professor was denied by a faculty committee.

Weaver's Work in South America and California

Weaver took an extended leave from teaching from 1919 to 1926 to work as a petroleum consultant in Central and South America. During the later part of this period, he made an exhaustive study of the Jurassic and Cretaceous rocks of Argentina. This work continued to occupy a central place in Weaver's life as he made two trips to Bonn, Germany, to compare his fossil material with collections in The Geologische-Paleontologische Institut at the University of Bonn (Barksdale, 1974,). In 1927, during his second trip to Germany, Weaver arranged to have hand-drawn illustrations of fossils made, at his own expense, for a monograph on the Argentinian material. Completing the text and 62 oversized plates for his monograph in 1928, Weaver then started a search for funds to underwrite what

Figure 7. J. C. Merriam and students, Berkeley, California, 1915. From left to right: J. P. Buwalda, E. L. Packard, Merriam, W.S.W. Kew, Jorgen Nomland, Chester Stock, and B. L. Clark.

would surely be a costly publication. About this time, an offer of a professorship at the University of Chicago was tendered him, and he seized on this as leverage to get his book published by the University of Washington. He agreed to stay on at Seattle provided that the University of Washington would underwrite publication. The monograph, a large format issue of 469 pages and the 62 plates of figured specimens, was published in 1931 at a cost of more than $10,000, surely a large sum of money during the height of the Depression (Barksdale, 1974). Weaver reportedly told that a copy of this publication (Weaver, 1931) was kept by the then governor of Washington, Roland Harley, in a prominent spot on his desk. Harley, a fiscal conservative who exercised almost direct fiscal control of the university from his office in Olympia, used to point to the massive Weaver monograph as "a horrible example of the waste of state funds in perilous times" (Barksdale, 1974, p. 27).

Following his return to the United States from Germany, Weaver's interests again turned to California, where he began a mapping program of the quadrangles north of San Francisco Bay. His summer field courses were held in California from 1928 to 1935 (Weaver, 1949). In later years, Weaver confided to one of his students that he only wanted to work in warm climates after his grueling pre-1920 work in the cold and damp forests of western Washington. Weaver's maps of nine Bay Area quadrangles and an extensive text that included much biostratigraphic data were completed in 1937 and submitted to the U.S. Geological Survey for publication as a Professional Paper; but publication was delayed for many years, and the work was finally released to the Geological Society of America (Weaver, 1949). Weaver returned to the study of the Tertiary rocks of western Washington and Oregon and completed a series of important stratigraphic and paleontologic syntheses (Weaver, 1937, 1942, 1945).

PALEOGENE STAGE AND ZONE BIOCHRONOLOGY

The culmination of Weaver's work on the Cenozoic of western Oregon and Washington came during World War II with publication of the Cenozoic correlation chart for the Pacific Coast States (Weaver and others, 1944). This cooperative venture between 21 West Coast stratigraphers brought about the first interpretation of marine microfaunal and macrofaunal zonations. Included in this massive correlation were 13 stratigraphic sections from Oregon to southern British Columbia, almost entirely Weaver's work. These sections provided a clear-cut frame of reference for provincial correlation and a point of departure for future biostratigraphic research. In many respects, this chart can be considered the principal milestone in Pacific Northwest biostratigraphy. It drew heavily on Weaver's (1937, 1942) syntheses of the Tertiary formations and their faunas. An important element was the newly developed late Eocene and Oligocene megafossil zonation of the northern Olympic Peninsula by his former student J. Wyatt Durham (1944). By this point in time, a better understanding of the time-stratigraphic classification of rocks (see Schenck and Muller, 1936, 1941) permitted a far more sophisticated stratigraphic synthesis than had heretofore been possible. The foraminiferal sequence was based entirely on California sections, whereas the molluscan sequence included three Paleogene stages based on Oregon and Washington sequences—the "Keasey," "Lincoln," and "Blakeley"—in addition to the seven faunal zones of Durham (1944).

The greater part of the marine Cenozoic record of Oregon, Washington, and British Columbia, unlike that of California, is Paleogene. With the publication of the faunal and stratigraphic studies of Weaver (1937, 1942) and Durham (1944), the basic framework of Pacific Northwest

Tertiary stratigraphy was well established. Parts of the Paleogene molluscan sequence have since been formally named (Armentrout, 1975; Addicott, 1976), and a complementary sequence of Neogene molluscan stages has been introduced for this region (Addicott, 1976).

RISE OF MICROPALEONTOLOGY

Soon after the close of World War I, the discovery that benthic foraminifers could be effectively utilized in biostratigraphy led to a new and important phase of research in California, one directly associated with the intensified search for new petroleum reserves. After a period of description and basic data gathering, the first foraminiferal biochronologies for the Pacific Coast were fashioned in the late 1930s (Schenck and Kleinpell, 1936; Kleinpell, 1938; Laiming, 1939, 1940). Foraminiferal studies in the Pacific Northwest lagged behind those in California, again because of the advancement of stratigraphic micropaleontology with oil exploration (see Kleinpell, 1971, for an account of commercial micropaleontology in California). The California biochronologies were utilized in Oregon and Washington but often with less precision owing to latitudinal change in faunal composition, particularly in Neogene rocks. Some 20 years after the California sequences were defined, Weldon Rau began a series of biostratigraphic studies in western Washington that resulted in a Tertiary zonation for the Pacific Northwest region (Rau, 1958, 1964, 1966, 1967, 1970, 1979).

In the United States, the first foraminifers described from bore-hold samples were from the West Coast (Chapman, 1900); yet it was J. A. Cushman's study of foraminifers from well samples from the Atlantic Coastal Plain (in Stephenson, 1914) that first demonstrated that age determinations and correlations could be made by these microorganisms. This, and a later address to a meeting of the Geological Society of America in 1921 (Cushman, 1922, p. 145) in which it was stated that foraminiferal "faunas of these [Gulf Coastal Plain] formations are easily recognizable" presumably led to the subsequent acceleration of foraminiferal biostratigraphy on the West Coast.

The person possibly most responsible for early development in this field on the Pacific Coast was G Dallas Hanna. A broadly versed naturalist, Hanna's primary paleontologic interest was in diatoms, work he started before his appointment to the California Academy of Sciences in 1919. In 1922, Hanna received a grant to study the possibility of utilizing foraminifers in subsurface stratigraphy. This work resulted in publication of a definitive report on operating a foraminiferal laboratory (Hanna and Driver, 1924) and his starting the first commercial laboratory on the Pacific Coast in May 1923. The following year, Hanna published the first paper on Pacific Northwest foraminifers with his brother Mark, a micropaleontologist who was at the University of Washington filling in for the absent C. E. Weaver. This report dealt with an assemblage from the Eocene Cowlitz Formation of southwest Washington (Hanna and Hanna, 1924).

During the ensuing 15 years, reports on foraminifers from Oregon and Washington were published by several workers (Cushman and Schenck, 1928; Frizzell and Blackwelder, 1933; Kleinpell in Packard and Kellogg, 1934; Frizzell, 1937; Kleinpell, 1938; Berthiaume, 1938). Descriptive accounts of foraminiferal assemblages increased during the forties (Detling, 1946; Cushman and others, 1947a–1947e, 1949a 1949b; Beck, 1943; Bandy, 1944; Rau, 1948, 1949), but larger marine invertebrates continued to be the primary means of age determination and correlation for Pacific Northwest Tertiary formations into the 1950s (see, for example, Warren and others, 1945, Snavely and Vokes, 1949, and Vokes and others, 1949). In the definitive Cenozoic correlation study of the West Coast (Weaver and others, 1944), all of the Pacific Northwest sections were correlated on the basis of megafossils. Only with the development of the regional foraminiferal zonations during the past two decades (see Rau, 1979) have foraminifers become widely utilized in Cenozoic biostratigraphy of the Pacific Northwest. Petroleum companies have refined their own biozonations, but only the most general outlines of these have been published (Rothwell, 1964; Billman, 1965; Bergen and Bird, 1972).

The appointment of micropaleontologists to the state surveys of Oregon and Washington following World War II gave considerable impetus to the development of regional foraminiferal biostratigraphy in the Pacific Northwest. Roscoe E. Stewart joined the Oregon Department of Geology and Mineral Industries in the mid-1940s, collaborating on many reports on Oregon foraminifers with his wife Katherine and J. A. Cushman. In 1960, Weldon Rau accepted a position with the Washington Division of Mines and Geology that resulted in the biostratigraphic reports and geologic maps that provided the basis for his Tertiary zonation of western Washington (Rau, 1979).

BENTHIC FORAMINIFERAL ZONES

The development of a series of provincial benthic foraminiferal zones from 1958 to 1967 (Rau, 1979) was another milestone in the biostratigraphy of Oregon and Washington. These zones provide the framework for geologic mapping of the heavily forested, predominantely siltstone and shale terrane of western Oregon and Washington in conjunction with determining the California Paleogene and Neogene benthic foraminiferal stages (Mallory, 1959; Schenck and Kleinpell, 1936; Kleinpell, 1938). Many, if not most, of the marine Cenozoic formations are discriminated on the basis of biostratigraphic data provided by benthic foraminifers, a fact that is not always clearly spelled out in map explanations.

PLANKTONIC CALIBRATIONS

The modern study of planktonic microfossils marks the most recent development in Cenozoic biostratigraphy of the West Coast. Generally disregarded in early foraminiferal zonations, the initial planktonic studies are only about 20 years old. From the first work on Paleogene examples (Loeblich, 1958), interest quickly spread to Neogene sections (Parker, 1964; Lipps, 1964, 1967; Fowler, 1956; Ingle, 1967). Studies of siliceous microfossils, initiated by Hanna in the 1920s, have taken on new interest during the past two decades (Mandra, 1960, 1963; Wornardt, 1967; Casey, 1972), and work on calcareous nannofossils was begun in the 1960s (Bramlette and Sullivan, 1961).

Interest in microfossils has been greatly stimulated by the Deep Sea Drilling Project and concomitant refinement of various planktonic biochronologies. Work on planktonic organisms in Pacific Northwest Cenozoic rocks has involved mainly qualitative studies yielding tie-points with the European series-epoch standard sections (Fowler, 1965; Orr and others, 1971; Orr, 1971; McKeel and Lipps, 1972, 1975; Worsley and Crecelius, 1972; Barron, 1979).

In California, several planktonic microfossil specialists (Milow and Ennis, 1961; Steineck and Gibson, 1971; Bukry and others, 1977; Poore, 1976; Brabb and others, 1977; Crouch and Bukry, 1979) have alleged that the benthic foraminiferal stages and zones of Kleinpell (1938), Laiming (1940), and Mallory (1959) have overlapping boundaries and are in part coeval. This California-based controversy has broad implications for Pacific Northwest biostratigraphy, as geologic mapping of marine Cenozoic formations commonly depends on biostratigraphic data from benthic foraminifers. Accordingly, the oft-repeated allegation that benthic foraminiferal stages and zones are "inherently time-transgressive" promises to be a focal point for biostratigraphic research during the next decade. Planktonic workers have raised this issue, but response by benthic specialists has been limited and has not adequately dealt with it. Before the two schemes can be brought into better agreement, benthic foraminiferal workers must reassess their data and respond. Whatever the final solution to this problem, it is clear that benthic foraminifers, by virtue of their consistent occurrence and great prevalence in Tertiary sections of the West Coast, will continue to be the principal biostratigraphic indices.

REFERENCES CITED

Addicott, W. O., 1972, Neogene molluscan paleontology along the West Coast of North America, 1840-1969—Trends and status: Journal of Paleontology, v. 46, p. 627-636.

——1976, Neogene molluscan stages of Oregon and Washington, *in* Neogene symposium: Society of Economic Paleontologists and Mineralogists, Pacific Section, San Francisco, California, April 1976, p. 95-115, Pls. 1-5.

——1979, The marine Cenozoic of the Pacific Coast States: An overview of the growth and development of biostratigraphic research, *in* Armentrout, J. M., and others, eds., Pacific Coast paleogeography symposium III, Cenozoic paleogeography of the western United States: Society of Economic Paleontologists and Mineralogists, Los Angeles, California, March 1979, p. 325.

Anderson, F. M., 1932, Pioneers in the geology of California: California Division of Mines Bulletin 104, p. 1-24.

Anderson, F. M., and Martin, Bruce, 1914, Neocene record in the Temblor Basin, California, and Neocene deposits of the San Juan district, San Luis Obispo County: California Academy of Sciences Proceedings, ser. 4, v. 4, no. 3, p. 15-112, Pls. 1-10.

Armentrout, J. M., 1975, Molluscan biostratigraphy of the Lincoln Creek Formation, southwest Washington, *in* Weaver, D. W., and others, eds., Paleogene symposium and selected technical papers: American Association of Petroleum Geologists-Society of Economic Paleontologists and Mineralogists-Society of Economic Geologists, Pacific Sections, 1975 Annual Meeting, Long Beach, California, p. 14-48.

Arnold, Ralph, 1905, Coal in Clallam County, Washington: U.S. Geological Survey Bulletin 260, p. 413-421.

——1906a, Geological reconnaissance of the coast of the Olympic Peninsula, Washington: Geological Society of America Bulletin, v. 17, p. 451-468, 1 map.

——1906b, The Tertiary and Quaternary pectens of California: U.S. Geological Survey Professional Paper 47, 264 p., 53 pls., 2 figs.

Arnold, Ralph, and Hannibal, Harold, 1913, The marine Tertiary stratigraphy of the north Pacific Coast of America: American Philosophical Society Proceedings, v. 52, no. 212, p. 559-605.

Bandy, O. L., 1944, Eocene foraminifera from Cape Blanco, Oregon: Journal of Paleontology, v. 18, p. 366-377.

Barksdale, J. D., 1974, Geology at the University of Washington 1895-1973: University of Washington [Seattle] Publications in Geological Science, no. 4, 109 p.

Barron, J. A., 1981, Marine diatom biostratigraphy of the Montesano Formation near Aberdeen, Washington: Geological Society of America Special Paper 184 (this volume).

Bartsch, Paul, Rehder, H. A., and Shields, B. E., 1946, A bibliography and short biogeographical sketch of William Healey Dall: Smithsonian Miscellaneous Collections, v. 104, no. 15, 96 p.

Beck, R. S., 1943, Eocene Foraminifera from Cowlitz River, Lewis County, Washington: Journal of Paleontology, v. 17, p. 584-614.

Bergen, F. W., and Bird, K. J., 1972, The biostratigraphy of the Ocean City area, Washington: Society of Economic Paleontologists and Mineralogists, 47th Annual Convention, Bakersfield, California, March 1972, Proceedings, p. 173-191.

Berthiaume, S. A., 1938, Orbitoids from the Crescent formation (Eocene) of Washington: Journal of Paleontology, v. 12, p. 494-497, 1 pl.

Billman, H. G., 1965, Review of Tertiary stratigraphy and foraminiferal zonations of western Washington State [abs.]: American Association of Petroleum Geologists Bulletin, v. 49, p. 1080-1081.

Brabb, E. E., Clark, J. C., and Throckmorton, C. B., 1977, Measured sections of Paleogene rocks from the California Coast Ranges: U.S. Geological Survey Open-File Report 77-714, 114 p.

Bramlette, M. N., and Sullivan, F. R., 1961, Cocolithophorids and related nannoplankton of the early Tertiary in California: Micropaleontology, v. 7, p. 129-188, Pls. 1-14.

Bukry, David, Brabb, E. E., and Vedder, J. G., 1977, Correlation of Tertiary nannoplankton assemblages from the Coast and Peninsular Ranges of California: Memoria Segundo Congreso Latinoamericano de Geología, T. 3, Venezuela Boletín de Geología, Publicación Especial no. 7, p. 1461-1483.

Casey, R. E., 1972, Neogene radiolarian biostratigraphy and paleotemperatures: Southern California, the experimental Mohole, Antarctic Core E14-8: Palaeogeography, Palaeoclimatology, Palaeoecology, v. 12, p. 115-130.

Chapman, Frederick, 1900, Foraminifera from the Tertiary of California: California Academy of Sciences Proceedings, ser. 3, Geology, v. 1, p. 241-258, Pls. 29-30.

Clark, B. L., 1925, Pelecypoda from the marine Oligocene of western North America: University of California Publications, Department of Geological Sciences Bulletin, v. 15, no. 4, p. 69-136, Pls. 8-22.

____1926, The Domengine horizon, middle Eocene of California: University of California Publications, Department of Geological Sciences Bulletin, v. 16, no. 5, p. 99-118.

____1929, Stratigraphy and faunal horizons of the Coast Ranges of California: (privately published), 132 p., 50 pls.; Ann Arbor, Michigan, University Microfilms, Inc., 30 p., 50 pls.

Clark, B. L., and Arnold, Ralph, 1923, Fauna of the Sooke Formation, Vancouver Island, with description of a new coral by T. Wayland Vaughan: University of California Publications, Department of Geological Sciences Bulletin, v. 14, no. 5, p. 123-234, Pls. 15-42.

Clark, B. L., and Vokes, H. E., 1936, Summary of marine Eocene sequence of western North America: Geological Society of America Bulletin, v. 47, p. 851-878, 2 pls.

Clarke, J. M., 1921, James Hall, of Albany, geologist and paleontologist, 1811-1898: Albany, N.Y., 565 p.

Conrad, T. A., 1848, Fossil shells from Tertiary deposits on Columbia River, near Astoria [Oregon]: American Journal of Science, ser. 2, v. 5, p. 432-433, 14 figs. [Reprinted in Dall, 1909, U.S. Geological Survey Professional Paper 59, p. 150-151.]

____1849, Fossils from northwestern America, *in* Dana, J. D., Geology, Volume X of the United States exploring expedition... 1838-1842, under the command of Charles Wilkes: Philadelphia, app. p. 722-728; atlas, Pls. 17-21. [Test reprinted *in* Dall, 1909, U.S. Geological Survey Professional Paper 59, p. 152-156.]

____1855a, Notes on the Miocene and post-Pliocene deposits of California with descriptions of two new fossil corals: Philadelphia Academy of Natural Sciences Proceedings, v. 7, p. 441.

____1855b, Report on the fossil shells collected in California by W. P. Blake, geologist of the expedition under the command of Lieutenant R. S. Williamson, U.S. Topographical Engineers, Pacific Railroad Explorations, 1853, *in* Appendix to the preliminary geological report of W. P. Blake: U.S. 33rd Congress, 1st Session, House Executive Document 129, p. 5-20. [Reprinted *in* Dall, 1909, U.S. Geological Survey Professional Paper 59, p. 163-171.]

____1857a, Descriptions of the fossil shells, *in* Williamson, R. S., Report of explorations in California for railroad routes: U.S. 33rd Congress, 2nd session, Senate Executive Document 78 and House Executive Document 91, v. 5, pt. 2, app. art. 2, p. 317-329, Pls. 2-9. [Pacific Railroad reports.]

____1857b, Report on the paleontology of the survey [Parke's survey], *in* Reports of explorations and surveys...railroad... from the Mississippi River to the Pacific Ocean: U.S. 33rd Congress, 2nd Session, Senate Executive Document 78 and House Executive Document 91, v. 7, pt. 2, p. 189-196, Pls. 1-10. [Reprinted *in* Dall, 1909, U.S. Geological Survey Professional Paper 59, p. 180-185.] [Pacific Railroad reports.]

____1857c, Description of three new genera; twenty-three new species middle Tertiary fossils from California and one from Texas: Philadelphia Academy of Natural Sciences Proceedings (for 1856), v. 8, p. 312-316. [Reprinted *in* Dall, 1909, U.S. Geological Survey Professional Paper 59, p. 173-175.]

____1857d,, Description of the Tertiary fossils collected on the survey [Williamson's survey], *in* Reports of explorations and surveys...railroad...from the Mississippi River to the Pacific Ocean: U.S. 33rd Congress, 2nd Session, Senate Executive Document 78 and House Executive Document 91, v. 6, pt. 2, p. 69-73, Pls. 2-5. [Reprinted *in* Dall, 1909, U.S. Geological Survey Professional Paper 59, p. 176-179.] [Pacific Railroad reports.]

____1865a, Catalogue of the older Eocene shells of Oregon: American Journal of Conchology, v. 1, p. 150-154.

____1865b, Observations of the Eocene Lignite formation of the United States: Philadelphia Academy of Natural Sciences Proceedings, v. 9, 2nd ser., p. 70-73.

Cope, E. D., 1880, Corrections of the geological maps of Oregon: American Naturalist, v. 14, p. 457-458.

Crouch, J. K., and Bukry, David, 1979, Comparison of Miocene provincial foraminiferal stages to coccolith zones in the California Continental Borderland: Geology, v. 7, p. 211-215.

Cushman, J. A., 1922, Use of Foraminifera in determining underground structure, especially in petroleum mining: Geological Society of America Bulletin, v. 33, p. 145-146.

Cushman, J. A., and Schenck, H. G., 1928, Two foraminiferal faunules from the Oregon Tertiary: University of California Publications, Department of Geological Sciences Bulletin, v. 17, p. 305-348.

Cushman, J. A., Stewart, R. E., and Stewart, K. E., 1947a, Astoria Miocene Foraminifera from the northwest corner of Tenth Street and Morrison Avenue, Astoria, Clatsop County, Oregon: Oregon Department of Geology and Mineral Industries Bulletin 36, pt. 1, p. 11-38.

____1947b, Astoria Miocene Foraminifera from Agate Beach, Lincoln County, Oregon: Oregon Department of Geology and Mineral Industries Bulletin 36, pt. 2, p. 41-54.

____1947c, Upper Coaledo (Upper Eocene) Foraminifera from Yokam Point, Coos County, Oregon: Oregon Department of Geology and Mineral Industries Bulletin 36, pt. 3, p. 57-68.

____1947d, Lower Coaledo (Upper Eocene) Foraminifera from Sunset Bay, Coos County, Oregon: Oregon Department of Geology and Mineral Industries Bulletin 36, pt. 4, p. 71-90.

____1947e, Eocene Foraminifera from Helmick Hill, Polk County, Oregon: Oregon Department of Geology and Mineral Industries Bulletin 36, pt. 5, p. 93-110.

____1949a, Upper Eocene Foraminifera from the Toledo formation, Toledo, Lincoln County, Oregon: Oregon Department of Geology and Mineral Industries Bulletin 36, pt. 6, p. 126-145.

____1949b, Quinault Pliocene Foraminifera from western Washington: Oregon Department of Geology and Mineral Industries Bulletin 36, pt. 7, p. 148-163.

Dall, W. H., 1909, Contributions to the Tertiary paleontology of the Pacific Coast, I. The Miocene of Astoria and Coos Bay, Oregon: U.S. Geological Survey Professional Paper 59, 278 p., 23 pls.

Detling, M. R., 1946, Foraminifera of the Coos Bay Lower Tertiary, Coos County, Oregon: Journal of Paleontology, v. 20, p. 348-361.

Diller, J. S., 1896, A geological reconnaissance in northwestern Oregon: U.S. Geological Survey 17th Annual Report, pt. 1, p. 441-520.

____1898, Description of the Roseburg quadrangle [Oregon]: U.S. Geological Survey Geologic Atlas, Roseburg Folio, no. 49, 4 p., maps.

____1899, The Coos Bay coal field, Oregon: U.S. Geological Survey 19th Annual Report, pt. 3, p. 309-379.

____1901, Description of the Coos Bay quadrangle [Oregon]: U.S. Geological Survey Geologic Atlas, Coos Bay Folio, no. 73, 5 p., maps.

____1902, Topographic development of the Klamath Mountains [California and Oregon]: U.S. Geological Survey Bulletin 196, p. 1-69, Pls. 1-13.

____1903, Description of the Port Orford quadrangle, Oregon: U.S.

Geological Survey Geologic Atlas, [Port Orford] Folio 89, 6 p.

____1905, Coal in Washington, near Portland, Oregon: U.S. Geological Survey Bulletin 260, p. 411-412.

Durham, J. W., 1937, Gastropods of the family Epitoniidae from Mesozoic and Cenozoic rocks of the West Coast of North America, including one species by F. E. Turner and one by R. A. Bramkamp: Journal of Paleontology, v. 16, p. 183-191, Pls. 56-57.

____1942, Notes on Pacific coast Galeodeas: Journal of Paleontology, v. 16, p. 183-191, Pls. 29-30.

____1944, Megafaunal zones of the Oligocene of northwestern Washington: University of California Publications, Department of Geological Sciences Bulletin, v. 27, no. 5, p. 101-211,Pls. 13-18.

Effinger, W. L., 1938, The Gries Ranch fauna (Oligocene) of western Washington: Journal of Paleontology, v, 12, p. 355-390, 3 pls.

Fowler, G. A., 1965, The stratigraphy, foraminifera, and paleoecology of the Montesano formation, Grays Harbor County, Washington [Ph.D. thesis]: Los Angeles, University of Southern California [Ann Arbor, Michigan, University Microfilms, Inc.], Inc.], 354 p.

Frizzell, D. L., 1937, Foraminifera of the type area of the Lincoln Formation of Washington [abs.]: Geological Society of America Proceedings, 1936, p. 383.

Frizzell, D. L., and Blackwelder, R. E., 1933, Preliminary analysis of the type Lincoln fauna (Oligocene) of Washington: Micropaleontology Bulletin, v. 4, no. 2, p. 53-63.

Hanna, G D., and Driver, H. L., 1924, the study of subsurface formations in California oil field development: California State Mining Bureau, Summary of Operations California Oil Fields, v. 10, no. 3, p. 5-26.

Hanna, G D., and Hanna, M. A., 1924, Foraminifera from the Eocene of Cowlitz River, Lewis County, Washington: University of Washington Publications in Geology, v. 1, no. 4, p. 57-62, 1 pl.

Hannibal, Harold, 1912, A synopsis of the recent and Tertiary freshwater Mollusca of the Californian province, based upon an ontogenetic classification: Malacological Society of London Proceedings, v. 10, p. 112-211, Pls. 5-8.

Hertlein, L. G., 1925, New Species of marine fossil Mollusca from western North America: Southern California Academy of Sciences Bulletin, v. 24, pt. 2, p. 39-46, Pls. 3-4.

Howe, H. V., 1926, Astoria: Mid-Tertic type of Pacific coast: Pan American Geologist, v. 45, p. 295-306.

Ingle, J. C., Jr., 1967, Foraminiferal biofacies variation and the Miocene-Pliocene boundary in southern California: American Paleontology Bulletins, v. 52, no. 236, p. 217-394.

Kleinpell, R. M., 1938, Miocene stratigraphy of California: Tulsa, Oklahoma, American Association of Petroleum Geologists, 450 p.

____1971, California's early "oil bug" profession: Journal of the West, v. 10, no. 1, p. 72-101.

Laiming, Boris, 1939, Some foraminiferal correlations in the Eocene of San Joaquin Valley, California: Sixth Pacific Science Congress, 1939, Proceedings, v. 2, p. 535-568 [1940].

____1940, Foraminiferal correlations in Eocene of San Joaquin Valley, California: American Association of Petroleum Geologists Bulletin, v. 24, p. 1923-1939.

Lipps, J. H., 1964, Miocene planktonic Foraminifera from Newport Bay, California: Tulane Studies in Geology, v. 2, p. 109-133, 4 pls.

____1967, Planktic Foraminifera, intercontinental correlation and age of California mid-Cenozoic microfaunal stages: Journal of Paleontology, v. 41, p. 994-999.

Loeblich, A. R., Jr., 1958, Danian Stage of Paleocene in California: American Association of Petroleum Geologists, v. 42, p. 2260-2261.

Mallory, V. S. 1959, Lower Tertiary biostratigraphy of the California Coast Ranges: Tulsa, Oklahoma, American Association of Petroleum Geologists, 416 p.

Mandra, Y. T., 1960, Fossil silicoflagellates from California, U.S.A.: International Geological Congress, 21st, Copenhagen, Report, pt. 6, Pre-Quaternary Micropaleontology, p. 77-89.

____1963, Buttle Diatomite, a new member of the Monterey Formation, Salinas Valley, California, *in* Guidebook to the geology of Salinas Valley and the San Andreas fault: American Association of Petroleum Geologists-Society of Economic Paleontologists and Mineralogist, Pacific Sections, Annual Trip, 1963, p. 104-105.

McCornack, E. C., 1928, Thomas Condon, pioneer geologist of Oregon: Eugene, University of Oregon Press, 355 p.

McKeel, D. R., and Lipps,J. H. 1972, Calcareous plankton from the Tertiary of Oregon: Palaeogeography, Palaeoclimatology, Palaeoecology, v. 12, p. 75-93.

____1975, Eocene and Oligocene planktonic foraminifera from the central and southern Oregon Coast Range: Journal of Foraminiferal Research, v. 5, p. 249-269, Pls. 1-5.

Meek, F. B., 1864, Check list of the invertebrate fossils of North America, Miocene: Smithsonian Miscellaneous Collections, v. 4, pub. 183, p. 1-32.

Merriam, C. W., 1941, Fossil Turritellas from the Pacific coast region of North America: University of California Publications, Department of Geological Sciences Bulletin, v. 26, no. 1, p. 1-214, Pls. 1-41.

Merrill, G. P., 1924, The first one hundred years of American geology: New Haven, Connecticut, Yale University Press, 773 p.

Milow, E. D., and Ennis, D. B., 1961, Guide to geologic field trip of southwestern San Diego County [California], *in* Field trip guidebook, San Diego County: Geological Society of America, 57th Annual Meeting, Cordilleran Section, p. 23-43.

Moore, E. J., 1963, Miocene mollusks from the Astoria Formation in Oregon: U.S. Geological Survey Professional Paper 419, 109 p., 33 pls. [1964].

____1971, Timothy Abbott Conrad, *in* Gillispie, C. C. ed., Dictionary of science biographies: New York, Charles Scribner's Sons, v. 3, p. 391-393.

Orr, W. N., 1971, Pacific Northwest siliceous phytoplankton: Palaeogeography, Palaeoclimatology, Palaeoecology, v. 12, p. 95-114.

Orr, W. N., Ehle, Judi, and Zaitzeff, J. B., 1971, A late Tertiary diatom flora from Oregon: California Academy of Sciences Proceedings, v. 37, no. 16, p. 489-500.

Packard, E. L., and Kellogg, Remington, 1934, A new cetothere from the Miocene Astroia formation of Newport, Oregon: Carnegie Institute of Washington, Contributions in Paleontology, Pub. 447, p. 1-62.

Parker, F. L., 1964, Foraminifera from the experimental MOHOLE drilling near Guadalupe Island, Mexico: Journal of Paleontology, v. 38, p. 617-636, Pls. 97-102.

Poore, R. Z., 1976, Microfossil correlation of California lower Tertiary section: A comparison: U.S. Geological Survey Professional Paper 743-F, 8 p.

Rau, W. W., 1948, Foraminifera from the Porter Shale (Lincoln Formation), Grays Harbor County, Washington: Journal of Paleontology, v. 22, p. 152-174.

____1949, Foraminifera from the Miocene Astoria formation in southwestern Washington: Journal of Paleontology, v. 22, p. 774-782.

____1958, Stratigraphy and foraminiferal zonation in some of the Tertiary rocks of southwestern Washington: U.S. Geological Survey oil and Gas Investigation Chart OC-57, 2 sheets.
____1964, Foraminifera from the northern Olympic Peninsula, Washington: U.S. Geological Survey Professional Paper 374-G, p. G1-G33.
____1966, Stratigraphy and foraminifera of the Satsop River area, southern Olympic Peninsula, Washington: Washington Division of Mines and Geology Bulletin 53, 66 p.
____1967, Geology of the Wynoochee Valley quadrangle, Grays Harbor County, Washington: Washington Division of Mines and Geology Bulletin 56, 51 p.
____1970, Foraminifera, stratigraphy, and paleoecology of the Quinault Formation, Point Grenville-Raft River coastal area, Washington: Washington Division of Mines and Geology Bulletin 62, 40 p.
____1979, An overview of state and federal studies in foraminiferal biostratigraphy of the Pacific northwest Tertiary province: Geological Society of America Special Paper 184 (this volume).
Rothwell, W. T., Jr., 1964, Western Washington microfaunal correlations [abs.]: Geological Society of America Special paper 62, p. 273-274 [1965].
Schenck, H. G., 1926, Cassididae of western America: University of California Publications, Department of Geological Sciences Bulletin, v. 16, no. 4, p. 69-98, Pls. 12-15.
____1928, Stratigraphic relations of western oligocene formations: University of California Publications, Department of Geological Sciences Bulletin, v. 18, no. 1, p. 1-50.
____1931, Cephalopods of the genus *Aturia* from western North America: University of California Publications, Department of Geological Sciences Bulletin, v. 19, no. 19, p. 435-490, Pls. 66-78.
____1936, Nuculid bivalves of the genus *Acila*: Geological Society of America Special paper 4, 149 p., 18 pls.
Schenck, H. G., and Kleinpell, R. M., 1936, Refugian Stage of Pacific coast Tertiary: American Association of Petroleum Geologists Bulletin, v. 20, p. 215-225.
Schenck, H. G., and Muller, S. W., 1936, Stratigraphic terminology: Geological Society of America Proceedings, 1935, p. 101-102.
____1941, Stratigraphic terminology: Geological Society of America Bulletin, v. 52, p. 1419-1426.
Shumard, B. F., 1858, Descriptions of new fossils from the Tertiary formations of Oregon and Washington Territories and the Cretaceous of Vancouver Island: St. Louis Academy of Sciences Transactions, v. 1, no. 2, p. 120-123.
Snavely, P. D., Jr., and Vokes, H. E., 1949, Geology of the Coastal area between Cape Kiwanda and Cape Foulweather, Oregon: U.S. Geological Survey Oil and Gas Investigations Preliminary Map 97.
Steineck, P. L., and Gibson, J. M., 1971, Age and correlation of the Eocene Ulatisian and Narizian Stages, California: Geological Society of America Bulletin, v.82, p. 447-480.
Stephenson, L. W., 1914, A deep well at Charleston, S.C.: U.S. Geological Survey Professional Paper 90, p. 69-90.
Taylor, D. W., and Smith, A. G., 1971, Harold Hannibal (1889-1965) with a review of his molluscan research: The Veliger, v. 13, p. 303-315.
Tegland, N. M., 1929, Correlation and affinities of certain species of *Pitaria*: University of California Publications, Department of Geological Sciences Bulletin, v. 18, no. 10, p. 275-290, 2 pls.
____1931, Gastropod genus *Galeodea* in the Oligocene of Washington: University of California Publications, Department of Geological Sciences Bulletin, v. 19, no. 18, p. 397-444, Pls. 59-65.
Townsend, J. I., 1839, Narrative of a journey across the Rocky Mountains to the Columbia River, and a visit to the Sandwich Islands, Chili, etc., with a scientific appendix: Philadelphia, Henry Perkins.
Turner, F. E., 1938, Stratigraphy and mollusca of the Eocene of western Oregon: Geological Society of America Special Paper 10, p. 1-130, 7 figs., 22 pls.
Vokes, H. E., Norbisrath, Hans, and Snavely, P. D., Jr., 1949, Geology of the Newport-Waldport area, Lincoln County, Oregon: U.S. Geological Survey Oil and Gas Investigations Preliminary Map 88.
Warren, W. C., Grivetti, R. M., and Norbisrath, Hans, 1945, Geology of northwestern Oregon, west of Willamette River and north of latitude 45°115′ N.: U.S. Geological Survey Oil and Gas Investigations Preliminary Map 42, scale 1 in. to 2¼ mi.
Washburne, C. W., 1914, Geology and oil prospects of northwestern Oregon: U.S. Geological Survey Bulletin 590, 111 p.
Weaver, C. E., 1912, A preliminary report on the Tertiary paleontology of western Washington: Washington Geological Bulletin 15, 80 p., 15 pls.
____1916a, Tertiary faunal horizons of western Washington: University of Washington [Seattle] Publications in Geology, v. 1, no. 1, p. 1-67, Pls. 1-5.
____1916b, The Tertiary formations of western Washington: Washington Geological Survey Bulletin 13, 327 p.
____1916c, The post-Eocene formations of western Washington: California Academy of Sciences Proceedings, ser. 4, v. 6, no. 2, p. 19-40.
____1931, Paleontology of the Jurassic and Cretaceous of west-central Argentina: University of Washington [Seattle Memoirs, v. 1, 469 p., 62 pls.
____1937, Tertiary stratigraphy of western Washington and northwestern Oregon: University of Washington [Seattle] Publications in Geology, v. 4, 266 p.
____1942, Paleontology of the marine Tertiary formations of Oregon and Washington: University of Washington [Seattle] Publications in Geology, v. 5, pts. 1-3, 789 p., 104 pls.
____1945, Stratigraphy and paleontology of the Tertiary formations at Coos Bay, Oregon: University of Washington [Seattle] Publications in Geology, v. 6, p. 31-62.
____1949, Geology of the Coast Ranges immediately north of the San Francisco Bay region, California: Geological Society of America Memoir 35, 242 p.
Weaver, C. E., chairman, and others (20), 1944, Correlation of the marine Cenozoic formations of western North America: Geological Society of America Bulletin, v. 55, p. 569-598.
Wheeler, H. E., 1935, Timothy Abbott Conrad, with particular reference to his work in Alabama one hundred years ago: American Paleontology Bulletins, v. 23, no. 77, p. 1-159.
White, C. A., 1889, On invertebrate fossils from the Pacific Coast: U.S. Geological Survey Bulletin 51, 102 p.
Woodring, W. P., 1958, William Healey Dall: National Academy of Sciences Biographical Memoirs, v. 31, p. 90-113.
Wornardt, W. W., Jr., 1967, Miocene and Pliocene marine diatoms from California: California Academy of Sciences Occasional Paper 63, p. 1-108.
Worsley, T. R., and Crecelius, E., 1972, Paleogene calcareous nannofossils from the Olympic Peninsula, Washington: Geological Society of America Bulletin, v. 83, p. 2859-2862.

MANUSCRIPT RECEIVED BY THE SOCIETY MARCH 14, 1980
MANUSCRIPT ACCEPTED MAY 7, 1980

Printed in U.S.A.

Geological Society of America
Special paper 184
1981

Significance of pectinids in Tertiary biochronology of the Pacific Northwest

WARREN O. ADDICOTT
U.S. Geological Survey, 345 Middlefield Road, MS 52, Menlo Park, California 94025

ABSTRACT

Range zones of pectinid bivalves provide a principal means of age determination and correlation of shallow-water, inshore facies of middle and late Tertiary marine basins of Oregon and Washington. During the Paleogene, pectinids were represented by a few species of *Chlamys*, *Vertipecten*, and some thin-shelled deep-water mud pectens of the *Eburneopecten* and *Amusium* groups. Neogene pectinids are much more numerous than Paleogene species in this region, although in each case, taxonomic diversity is far lower than in the warmer-water Neogene faunas of coastal California to the south. Giant pectinids, species that regularly attain at least 90 mm in diameter, appear near the Oligocene-Miocene boundary. Range zones of giant pectinids are useful in provincial age determination and correlation of marine Neogene strata in Oregon, Washington, and coastal British Columbia: *Vertipecten fucanus* (Dall)—lower lower Miocene (Pillarian Stage); *Patinopecten propatulus* (Conrad)—upper lower Miocene to middle Miocene (Newportian Stage); *Patinopecten oregonensis* (Howe)—lower upper Miocene (Wishkahan Stage); *Yabepecten condoni* (Hertlein)—upper upper Miocene (Graysian Stage); and *Lituyapecten dilleri* (Dall)—Pliocene (Moclipsian Stage). Certain widespread species facilitate interprovincial correlation between the California Neogene molluscan province, to the south, and the Gulf of Alaska Neogene molluscan province, to the north. *Vertipecten fucanus*, an early Miocene species, ranged southward to southern California and northward to the head of the Gulf of Alaska. *Leptopecten andersoni* (Arnold), a late early Miocene to middle Miocene species, ranged southward to Baja California Sur. *Lituyapecten dilleri* (Dall) ranged northward to near the head of the Gulf of Alaska and southward to Cedros Island, off Baja California, Mexico, during the Pliocene.

INTRODUCTION

Range zones of pectinids provide one of the best means of provincial age determination and correlation of shallow-water, nearshore facies of marine Tertiary basins of the eastern North Pacific margin. In Oregon and Washington, USA, and British Columbia, Canada, these bivalve mollusks reached their peak diversity and abundance near the Oligocene-Miocene boundary and serve as particularly useful indices in the biochronologic subdivision of upper Tertiary, or Neogene, rock units. The purpose of this report is to examine the Tertiary record of the family Pectinidae in this sector of the eastern North Pacific with particular reference to those species and genera that are of biochronologic and biogeographic significance.

Small, thin-shelled pectinids characteristic of deep-water muddy substrates occur throughout the marine Tertiary of the Pacific Northwest States and are known from isolated parts of western Vancouver Island, British Columbia. These fragile so-called mud pectens make up about 25% of the known species in this region. They tend to be most abundant and diverse in lower Tertiary, or Paleogene, formations. Most of the late Paleogene and Neogene pectinids, however, are larger, thick-shelled species that are characteristic of middle and inner sublittoral environments and relatively coarse-textured substrates. They are quite useful in provincial biochronology and are the principal subject of this paper.

More than 35 described species of pectinids occur in Eocene to Pliocene formations in western Oregon, Washington, and British Columbia. Only a few of these are treated in Arnold's (1906) early monograph of Cenozoic pectinids from California, and fewer than 60% are included in a later monograph of the Tertiary mollusks of Oregon and Washington (Weaver, 1942). The taxonomy of the large-shelled pectinids is refined by recent work of North

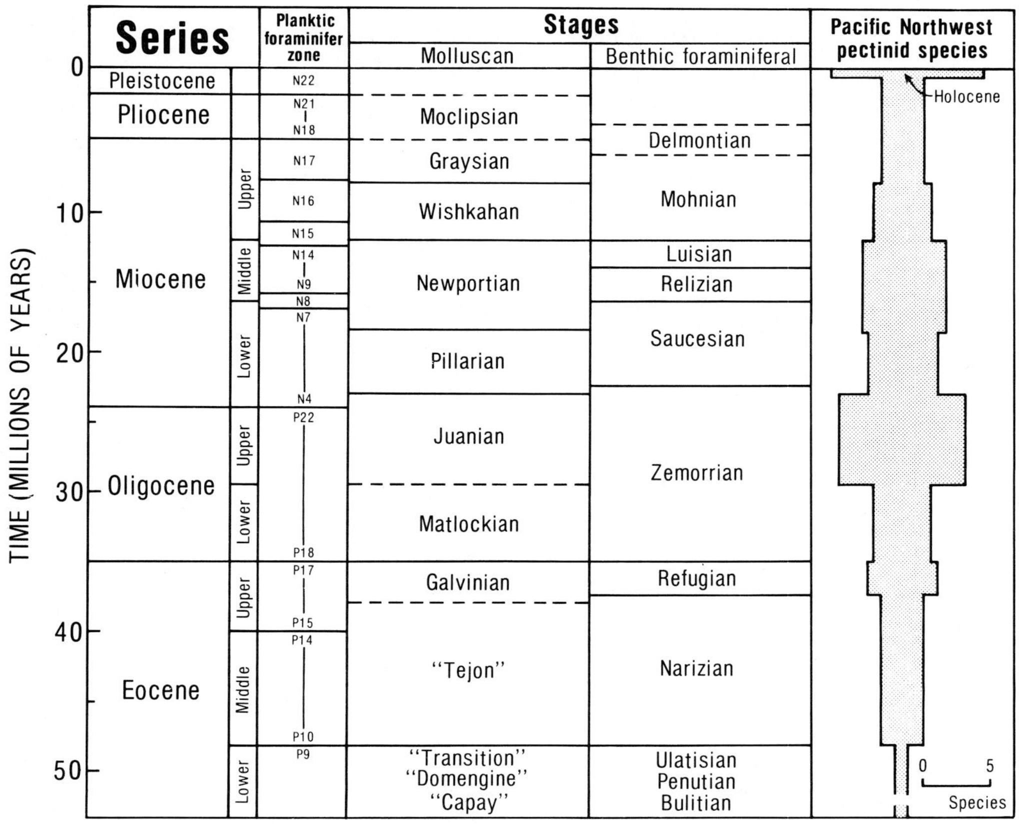

Figure 1. Correlation of the Pacific Northwest provincial molluscan chronology with foraminiferal stages and stage-by-stage taxonomic diversity of the Pectinidae. Positioning of planktonic foraminifera zones after Berggren and Van Couvering (1974), Cavelier and Pomerol (1977), and Ryan and others (1974).

Pacific molluscan specialists, notably MacNeil (1961, 1967), Hertlein and Grant (1972), Masuda (1962, 1971, 1972, 1973), and Sinelnikova (1975). Taxonomy of the mud pectens is still poorly developed, although when better known these small species may provide a complementary basis for correlation of the deep-water facies in this region.

GENERAL CONSIDERATIONS

It is appropriate, at the outset, to consider why the pectinids are such good biochronologic indicators in the marine Tertiary rocks of this area. Perhaps the most significant reason is the relatively rapid rate of evolution of the large-shelled Pectinidae during the Tertiary. This, of course, resulted in species with relatively brief geologic ranges—one of the principal criteria for biochronologic significance. Secondly, many of the species have relatively broad dispersal patterns, a factor that further enhances their usefulness in provincial age determination and correlation. In addition, two properties of the pectinid shell make these bivalves more amenable to preservation and to identification than the vast majority of molluscan species: (1) the calcitic shell composition and (2) the usually distinctive, intricate surface sculpture. In coarse sandstone, the valves of aragonitic-shelled species are selectively leached by ground water so that as a rule, only the small proportion of larger calcitic-shelled invertebrates is preserved. In relatively coarse-grained sandstone there is usually insufficient clay and silt to record the shape and sculpture of the dissolution-prone aragonitic-shelled mollusks as external molds. A good example is some of the coarse-grained pectinid-bearing sandstone of the Clallam Formation in northwestern Washington; another is the Santa Margarita Sandstone of central California with its characteristic low-diversity assemblages of calcitic-shelled pectinids, ostreids, muricid gastropods, barnacles, and echinoids. Finally, the relatively refined taxonomic knowledge of this family, already alluded to, is an important factor in its effectiveness in Cenozoic biochronology.

PROVINCIAL TIME SCALE

The provincial time scale for the Pacific Northwest States and adjacent coastal British Columbia (Fig. 1) is based on a progressive refinement of the early molluscan chronologies developed by Arnold (1906), Arnold and Hannibal (1913), Weaver (1916), and Weaver and others (1944). It has been recently formalized through the description and naming of stages (Armentrout, 1975; Addicott, 1976a). The status of the provincial molluscan chronology is reviewed elsewhere in this volume (Armentrout). The molluscan chronology, being based on nearshore marine sequences, is useful in the coarse-grained clastic, inner sublittoral facies of marine Tertiary basins. It complements the benthic foraminiferal chronology (Kleinpell, 1938; Schenck and Kleinpell, 1936; Mallory, 1959) that

Figure 2. Species diversity of Tertiary Pectinidae for California, the Pacific Northwest, and Alaska. Subdivisions of Tertiary series are based on the provincial molluscan stages shown in Figure 1.

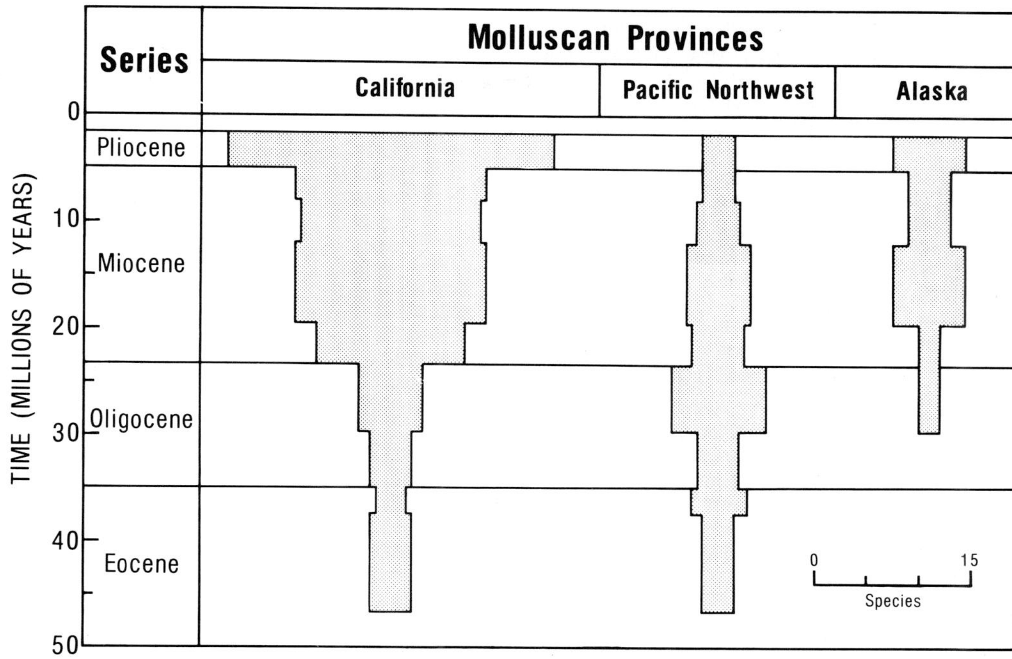

is most effectively utilized in the outer sublittoral and bathyal facies. The use of mollusks in age determination and refined correlation of the deep-water facies has been far less successful than benthic foraminifers, although Turrid gastropods are useful in lower Tertiary biostratigraphy (Hickman, 1977).

Generally speaking, the Paleogene was a period of widespread marine deposition and extensive development of moderately deep to deep-water facies in the Pacific Northwest region. Consequently, foraminiferal zonations are more useful than mollusks in biostratigraphic correlation of Paleogene formations. During the Neogene, marine embayments became more localized; deposition was predominantly of shallow-water (sublittoral) aspect. Mollusks, including pectinid bivalves, are especially useful in reginal biochronology of the Neogene.

PECTINID DIVERSITY

During the early and middle Eocene of the Pacific Northwest province, pectinid diversity was relatively low (Fig. 1). The low diversity during the early Cenozoic is a reflection of a general worldwide tendency for much lower levels of pectinid diversity before the late Oligocene or early Miocene. In Oregon and Washington, both deep- and shallow-water depositional facies are well represented in the Paleogene. The greatest diversity of pectinids occurred during the late Oligocene and again during the early and middle Miocene (Fig. 1). These apparent flourishings are atypical in comparison with the data from other regions in the Northern Hemisphere, which show a Miocene peak in species diversity following very low levels during the Paleogene (Fig. 2; Masuda, 1962, 1976). These two peaks can seemingly be explained by the fact that both deep-water muddy substrates and shallow-water sandy substrates, each with their distinctive pectinid species, were widely developed during these time periods. The uniquely low diversity of species in the post-middle Miocene stages of the marine Tertiary of this province also could be attributed to the very limited distribution of marine strata of this age in onshore marine basins of coastal Oregon and Washington and the resulting problem of obtaining fully representative samples.

Examination of diversity trends in California, to the south, and on the Pacific coast of Alaska, to the north (Fig. 2), may be helpful in interpreting the Tertiary record of the Pectinidae in coastal Oregon, Washington, and British Columbia. The somewhat higher level of diversity during the late Oligocene in the Pacific Northwest province probably is due to the widespread development of marine facies of this age. In California, marine strata of this age are scant, the principal deposition having been of non-marine units (Sespe Formation and a few other units). During the Miocene and Pliocene, however, the gamut of facies was widely deposited in California. This, coupled with a general worldwide tendency for markedly higher species diversity in Neogene rocks (Csepreghy-Meznerics, 1960; Masuda, 1962; Ctyroky, 1966; Demarcq and Barbillat, 1971; Addicott, 1974; Steininger and others, 1976), seems to be responsible for the sharp change at or near the Oligocene/Miocene boundary. A similar change is noted in the pectinid faunas of Alaska (Fig. 2). This change in diversity is not apparent in the Pacific Northwest, for reasons already indicated.

In comparing levels of specific diversity between adjacent molluscan provinces along the Pacific Coast of North America, it is important to keep in mind the modern tendency for high levels of taxonomic diversity within the Mollusca to be correlated with low-latitude warm-water faunas (Stehli and others, 1967). In the Pectinidae, for example, there are approximately 30 species now living in the Panamic molluscan province, 17 living in the northern part of the Californian molluscan province off southern California, and 11 in the Oregonian molluscan province along the Oregon and Washington coasts (Dall, 1921; Grau, 1959; Bernard, 1970; Keen, 1971). This relationship helps to explain the higher diversity of pectinids in the warm-water Miocene faunas of California as contrasted with the much lower diversity in the cooler water Miocene faunas of the Pacific Northwest and coastal Alaska (Fig. 2).

PECTINID BIOGEOGRAPHY

Paleogene Occurrences

Eocene mollusks of the middle latitudes of the eastern North Pacific margin have, for the most part, such widespread distributions that a single molluscan province extending from California to the Gulf of Alaska has been recognized (Addicott, 1977). Foraminiferal distributions during the Paleogene were also quite broad; Tipton (1976), for example, recognized a single benthic foraminiferal province extending from southern California to northwestern Washington.

Pre-late Oligocene molluscan faunas usually are devoid of pectinids. When present, the pectinids normally are represented by species of *Chlamys*, *Propeamussium*, or *Delectopecten* (Fig. 3). Aside from certain species of the mud pecten *Propeamussium—P. vacavillensis* (Palmer) and *P. stanfordensis* (Arnold)—the widely scattered Paleogene pectinids of this region tend to be unique species with limited distribution, often known only from their type areas. The large, irregularly ribbed genus *Vertipecten*—characterized by a convex left valve and a flat, or nearly flat, right valve—apparently evolved from *Chlamys* during the late Eocene (MacNeil, 1967). This genus continued into the early Miocene but became extinct in the Pacific Northwest before latest early Miocene time (Newportian). During the late Oligocene, these genera are joined by *Leochlamys* and *Leptopecten* (Fig. 3).

Neogene

During the late Oligocene, the widespread Paleogene molluscan province began to break up. By latest Oligocene and early Miocene time, molluscan faunas had developed geographically limited distributional patterns owing to an intensification of the latitudinal thermal gradient that began at or near the Oligocene-Miocene boundary (Addicott, 1970). As a consequence, three Neogene molluscan provinces can be recognized in the eastern North Pacific region: the California, Pacific Northwest, and Gulf of Alaska molluscan provinces (Fig. 4) (Addicott, 1976a). This breakup is manifested by the greatly increased difficulty of faunal correlation between the respective geographic areas during the latest part of the Oligocene. The Pacific Northwest Neogene molluscan faunal province extended roughly from the northwestern border of Canada (lat 55°N) to northernmost California (lat 40.5°N) (Fig. 4).

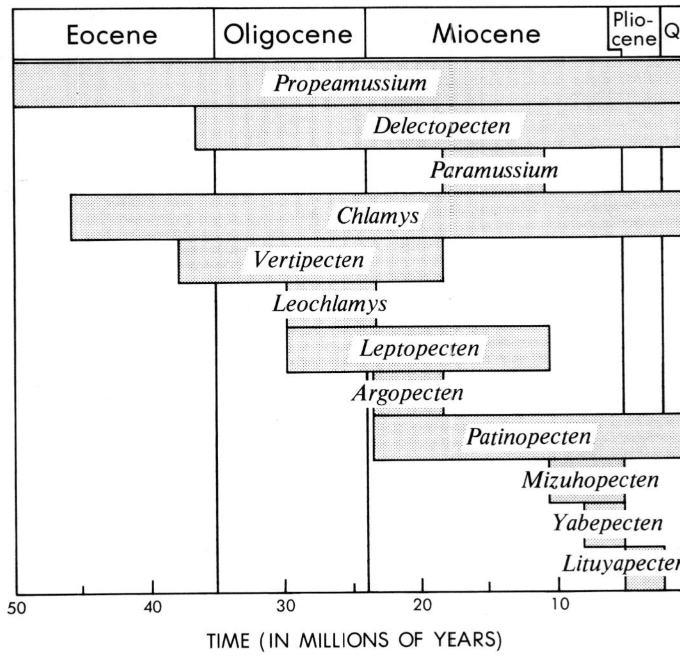

Figure 3. Stratigraphic ranges of pectinid genera in the Cenozoic rocks of the Pacific Northwest.

Probably the most significant evolutionary and biogeographic event among the Pectinidae during the Neogene was the appearance and diversification of giant pectinids. These species attain at least 90 mm in diameter as adult speciments (Addicott, 1974). They appear at or near the Oligocene/Miocene boundary in the Pacific Northwest, as exemplified by the occurrence of *Vertipecten fucanus* (Dall) in the lower part of the Pillarian Stage (Addicott, 1976a). Most of the giant pectinids are warm-water taxa that are limited in distribution to California. There are, however, two genera that are found in warm-temperate to cool-water fossil faunas: *Vertipecten* and *Patinopecten* (*Patinopecten* s.s. and *Lituyapecten*). Species of these genera serve as useful biochronologic indices of the Neogene molluscan stages of this province (Fig. 5). Interestingly, these indicator species are not closely related phylogenetically; their mutually exclusive stratigraphic ranges are fortuitous as is the case with the upper Eocene to lower Miocene *Acila* biozones of Schenck (1936): *A. shumardi* and A. *gettysburgensis*.

Another important biogeographic event in western Washington was the arrival, during the late Miocene, of two Japanese pectinid genera, *Mizuhopecten* and *Yabepecten* (Fig. 3). These genera permit correlation with the western North Pacific and, possibly in the case of *Yabepecten*, the Gulf of Alaska (Masuda and Addicott, 1970; Masuda, 1976).

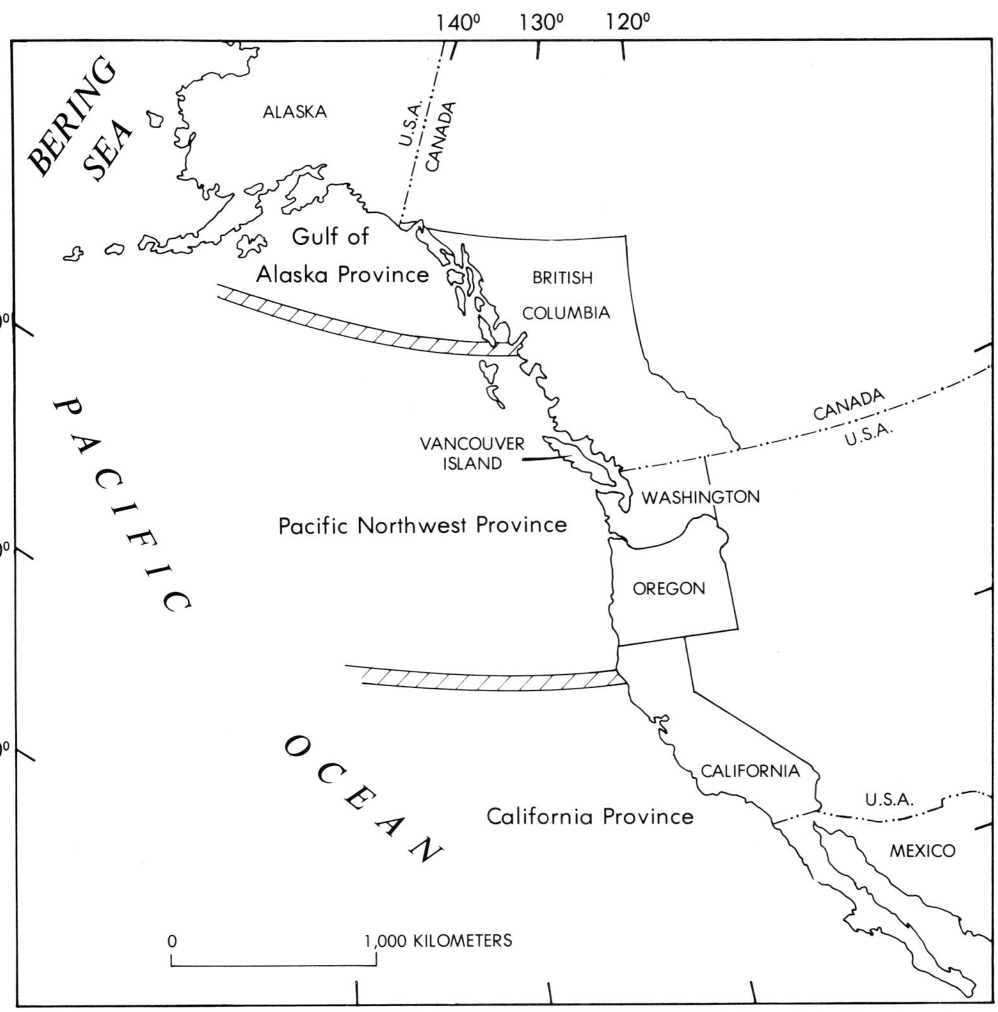

Figure 4. Neogene shallow-water molluscan provinces of the eastern North Pacific Ocean.

STRATIGRAPHIC OCCURRENCE OF PECTINIDS

This discussion is made with reference to the Tertiary molluscan stages indicated in Figure 1. It is mostly based on those species that are characteristic of the shelly, inner sublittoral facies of Tertiary rocks in the Pacific Northwest. The small mud pectens characteristic of deep-water environments may prove useful in biostratigraphic zonation, but at present their stratigraphic occurrences are poorly documented. At least a few of the species have substantially longer geologic ranges than their shallow-water counterparts: *Delectopecten peckhami* (Gabb), Juanian to Graysian Stages; *D. vanwinklae* (Clark), Galvinian and Matlockian Stages of Armentrout (1975 and this volume); *Propeamussium clallamensis* (Arnold), Matlockian to Pillarian Stages; and *P. waylandi* (Arnold), Matlockian and Juanian Stages. [The Matlockian Stage as now recognized by Armentrout (this volume) is restricted to the lower part of his originally defined unit (Armentrout, 1975). Thus the Matlockian (early Oligocene) and Juanian (late Oligocene and earliest Miocene) Stages are now successive, nonoverlapping units (see Armentrout, this volume).] The mud pectens are most commonly represented in Paleogene rocks, possibly in part owing to the prevalence of deep-water fine-grained clastic facies. Many species have been recognized in strata of late Eocene and Oligocene age (Arnold, 1906; Clark and Arnold, 1923; Clark, 1925; Slodkevich, 1938; Durham, 1944; Warren and others, 1945; Vokes and others, 1949; Snavely and Vokes, 1949; Snavely and others, 1958; Moore, 1963; Moore and Vokes, 1963; Wolfe and McKee, 1972; Jeletsky, 1975a).

"Tejon" Stage

Shallow-water pectinids have their lowest occurrence in strata referable to the middle and lower upper Eocene "Tejon" Stage (Fig. 1) in southwestern Washington. The three known species are referable to the genus *Chlamys*: *C. cowlitzensis* (Weaver), *C. landesi* (Arnold), and *C.* n. sp.

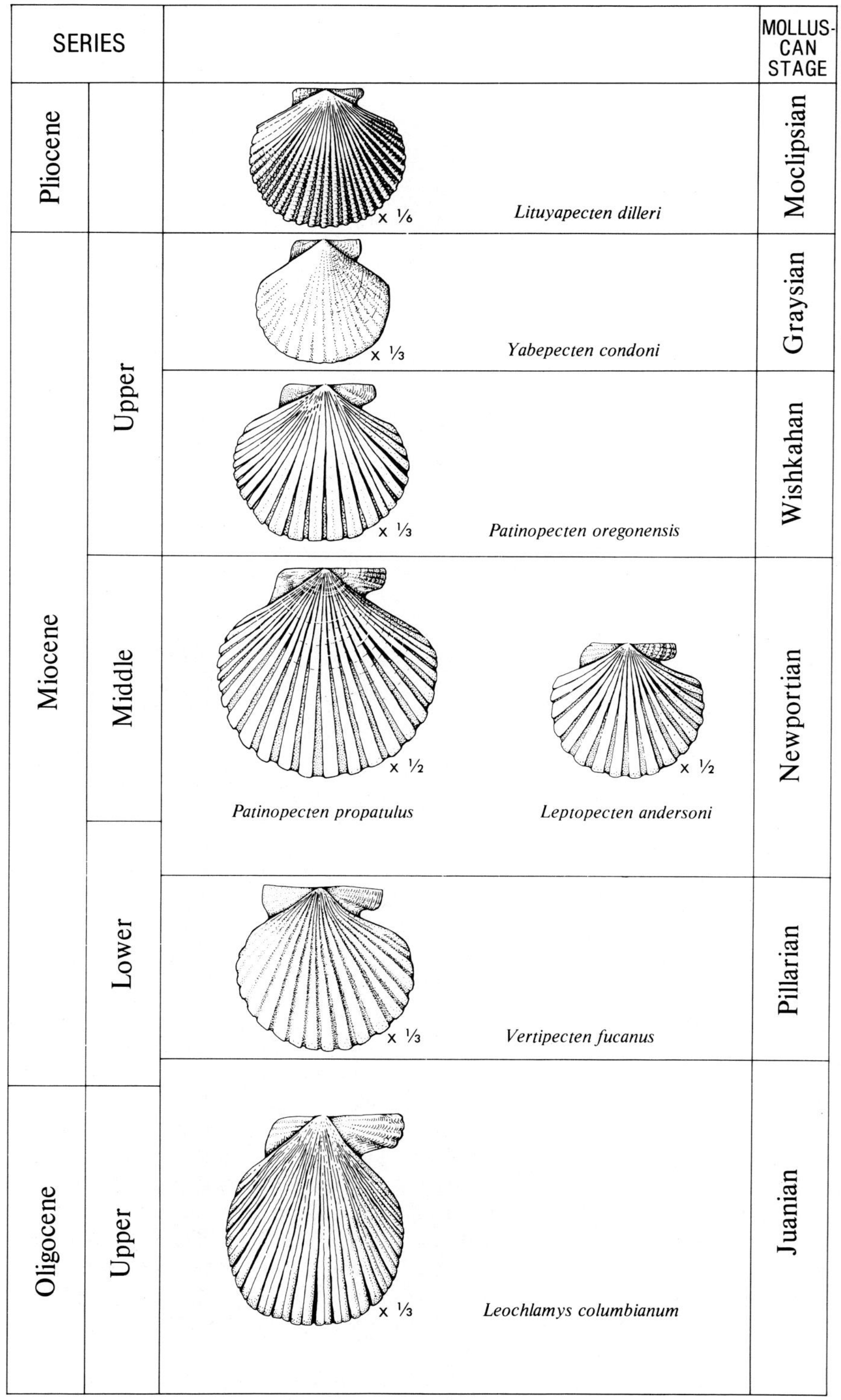

Figure 5. Range zones of some late Paleogene and Neogene pectininds that are useful in provincial biochronology.

Vokes (in Snavely and others 1958). These species have very limited geographic distribution, being known only from their type areas.

Galvinian Stage of Armentrout (1975)

The Galvinian Stage (= the "Keasey" and "Lincoln Stages of earlier reports) contains four species of pectinids: *Chlamys washburnei* (Arnold), *C. grunskyi* (Hertlein), *C.* n. sp. A, and *Vertipecten porterensis* (Weaver).

Chlamys washburnei, a large species, is known from the basal part of an unnamed basaltic sandstone of early Refugian age (Snavely and MacLeod, 1974) that overlies the Yachats Basalt on the central Oregon coast. The locality was once thought to be of Pliocene age (Arnold, 1906) but subsequently was shown to be of Paleogene age (Snavely and Vokes, 1949; Moore and Vokes, 1963). According to Snavely (June 1977, oral commun.), the underlying basalt has a radiometric age of about 36 to 38 m.y. Previously unreported localities for *C. washburnei* include basaltic sandstone near Heceta Head, about 12 mi south of Yachats (USGS locs. M1659, M2125)[1] and exposures of the Nestucca Formation near the mouth of the Nestucca River, northwestern Oregon (USGS loc. M2128). Smaller, incomplete specimens with fewer radial ribs than the type of *C. washburnei* (Pl. 1, fig. 4) occur in a basal conglomerate of the Twin River Formation near Sequim, northwest Washington (USGS loc. M2299). These specimens, and the material from USGS locality M1659 on the Oregon coast, suggest that the medial ribbing in *C. washburnei* may be highly variable.

Chlamys n. sp. A (Pl. 1, figs. 1, 6, 7) is from Pilchuck Creek, a few miles north of Arlington, Skagit County, Washington (USGS loc. M6523). This large *Chlamys* has scaly ribs that are much higher than those of *C. washburnei*. It occurs with *Acila shumardi* (Dall), *Echinophoria dalli* (Dickerson) of Durham (1942), *Terebratalia transversa* Sowerby, *Trochita sookensis* (Clark and Arnold), and other species that support a latest "Keasey" or earliest "Lincoln" Age, that is, a position in the middle part of the Galvinian Stage of Armentrout (1975).

Vertipecten porterensis occurs in strata referable to the upper part of the Galvinian Stage exposed near Oakville and Porter, southeastern Grays Harbor County, Washington. This species occurs in the Escalante Formation of the west coast of Vancouver Island (lat 49.5°N) (Jeletsky, 1975a), and it has been doubtfully identified from the Hesquiat Formation of the same area (Jeletsky, 1975b).

Matlockian Stage

Pectinids from the Matlockian Stage (= lower part of the "Blakeley" Stage) are principally small species of the deep-water genera *Delectopecten* and *Propeamussium*. Only one larger species is known from the Matlockian—*Pecten* sp., "a fragment of a valve of a large Pecten" (Durham, 1944, p. 138) from the Blakeley Formation of Weaver (1912) near Seattle, Washington. The paucity of larger species of pectinids in the Matlockian can be ascribed to the predominance of deep-water facies.

Juanian Stage

Pectinids from the Juanian Stage (late Oligocene) include *Leochlamys columbianum* (Clark and Arnold) and *Chlamys cornwalli* (Clark and Arnold), both from the Sooke Formation of the southwest coast of Vancouver Island, British Columbia. An incomplete specimen of *Leptopecten* is known from the upper part of the Twin River Formation, northern Olympic Peninsula, Washington (Addicott, 1976b). A small *Chlamys* from an unnamed marine tuff and sandstone unit near Scotts Mills (Peck and others, 1964), about 50 km south of Portland, Oregon, has been variously identified as *Pecten sespeensis* (Arnold) (Durham and others, 1942) and *Chlamys* aff. *C. columbianum* (Clark and Arnold) (Vokes in Peck and others, 1964). This pectinid (Pl. 1, figs. 3, 5) is here regarded as an undescribed species of *Chlamys*; it is represented in collections from two USGS localities (21347 and M2037). It occurs with *Bruclarkia acuminata* (Anderson and Martin) and *Liracassis* cf. *L. apta* (Tegland) in the collection from USGS locality 21347 (Vokes in Peck and others, 1964).

The peak in taxonomic diversity registered during the Juanian Stage (fig. 1) is an indication of the occurrence of both shallow and moderately deep to deep-water facies, each with its distinctive pectinid associations. In addition to the species listed above, five species of mud pectens are described or recorded from this stage.

Pillarian Stage

Vertipecten fucanus (Dall), perhaps the best index species for the Pillarian Stage, ranged from the head of the Gulf of Alaska through western Washington and Oregon to the southern part of the San Joaquin basin, California, as discussed elsewhere in this paper. This species (Pl. 2, fig. 1) marks the highest stratigraphic occurrence of *Vertipecten* in the Tertiary rocks of the Pacific Northwest (Fig. 3). Pre-Miocene species of *Vertipecten* are not included with the giant pectinids owing to their relatively small adult size (Addicott, 1974). *Argopecten neahensis* (Arnold), a distinctive but rare species, is known only from the type area in the Clallam Formation of the northern margin of the Olympic Peninsula, Washington (Addicott, 1976c).

Newportian Stage

Shallow-water pectinids are most abundant in provincial upper lower and middle Miocene rocks. The most commonly occurring species is *Patinopecten propatulus* (Conrad). This pectinid (Pl. 3, fig. 1) ranges from western Washington (Wolfe and McKee, 1972) to the San Joaquin basin of central California (Adegoke, 1969). Another pectinid restricted to this stage is *Leptopecten andersoni*

[1]U.S. Geological Survey locality descriptions are on file at Menlo Park, California.

(Arnold). This species (Pl. 2, fig. 2) ranges from the Grays Harbor basin of western Washington (Etherington, 1931; as *Pecten andersoni clemonensis*) to the Los Angeles basin, southern California (Vedder in Headlee and others, 1970) in the Newportian and the coeval "Temblor" Stage (Addicott, 1972) of California. Other species from the Newportian Stage include the giant pectinid *Patinopecten oregonensis cancellosus* Moore (Pl. 2, fig. 7) from unnamed Miocene rocks dredged from Coos Bay, southwestern Oregon, and *Leptopecten alockmanensis* (Weaver) from the Astoria (?) Formation of southwestern Washington. Both of these distinctive pectinids, however, are known only from the type localities and are therefore not particularly useful in provincial biochronology.

Wishkahan Stage

Patinopecten oregonensis (Howe) (Pl. 3, fig. 3), the giant pectinid that serves as the index species for the Wishkahan Stage, is abundant at localities in the Empire Formation at Coos Bay. It was reported from the Eel River basin of northwestern California (Howe, 1922) but was not found in subsequent studies of that basin (Stewart and Stewart, 1949; Faustman, 1964). It ranges northward from Coos Bay to the Grays Harbor basin of western Washington where it occurs in the Montesano Formation of Weaver (1912) as defined by Fowler (1965) (Addicott, 1976a). *Chlamys hodgei* (Hertlein) (Pl. 3, fig. 2; Pl. 4, figs. 3, 4) occurs in the basal part of the Empire Formation at the southwest tip of Cape Blanco (USGS loc. M3965). This species occurs in both the "Margaritan" and "Jacalitos" Stages, successive upper Miocene units in the California Coast Ranges. It was recorded from the Pancho Rico Formations of the Salinas Valley, central California, as *Chlamys* n. sp. aff. *C. nipponensis* Kuroda by Durham and Addicott (1965) (Pl. 3, figs. 4, 5).

Another species from the Empire Formation at Coos Bay, *Patinopecten coosensis* (Shumard) (Pl. 4, fig. 1), is characterized by numerous radial ribs (as many as 30 per adult valve) that are deeply undercut or somewhat T-shaped in cross section (Trumbull, 1958). Weaver (1942) reported this species from his Montesano Formation in southwest Washington, but subsequent study of this formation and its molluscan fauna (Addicott, 1976a) has not confirmed this occurrence. There is another unverified report from near Scammons Lagoon, on the west coast of Baja California (Jordan and Hertlein, 1926). This species was doubtfully assigned to the northern genus *Lituyapecten* by MacNeil (1961) and was unequivocally assigned to that genus by Hertlein and Grant (1972). The species, however, lacks the frills on the ribs of the left valve that characterize *Lituyapecten*, which probably led MacNeil (1961) to regard it as a possibly aberrant member of that genus.

Mizuhopecten warreni Masuda (Pl. 4, figs. 2, 5), a moderately large species closely allied to *Patinopecten*, but lacking the auricular crurae of that genus (Masada, 1971), appears in the Wishkahan Stage (USGS loc. M2978 and Stanford University loc. NP 243) and ranges upward into the lower part of the Graysian Stage (USGS loc. M1551). The right valve of this species resembles *Yabepecten condoni* (Hertlein) in having a similar rib count. The two can be differentiated by the more strongly developed ribbing on right valves of *M. warreni*. Moreover, the posterior segment of the growth lines on the right anterior ear of *M. warreni* are recurved toward the beak whereas on *Y. condoni* they are more nearly vertical and meet the hinge line at almost 90°.

Graysian Stage

Exposures of the Graysian Stage are limited to the Grays Harbor basin of western Washington. *Yabepecten condoni* (Hertlein) (Pl. 2, figs. 3, 4) is one of the principal index fossils for this stage. This distinctive, almost smooth species occurs commonly in the middle part of the Montesano Formation as recognized by Fowler (1965) in western Washington. This species was doubtfully recognized from the upper part of the Yakataga Formation of the Malaspina District, Alaska (Masuda and Addicott, 1970). *Mizuhopecten warreni* Masada, another sublittoral pectinid, occurs in strata at Hoquiam, Washington (USGS loc. M1551), referable to the Graysian Stage. This species (Pl. 4, fig. 2) appears in the underlying Wishkahan Stage in southwestern Washington.

Moclipsian Stage

The Pliocene species *Lituyapecten dilleri* (Dall) (Pl. 5), discussed elsewhere in this paper, occurs in the Quinault Formation of coastal Washington (Addicott, 1976a). Also occurring in this formation are undetermined species of *Patinopecten* and *Chlamys* (unpub. records from USGS locs. 4043 and M3019). The few species contrast markedly with the extremely diverse pectinid fauna of California Pliocene rocks (Fig. 2) and, apparently, of Japan (Masuda, 1976). The abnormally low number of pectinid taxa presumably is a manifestation of the very limited distribution of marine Pliocene strata in the Pacific Northwest.

INTERPROVINCIAL CORRELATION

A few widely distributed species within well-known plylogenetic lineages can be utilized in interprovincial correlation. These kinds of pectinids offer perhaps the best hope for extending the provincial molluscan biochronologies of the eastern North Pacific westward to Kamchatka, Sakhalin, and Japan in the western North Pacific, and southward from California to the equatorial latitudes of Central and South America. Some of the possibilities of finding and developing long-range faunal correlations based on the Pectinidae are considered by Masuda (1976). The Tertiary molluscan paleontology of the Pacific coast of Alaska is, as yet, inadequately known, particularly in comparison with the status of knowledge of the Japanese and conterminous United States sequences. Fortunately, the richness of these essentially undocumented marine

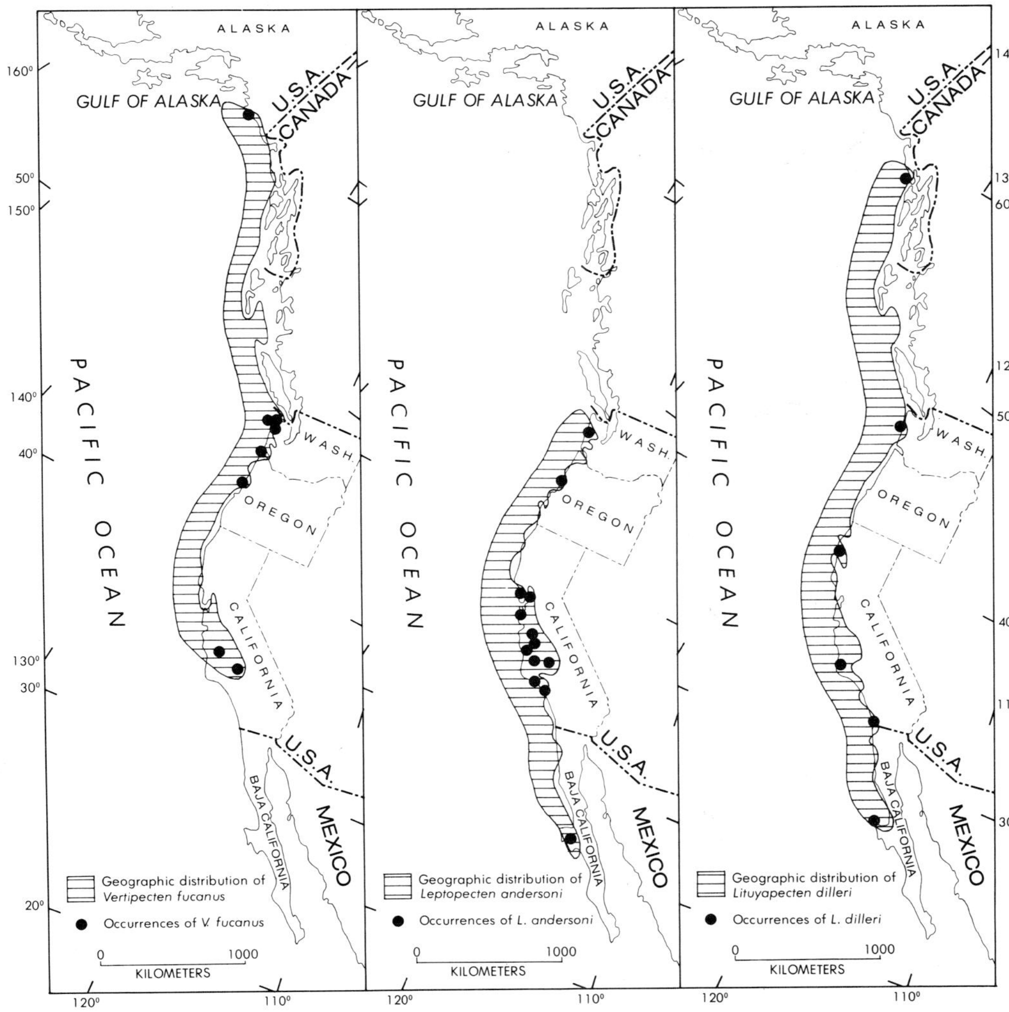

Figure 6. Early Miocene biogeography of *Vertipecten fucanus* (Dall), late early and middle Miocene biogeography of *Leptopecten andersoni* (Arnold), and Pliocene biogeography of *Lituyapecten dilleri* (Dall).

faunas and their geographic significance has led to an intensification of biostratigraphic research (see, for example, Allison, 1977a, 1977b, and Marincovich, 1976). Thus, this weak link in the circum-North Pacific chain of Cenozoic molluscan biostratigraphic sequences is being strengthened, and significant advances in interregional correlation may soon be realized. Correlation with South America is likewise impeded by the very incomplete knowledge of Tertiary faunas of western Mexico and Central America.

Three examples of species with broad geographic ranges and restricted geologic records can be given. These species are useful in interprovincial correlation along the Pacific coast of North America.

Vertipecten fucanus (Dall), an early Miocene species that ranges from southern California to the Gulf of Alaska (Fig. 6), marks the appearance of giant pectinids in the Tertiary of the Pacific Northwest and Alaska. This pectinid is characteristic of the Clallam Formation of northwest Washington, but it also occurs in the lower part of the Astoria Formation of southwestern Washington (Wolfe and McKee, 1972), the Hoh rock assemblage of Rau (1973) of the western Olympic Peninsula, the Nye Mudstone of northwestern Oregon (Addicott, 1976c), and localities in the Temblor Formation (Stewart, 1946) and Jewett Sand (J. T. Smith, 1976, oral commun. based on material in U.S. Geological Survey collections) in the San Joaquin basin, California. It has been found as far north as the head of the Gulf of Alaska in the upper part of the Poul Creek Formation (Addicott, 1977).

Another Miocene pectinid which extends southward into the California Neogene molluscan province is *Leptopecten andersoni* (Arnold) (Fig. 6). This species has a well-known record in southern and central California (Arnold, 1906;

1
2
3
4
5
6
7

PLATE 1 (facing page)

Eocene and Oligocene Pectinids

Figure

1. ***Chlamys* n. sp. A. USNM (U.S. National Museum) 252499, right valve, ×1. Unnamed strata referable to the Galvinian Stage (upper Eocene), Pilchuck Creek, Skagit County, Washington, USGS loc. M6523.**

2. ***Chlamys cornwalli* (Clark and Arnold). USNM 252500, right valve, ×1. Sooke Formation, Juanian Stage (upper Oligocene), near mouth of Kirby Creek, southwest Vancouver Island, British Columbia, USGS loc. M4060.**

3. ***Chlamys* n. sp. B. USNM 252501, right valve, ×1.5. Unnamed marine tuff and sandstone referable to the Juanian Stage (upper Oligocene), Butte Creek, Marion County, Oregon, USGS loc. M2037.**

4. ***Chlamys* cf. *C. washburnei* (Arnold). USNM 252502, ×1. Basal conglomerate and sandstone of the Twin River Formation, Galvinian Stage (upper Eocene), Sequim quadrangle, Clallam County, Washington, USGS loc. M2299.**

6. ***Chlamys* n. sp. A. USNM 252504, right valve, ×1. Unnamed strata referable to the Galvinian Stage (upper Eocene), Pilchuck Creek, Skagit County, Washington, USGS loc. M6523.**

7. ***Chlamys* n. sp. A. USNM 252505, left valve, ×1. Unnamed strata referable to the Galvinian Stage (upper Eocene), Pilchuck Creek, Skagit County, Washington, USGS loc. M6523.**

1
2
3
4
5
6
7

PLATE 2 (facing page)

Miocene Pectinids

Figure

1. *Vertipecten fucanus* (Dall). Latex cast of holotype, USNM 107790, left valve, ×1. Clallam Formation, Pillarian Stage (lower Miocene). On coast 11,000 ft southeast of Slip Point, Clallam County, Washington, USGS loc. 2464.

2. *Leptopecten andersoni* (Arnold). USNM 252506, right valve, ×1. Astoria Formation, Newportian Stage (lower and middle Miocene), Boiler Bay, Lincoln County, Oregon, USGS loc. 16005.

3. *Yabepecten condoni* (Hertlein). USNM 646455, right valve, ×1. Upper part of the lower member of the Montesano Formation of Fowler (1965), Graysian Stage (upper Miocene), Middle Fork of the Wishkah River, Grays Harbor County, Washington, USGS loc. M3039.

4. *Yabepecten condoni* (Hertlein). Stanford University (SU) Paleontological Type Collection 22340, left valve, ×1. Upper part of the lower member of the Montesano Formation of Fowler (1965), Graysian Stage (upper Miocene), West Fork of the Wishkah River, Grays Harbor County, Washington, SU loc. NP244.

5. *Delectopecten peckhami* (Gabb) Arnold. USNM 563247 left valve, ×1.5. Astoria Formation, Newportian Stage (lower and middle Miocene), Tillamook quadrangle, Tillamook County, Oregon, USGS loc. 15290.

6. *Delectopecten peckhami* (Gabb) Arnold. USNM 563244, right valve, ×2. Astoria Formation, Pillarian Stage (lower Miocene), no locality number.

7. *Patinopecten oregonensis cancellosus* Moore. USNM 252507, ×1. Astoria Formation, Newportian Stage (lower and middle Miocene), Coos Bay, Coos County, Oregon, USGS loc. 18284.

1

2

3

4

5

PLATE 3 (facing page)

Miocene Pectinids

Figure

1. *Patinopecten propatulus* (Conrad). USNM 563236, right valve, ×2. Astoria Formation, Newportian Stage (lower and middle Miocene), seacliff near mouth of Spencer Creek, Lincoln County, Oregon, USGS loc. 18885.

2. *Chlamys hodgei* (Hertlein). USNM 252508, right valve, ×1. Basal part of the Empire Formation, Wishkahan Stage (upper Miocene), south side of Cape Blanco, Curry County, USGS loc. M3965.

3. *Patinopecten oregonensis* (Howe). USNM 252509, right valve, ×1. Empire Formation, Wishkahan Stage (upper Miocene), Coos Head, Coos County, Oregon, USGS loc. M2137.

4. *Chlamys hodgei* (Hertlein). University of California Museum of Paleontology (UCMP) 14585, right valve, ×1. Pancho Rico Formation, "Jacalitos" Stage (upper Miocene), UCMP loc. A906.

5. *Chlamys hodgei* (Hertlein). USNM 254388, right valve, ×1. Pancho Rico Formation, "Jacalitos" Stage (upper Miocene), USGS loc. M902.

1
2
3
4
5

PLATE 4 (facing page)

Miocene and Pliocene Pectinids

Figure

1. ***Patinopecten coosensis*** **(Shumard). USNM 252510, right valve, ×1. Empire Formation, Wishkahan Stage (upper Miocene), Fossil Point, Coos County, Oregon, USGS loc. M2134.**

2. ***Mizuhopecten warreni*** **(Masuda). USNM 649124, left valve, ×1. Upper part of the lower member of the Montesano Formation of Fowler (1965), Graysian Stage (upper Miocene), Hoquiam, Grays Harbor County, Washington, USGS loc. M1551.**

3. ***Chlamys hodgei*** **(Hertlein). USNM 252511, right valve, ×1. Empire Formation, Wishkahan Stage (upper Miocene), inland exposure near Cape Blanco, Curry County, Oregon, USGS loc. M3961.**

4. ***Chlamys hodgei*** **(Hertlein). USNM 252512, left valve, ×1. Basal part of the Empire Formation, Wishkahan Stage (upper Miocene), south side of Cape Blanco, Curry County, Oregon, USGS loc. M3965.**

5. ***Mizuhopecten warreni*** **Masuda. Holotype. USNM 646491, right valve, ×1. Lower part of the lower member of the Montesano Formation of Fowler (1965), Wishkahan Stage, West Fork of the Wishkah River, Grays Harbor County, Washington, USGS loc. M2978.**

1

PLATE 5. Pliocene Pectinid

***Lituyapecten dilleri* (Dall). USNM 252514, right valve, ×1. Wildcat Group, Moclipsian Stage (Pliocene), Thompkins Hill, Humboldt County, California, USGS loc. M7258.**

Loel and Corey, 1932; Adegoke, 1969). It is also known from Baja California Sur (Minch and others, 1976). This species is less well known in the Pacific Northwest but has been recorded from localities in southwestern Washington (Etherington, 1931) and coastal Oregon (Moore, 1963) as *L. andersoni clemonensis* (Etherington). Because this small species belongs to a warm-water genus, it seems possible that it may have extended ever farther southward in Central America during the late early and middle Miocene.

Lituyapecten dilleri (Dall), a Pliocene species, has the widest distribution of these three examples, extending from the head of the Gulf of Alaska (MacNeil, 1961) to the West Coast of Mexico (lat 26°N) (Arnold, 1906) (Fig. 6). Its northernmost occurrence of the Washington Coast is in the Quinault Formation (Addicott, 1976a). The biogeographic distribution of *Lituyapecten* suggests that it was a temperate to cold-water genus (Addicott, 1974). Thus, its occurrence at about lat 26°N, an area that today lies at the boundary between the subtropical and warm temperate molluscan provinces (Hall, 1964), indicates that cool-water conditions extended into low latitudes during the Pliocene. The genus became extinct before the close of the Pliocene, but its modern counterpart, *Patinopecten*, is limited today to cool-temperate and cold-water provinces of the eastern North Pacific north of about lat 38°N.

Other pectinid genera and species are potentially useful

in interregional correlations with other sectors of the Pacific. Masuda (1976), for example, has demonstrated possibilities for trans-North Pacific correlation with the late Miocene genera *Yabepecten* and *Mizuhopecten*, as well as with other pectinids that occur in southeastern Alaska that have not, as yet, been found as far south as Oregon, Washington, or British Columbia.

SUMMARY

Pectinid bivalves are useful in age determination and correlation of nearshore marine strata of Tertiary age in the Pacific Northwest region. They are especially helpful in biochronologic subdivision of the Neogene owing to the prevalence of shallow-water deposition and the relatively rapid evolutionary diversification at that time. There are two basic kinds of Cenozoic pectinids—relatively large, thick-shelled, strongly sculptured species characteristic of inner sublittoral environments and small, relatively smooth, thin-shelled species characteristic of outer sublittoral and bathyal environments. Shallow-water species, which make up about 75% of the known pectinids of this region, are relatively well studied, whereas the small mud pectens are still poorly known taxonomically, and data on their stratigraphic distribution are likewise very limited.

During the early and middle parts of the Paleogene, pectinid diversity was relatively low. The known taxa were almost exclusively species of *Chlamys* and various mud pectens. Taxonomic diversification and the widespread development of both shallow- and deep-water marine facies during the latest part of the Paleogene led to a peak in species diversity.

The Neogene is characterized by the appearance of giant pectinids which form a series of consecutive range zones that are valuable in provincial biochronology: *Vertipecten fucanus* (Dall), Pillarian Stage; *Patinopecten propatulus* (Conrad), Newportian Stage; *P. oregonensis* (Howe), Wishkahan Stage; *Yabepecten condoni* (Hertlein), Graysian Stage; and *Lituyapecten dilleri* (Dall), Moclipsian Stage. Some of these species are important in interprovincial correlation owing to their occurrence in two or more eastern North Pacific Neogene faunal provinces. The appearance of the Japanese genera *Mizuhopecten* and *Yabepecten* during the late Neogene provides a means of interregional correlation with the western North Pacific.

ACKNOWLEDGMENTS

The critical reading of early versions of this report by John M. Armentrout, Carole S. Hickman, Scott McCoy, and C. A. Repenning is gratefully acknowledged. Fossil photography is by Kenji Sakamoto. Line drawings of fossil specimens are by Sarah Boore.

REFERENCES CITED

Addicott, W. O., 1970, Latitudinal gradients in Tertiary molluscan faunas of the Pacific coast: Palaeogeography, Palaeoclimatology, Palaeoecology, v. 8, p. 287-312.

——1972, Provincial middle and late Tertiary molluscan stages, Temblor Range, California, *in* Symposium on Miocene biostratigraphy of California: Society of Econonic Paleontologists and Mineralogists, Pacific Section, Bakersfield, California, March 1972, p. 1-26, Pls. 1-4.

——1974, Giant pectinids of the eastern North Pacific margin: Significance in Neogene zoogeography and chronostratigraphy: Journal of Paleontology, v. 48, p. 180-194, 2 pls.

——1976a, Neogene molluscan stages of Oregon and Washington, *in* Neogene symposium: Society of Economic Paleontologists and Mineralogists, Pacific Section, San Francisco, California, April 1976, p. 95-115, Pls. 1-5.

——1976b, New molluscan assemblages from the upper member of the Twin River Formation, western Washington: Significance in Neogene biostratigraphy: U.S. Geological Survey Journal of Research, v. 4, p. 437-447, 7 figs.

——1976c, Molluscan paleontology of the early Miocene Clallam Formation, northwestern Washington: U.S. Geological Survey Professional Paper 976, 44 p., 9 pls.

——1977, Neogene chronostratigraphy of nearshore marine basins of the eastern North Pacific: Northwestern Mexico to Canada: International Congress on Pacific Neogene Stratigraphy, First, Japan, 1976, Proceedings, p. 151-175.

Adegoke, O. S., 1969, Stratigraphy and paleontology of the marine Neogene formations of the Coalinga region, California: University of California Publications in Geological Sciences, v. 80, 241 p., 13 pls.

Allison, R. C., 1977a, Late Oligocene through Pleistocene molluscan faunas in the Gulf of Alaska region: International Congress on Pacific Neogene Stratigraphy, First, Japan, 1976, Proceedings, p. 313-316.

——1977b, Late Paleogene and Neogene molluscan faunas of the Gulf of Alaska region: Correlation, paleoclimatology, and paleozoogeography: Geological Society of America Abstracts with Programs, v. 9, p. 876.

Armentrout, J. M., 1975, Molluscan biostratigraphy of the Lincoln Creek Formation, southwest Washington, *in* Weaver, D. W., and others, eds., Paleogene symposium and selected technical papers: American Association of Petroleum Geologists-Society of Economic Paleontologists and Mineralogists-Society of Economic Geologists, Pacific Sections, Annual Meeting, Long Beach, California, p. 14-48.

——1981, Cenozoic geologic timetable for Oregon and Washington: Geological Society of America Special Paper 184 (this volume).

Arnold, Ralph, 1906, The Tertiary and Quaternary pectens of California: U.S. Geological Survey Professional Paper 47, 264 p., 53 pls., 2 figs.

Arnold, Ralph, and Hannibal, Harold, 1913, The marine Tertiary stratigraphy of the north Pacific Coast of America: American Philosophical Society Proceedings, v. 52, no. 212, p. 559-605.

Berggren, W. A., and Van Couvering, J. A., 1974, The late Neogene: Palaeogeography, Palaeoclimatology, Palaeoecology, v. 16, no. 1/2, p. 1-216.

Bernard, F. R., 1970, A distributional checklist of marine mollusks of British Columbia [Canada]: Based on faunistic surveys since 1950: Syesis, v. 3, p. 75-94.

Cavelier, Claude, and Pomerol, Charles, 1977, Proposition d'une echelle stratigraphique standard pour le Paleogene: Newsletter Stratigraphy, v. 6, no. 1, p. 56-65.

Clark, B. L., 1925, Pelecypoda from the marine Oligocene of western North America: University of California Publications, Department of Geological Sciences Bulletin, v. 15, no. 4, p. 69-136, Pls. 8-22.

Clark, B. L., and Arnold, Ralph, 1923, Fauna of the Sooke Formation, Vancouver Island, with description of a new coral by T. Wayland Vaughan: University of California Publications, Department of Geological Sciences Bulletin, v. 14, no. 5, p. 123-234, Pls. 15-42.

Csepreghy-Meznerics, Ilona, 1960, Pectinides du Neogene de la Hongrie et leur importance biostratigraphicque: Société Géologique de France, Mémoire 92, 56 p.

Ctyroky, P., 1966, The family Pectinidae in the lower Miocene of the Carpathian region: Committee on Mediterranean Neogene Stratigraphy, Session 3, Berne, 1966, p. 176-179.

Dall, W. H., 1921, Summary of the marine shell-bearing mollusks of the northwest coast of America: U.S. National Museum Bulletin 112, 217 p., 22 pls.

Demarcq, G., and Barbillat, R., 1971, Les Pectinides neogenes du Bassin Rhodanien: University of Lyon, Laboratoire Geologique Documents hors Serie 1971, v. 1, p. 45-59.

Durham, D. L., and Addicott, W. O., 1965, Pancho Rico Formation, Salinas Valley, California: U.S. Geological Survey Professional Paper 524-A, 22 p., 5 pls.

Durham, D. L., and Addicott, W. O., 1965, Pancho Rico Formation, Salinas Valley, California: U.S. Geological Survey Professional Paper 524-A, 22 p., 5 pls.

Durham, J. W., 1942, Notes on Pacific coast Galeodeas: Journal of Paleontology, v. 16, no. 2, p. 183-191, Pls. 29-30.

Durham, J. W., 1944, Megafaunal zones of the Oligocene of northwestern Washington: University of California Publications, Department of Geological Sciences Bulletin, v. 27, no. 5, p. 101-211, Pls. 13-18.

Durham, J. W., Harper, Herbert, and Wilder, Beverly, Jr., 1942, Lower Miocene in the Willamette Valley, Oregon [abs.]: Geological Society of America Bulletin, v. 53, p. 1817.

Etherington, T. J., 1931, Stratigraphy and fauna of the Astoria Miocene of southwest Washington: University of California Publications, Department of Geological Sciences Bulletin, v. 20, no. 5, p. 31-142, 14 pls.

Faustman, W. F., 1964, Paleontology of the Wildcat Group at Scotia and Centerville Beach, California: University of California Publications in Geological Sciences, v. 41, no. 2, p. 97-160, 3 pls.

Fowler, G. A., 1965, The stratigraphy, foraminifera, and paleoecology of the Montesano formation, Grays Harbor County, Washington [Ph.D. thesis]: Los Angeles, University of Southern California, 354 p.

Grau, G., 1959, Pectinidae of the eastern Pacific: Allan Hancock Pacific Expeditions, v. 23, p. 1-308, Pls. 1-57.

Hall, C. A., Jr., 1964, Shallow-water marine Climates and molluscan provinces: Ecology, v. 45, no. 2, p. 226-234.

Headlee, L. A., Warren, A. D., and Wildharber, T. L., 1970, Southestern rim of the Los Angeles basin, Orange County, California, *in* Newport Lagoon-San Joaquin Hills-Santa Ana Mountains geologic guidebook: American Association of Petroleum Geologists-Society of Economic Paleontologists and Mineralogists-Society of Economic Geologists, Pacific Sections, 49 p., illus.

Hertlein, L. G., and Grant, U. S., IV, 1972, The geology and paleontology of the marine Pliocene of San Diego, California (Paleontology: Pelecypods): San Diego Society of Natural History Memoir 2, pt. 2B, p. 135-411, Pls. 27-57.

Hickman, C. J. S., 1977, Molluscan paleoecology and biostratigraphy of Cenozoic deep-water facies in the Pacific Northwest: Geological Society of America Abstracts with Programs, v. 9, no. 7, p. 1014.

Howe, H. V., 1922, Faunal and stratigraphic relationships of the Empire Formation, Coos Bay, Oregon: University of California Publications, Department of Geological Sciences Bulletin, v. 14, no. 3, p. 85-114, Pls. 7-12.

Jeletsky, J. A., 1975a, Hesquiat Formation (new): A neritic channel and interchannel deposit of Oligocene age, western Vancouver Island, British Columbia: Canada Geological Survey Paper 75-32, 55 p., 6 pls.

——1975b, Age and depositional environment of the lower part of the Escalante Formation, western Vancouver Island, British Columbia (92E): Canada Geological Survey Paper 75-1C, p. 9-16, 1 pl.

Jordan, E. K., and Hertlein, L. G., 1926, Contribution to the geology and paleontology of the Tertiary of Cedros Island and adjacent parts of Lower California, *in* Expedition to the Revillagigedo Islands, Mexico, in 1925: California Academy of Sciences Proceedings, ser. 4, v. 15, no. 14, p. 409-464, Pls. 27-34.

Keen, A. M., with the assistance of McLean, J. H., 1971, Sea shells of tropical west America, marine mollusks from Baja California to Peru, second edition: Stanford, California, Stanford University Press, 1 064 p., illus.

Kleinpell, R. M., 1938, Miocene stratigraphy of California: Tulsa, Oklahoma, American Association of Petroleum Geologists, 450 p.

MacNeil, F. S., 1961, *Lituyapecten* (new subgenus of *Patinopecten*) from Alaska and California: U.S. Geological Survey Professional Paper 354-J, p. 225-237, Pls. 35-46.

——1967, Cenozoic pectinids of Alaska, Iceland, and other northern regions: U.S. Geological Survey Professional Paper 553, 57 p., 25 pls.

Mallory, V. S., 1959, Lower Tertiary biostratigraphy of the California Coast Ranges: Tulsa, Oklahoma, American Association of Petroleum Geologists, 416 p.

Marincovich, L. N., 1976, Miocene mollusks from the Topsy Formation, Lituya District [Alaska]: U.S. Geological Survey Circular 733, p. 58.

Masuda, Koichiro, 1962, Tertiary Pectinidae of Japan: University of Tohoku Science Reports, ser. 2, v. 33, no. 2, 238 p., 27 pls.

——1971, On some *Patinopecten* from North America: Palaeontological Society of Japan Transactions, n.s., no. 83, p. 166-178, Pls. 19-21.

——1972, *Swiftopecten* of the Northern Pacific: Palaeontological Society of Japan Transactions, n.s., no. 87, p. 395-408, Pls. 48, 49.

——1973, *Chlamys cosibensis* (Yokoyama) of the northern Pacific: University of Tohoku Science Reports, Series 2 (Geology), Special Volume 6 (Hatai Memorial Volume), p. 109-116, Pls. 8, 9.

——1976, Neogene Pectinidae of the Northern Pacific: International Congress on Pacific Neogene Stratigraphy, First, Japan, 1976, Abstracts of Papers, p. 140-143.

Masuda, Koichiro, and Addicott, W. O., 1970, On *Pecten* (*Amusium*) *condoni* Hertlein from the west coast of North America: Veliger, v. 13, no. 2, p. 153-156, 1 pl.

Minch, J. C., and others, 1976, Geology of the Vizcaino Peninsula [Baja California, Mexico], *in* Howell, D. G., ed., Aspects of the geologic history of the California Continental Borderland: American Association of Petroleum Geologists, Pacific Section, Miscellaneous Publication 24, p. 136-195.

Moore, E. J., 1963, Miocene mollusks from the Astoria Formation in Oregon: U.S. Geological Survey Professional Paper 419, 109 p., 33 pls. [1964].

Moore, R. C., and Vokes, H. E., 1963, Lower Tertiary crinoids from northwestern Oregon: U.S. Geological Survey Professional Paper 233-E, p. 113-148. Pls. 14-24.

Peck, D. L., and others, 1964, Geology of the central and northern parts of the Western Cascade Range in Oregon: U.S. Geological Survey Professional Paper 449, 65 p.

Rau, W. W., 1973, Geology of the Washington coast between Point Grenville and the Hoh River: Washington Department of Natural Resources, Geology and Earth Resources Division Bulletin 66, 58 p.

Ryan, W.B.F., and others, 1974, A paleomagnetic assignment of Neogene stage boundaries and the development of isochronous datum planes between the Mediterranean, the Pacific and Indian oceans in order to investigate the response of the world ocean to the Mediterranean "salinity crisis": Rivista Italiana di Paleontologia e Stratigrafia, v. 80, p. 631-688.

Schenck, H. G., 1936, Nuculid bivalves of the genus *Acila*: Geological Society of America Special Paper 4, 149 p., 18 pls.

Schenck, H. G., and Kleinpell, R. M., 1936, Refugian stage of the Pacific coast Tertiary: American Association of Petroleum Geologists Bulletin, v. 20, no. 2, p. 215-225.

Sinelnikova, V. N., 1975, Mio-Pliocene Pectinidae of Kamchatka: Academy of Sciences of the USSR, Geological Institute Transactions, v. 229, 140 p., 25 pls. [in Russian].

Slodkevich, W. S., 1938, Tertiary Pelecypoda from the Far East, v. 10, pt. 2: Academy of Sciences of the USSR, Paleontological Institute, Paleontology of USSR, v. 10, pt. 3, fasc. 19, 275 p., 106 pls.

Snavely, P. D., Jr., and others 1958, Geology and coal resources of the Centralia-Chehalis district, Washington: U.S. Geological Survey Bulletin 1053, 159 p.

Snavely, P. D., Jr., and MacLeod, N. S., 1974, Yachats Basalt—An upper Eocene differentiated volcanic sequence in the Oregon Coast Range: U.S. Geological Survey Journal of Research, v. 2, no. 4, p. 395-403.

Snavely, P. D., Jr., and Vokes, H. E., 1949, Geology of the coastal area between Cape Kiwanda and Cape Foulweather, Oregon: U.S. Geological Survey Oil and Gas Investigations Preliminary Map 97, Scale 1:62,500.

Stehli, F. G., McAlester, A. L., and Helsley, C. E., 1967, Taxominic diversity of Recent bivalves and some implications for geology: Geological Society of America Bulletin, v. 78, no. 4, p. 455-466.

Steininger, F. F., Rogl, F., and Marini, E., 1976, Current Oligocene/Miocene biostratigraphic concept of the central Paratethys (middle Europe): Newsletter Stratigraphy, v. 4, no. 3, p. 174-202.

Stewart, R. B., 1946, Geology of Reef Ridge, Coalinga district, California: U.S. Geological Survey Professional Paper 205-C, p. 81-115, Pls. 9-17.

Stewart, R. E., and Stewart, K. C., 1949, Local relationships of the Mollusca of the Wildcat coast section, Humboldt County, California, with related data on the foraminifera and ostracoda: Oregon Department of Geology and Mineral Industries Bulletin 37, pt. 8, p. 165-208, Pls. 19-22.

Tipton, Ann, 1976, Foraminiferal zonation of Refugian Stage, uppermost Eocene of California: American Association of Petroleum Geologist Bulletin, v. 60, p. 2190-2191.

Trumbull, E. J., 1958, Shumard's type specimens of Tertiary mollusks from Oregon and other types at Washington University, St. Louis: Journal of Paleontology, v. 32, no. 5, p. 893-906, Pls. 115-117.

Vokes, H. E., Norbisrath, Hans, and Snavely, P. D., Jr., 1949, Geology of the Newport-Waldport area, Lincoln County, Oregon: U.S. Geological Survey Oil and Gas Investigations (Preliminary) Map 88, scale 1:62,500.

Warren, W. C., Grivetti, R. M., and Norbisrath, Hans, 1945, Geology of northwestern Oregon, west of Willamette River and north of latitude 45°15′: U.S. Geological Survey Oil and Gas Investigations Preliminary Map 42, scale 1:145,728.

Weaver, C. E., 1912, A preliminary report on the Tertiary paleontology of western Washington: Washington Geological Survey Bulletin 15, 80 p., 15 pls.

——1916, The Tertiary formations of western Washington: Washington Geological Survey Bulletin 13, 327 p.

——1942, Paleontology of the marine Tertiary formations of Oregon and Washington: University of Washington [Seattle], Publications in Geology, v. 5, pts. 1-3, 789 p., 104 pls.

Weaver, C. E., chairman, and others, 1944, Correlation of the marine Cenozoic formations of western North America: Geological Society of America Bulletin, v. 55, p. 569-598.

Wolfe, E. W., and McKee, E. H., 1972, Sedimentary and igneous rocks of the Grays River quadrangle, Washington: U.S. Geological Survey Bulletin 1335, 70 p.

MANUSCRIPT RECEIVED BY THE SOCIETY MARCH 14, 1980
MANUSCRIPT ACCEPTED MAY 7, 1980

Printed in U.S.A.

Geological Society of America
Special Paper 184
1981

A chronologic framework for Cenozoic megafossil floras of northwestern North America and its relation to marine geochronology

JACK A. WOLFE
U.S. Geological Survey, 345 Middlefield Road, M.S. 95, Menlo Park, California 94025

ABSTRACT

Collection and analysis of fossil plants in several stratigraphic sections have produced a chronologic framework for plant-bearing rocks of the northwestern North American province. The geographic extent of this province varied during the Cenozoic, but the stage framework can generally be applied to the region west of the Rocky Mountains (including Alaska) during the Paleogene and Miocene. The occurrence of some plant megafossils in dominantly marine sections of Paleogene age provides only a general concept of relation to the marine geochronology. Numerous Neogene floras have been radiometrically dated and can thus be more accurately related to the marine geochronology. The changes in the floral assemblages are largely the result of major climatic changes. If these changes are assumed to represent secular phenomena, the provincial chronology can also be related to the Cenozoic terminology in those Northern Hemisphere regions where paleobotanical data indicate that a similar series of climatic changes occurred. Inferences based on a combination of radiometric ages, correlations of West Coast benthic foraminiferal stages to the planktonic chronologies, and marine-nonmarine correlations of West Coast units indicate that the Eocene/Oligocene boundary may be as young as 32 m.y.

INTRODUCTION

A traditional approach to plant biostratigraphy as it pertains to Tertiary rocks is that plants are far too subject to climate and other ecologic factors to be of use in geochronology. Coupled with this attitude is a nonevolutionary concept held by many paleobotanists: species and genera are too long-ranging to apply traditional biostratigraphic models. In fact, only in the last two decades have serious attempts been made to apply such traditional models to Tertiary paleobotany.

I do not mean to suggest that plants are not subject to ecologic controls and to the inherent problems such controls create relative to biostratigraphy—as is true with all other organisms. Two apparently dissimilar floras could live within a short distance of one another, and apparently similar floras could, because of convergence to a similar environment, be highly nonsynchronous.

Species that today occupy primarily streamside habitats—members, for example, of Salicaceae and Betulaceae—may range through regional vegetational types as varied as coniferous forests to broad-leaved evergreen forests to broad-leaved deciduous forests. Latitudinal ranges of 20° to 30° are not uncommon for streamside species. Fortuitously, it is exactly streamside species that are most commonly represented as plant megafossils (MacGinitie, 1953).

Gross similarities on the generic or familial levels have not provided a firm basis for biostratigraphic studies. For decades, it was not recognized that climatic fluctuations in the Eocene produced floras that were deceptively similar to floras of Oligocene or younger age, and thus a number of Eocene floras were assigned to the Neogene. In these instances, however, detailed morphologic studies indicate that markedly different species occur in Eocene versus Neogene floras.

I have, therefore, attempted to apply traditional biostratigraphic techniques to Tertiary plant megafossils (largely streamside). The result is the definition of several formal stages (Wolfe and others, 1966; Wolfe, 1968, 1977) that can be applied to plant-bearing rocks in the Pacific Coast States and Alaska and, during some periods, to the Cordilleran region. Although some controversy developed concerning synchroneity of these stages in Alaska with those in the conterminous United States (see Chaney, 1967), recent radiometric ages have substantiated the original approximate correlations of stages in Alaska with plant-bearing beds in the conterminous United States (Triplehorn and others, 1977; Turner and others, 1980).

Most of the original assignments of these stages to Tertiary series were based on the correlations of the Pacific Coast molluscan biostratigraphy with the European sections. It is now realized that these correlations were not entirely valid. Because various local chronologies are now being related to chronologies based on planktonic organisms abundant in DSDP samples (see Berggren, 1972), and because much of the local European classic sequence is being related to the planktonic chronologies, it seems most desirable to attempt a calibration of the western North American plant biostratigraphy to the planktonic chronologies.

Radiometric ages are of considerable value in attempts to calibrate one biostratigraphic scheme with another. Ages based on glauconite are traditionally suspect, and most radiometric ages from the European and Gulf Coast Tertiary rocks are based on glauconite. Ages based on other volcanic minerals (for example, sanidine, biotite) or whole-rock basalts are generally considered to be more trustworthy, but even such ages may be highly anomalous, particularly if a single analysis is relied on.

Evernden and James (1964) published a number of ages that purported to date plant-bearing beds in the western conterminous United States. Some of the "floras" dated by these workers are nonexistent; for example, no one has yet collected fossil plants on which to base the "Pinole flora," "Siesta flora," or "Orinda flora," and these floral names have never appeared in paleobotanical literature. In other cases, the samples dated were of uncertain relationship to the plant-bearing beds (see Gray and Kittleman, 1967). In still other cases, analyses of additional samples gave different ages (in the instance of the Molalla flora, 10.8 versus 12.9 m.y.). I have attempted to cull the ages presented by Evernden and James (1964) for those that have a proper stratigraphic relation to plant-bearing beds.

PLANT STAGES

The several formal stages that have been proposed are discussed below. As of this time, no formal stages have been proposed for certain parts of the geologic column. Having neither the space nor in some instances the basic data necessary to formally propose additional stages, I will refer to these unnamed stages in terms of one of their characteristic floras.

Denver and Fort Union Types

As defined by mammalian paleontologists, the basal Paleocene of the Rocky Mountain region contains a distinctive floral assemblage, as exemplified by the Raton and Denver floras (Knowlton, 1917, 1930) (see Fig. 1). Brown (1962) discussed and listed a large number of species that occur in these two floras but not in floras such as the Coalmont, Middle Park, and Fort Union. The absence of such species in the Coalmont and Middle Park is particularly significant because these units are at about the same latitude as the Denver flora. Near Whetstone Falls in southern Wyoming, an unpublished flora similar to the Denver also occurs stratigraphically below a typical Fort Union assemblage.

Whether the Denver type of flora (and the Puercan mammalian age associated with this flora) is in fact of Paleocene rather than of Cretaceous age has yet to be demonstrated. The 64.8-m.y. age obtained from the Denver flora and fauna (Evernden and James, 1964) is considerably older than any of the radiometric ages (all on glauconite) obtained from lower Paleocene marine rocks in either the Atlantic Coastal Plain (Owens and Sohl, 1973) or in the Mississippi Embayment (Ghosh, 1972). If these ages are valid, the Denver type of flora may, at least in part, be of latest Cretaceous age; on the other hand, the work of Butler and others (1977) indicates that the Puercan is indeed of Paleocene age, which indicates that the 64.8-m.y. age might be anomalous and should be reinvestigated.

Elsewhere in the western conterminous United States, Paleocene floras are apparently rare. Neither the Elsinore flora in southern California nor the Chuckanut flora of western Washington has been thoroughly studied. Pabst's (1968) work on the conifers and ferns of the Chuckanut Formation does indicate a general age equivalency to the Fort Union.

In Alaska (see Fig. 2), Paleocene floras of the Fort Union type are of wide occurrence (Wolfe, 1972). Whether an age equivalent of the Denver type of flora is represented in Alaska is not known; most of the Alaskan floras (for example, the Chickaloon) appear to be similar to the Fort Union.

Ivanof Type

Only one floral assemblage was previously described (Hickey, 1977) for this unnamed floral stage. Several assemblages of approximately the same age as Hickey's Camels Butte flora in North Dakota are, however, known from the Cook Inlet region (West Foreland flora) and the Alaska Peninsula (Tolstoi flora). These floras are all of more modern aspect than the typical Fort Union–Chickaloon assemblages and contain "Eocene" indices such as *Platycarya* and *Liquidambar*. At the same time, the Ivanof-type floras contain Paleocene indices such as *Thuja interrupta, Protophyllum,* and *Chaetoptelea microphylla.* In North Dakota, the Camels Butte flora occurs above a typical Fort Union flora, and on Ivanof Bay (Alaska Peninsula) several floras of the Fort Union–Chickaloon type occur stratigraphically below the Ivanof-type assemblages.

No demonstrable relations of any Ivanof-type flora to marine rocks are known. No Ivanof-type floras are radiometrically dated. Hickey (1977) has shown that the Camels Butte flora can be equated to the earliest part of the Wasatchian mammalian age ("Graybull"). Although the Wasatchian has been traditionally assigned to the lower Eocene, Berggren, McKenna, and Hardenbol (1978) have suggested that the lower part of the Wasatchian is of latest Paleocene (Thanetian) age. If this suggestion is accepted, then insofar as known, the Ivanof type is of latest Paleocene age.

Ivanof Type to Franklinian Stage

If the Ivanof type is entirely of late Paleocene age (no younger than planktonic foraminiferal Zone P6a) and the Franklinian is no older than Zone P8 (or possibly P7), then a hiatus exists between the known floral assemblages in western North America. Collections have been made from middle Wasatchian beds in Wyoming (Hickey, 1977), but these have not been sufficiently studied to indicate whether it would be best to extend the concept of Ivanof type upward to include such assemblages or to formulate a new stage concept.

Franklinian Stage

The Franklinian Stage was defined on the basis of plants found in the lower part of the Green River Canyon section of the Puget Group (Wolfe, 1968). Assemblages of this age are unknown in Alaskan Tertiary rocks but have been found in California and Wyoming. In California, the Chalk Bluffs flora (MacGinitie, 1941) occurs in the gravels of the Ione Formation of the Sierra Nevada; these gravels extend westward into the Sacramento Valley, where they occur immediately above marine beds that have been assigned to the benthic foraminiferal Penutian Stage (Mallory, 1959). Poore (1976) has shown that the Penutian is typically equivalent to the *Discoaster lodoensis* Zone of Bukry (1973), although some strata placed in the Penutian are assignable to either subjacent or superjacent zones. In terms of the planktonic foraminiferal zonation, therefore, the Ione gravels are most probably P8, that is, in the upper part of the lower Eocene.

MacGinitie (1941, 1974) has correlated the Chalk Bluffs assemblage with the assemblages found in the early acid breccia of Yellowstone National Park, although an almost equally strong resemblance between the Chalk Bluffs and the assemblages from the early basic breccia indicates that the Chalk Bluffs could be correlated with either group of assemblages. In terms of local correlations and radiometric ages in Wyoming, this interval is from about 49 to 52 m.y. B.P. (MacGinitie, 1974). Berggren (1972) has indicated that this time interval corresponds to planktonic foraminiferal Zones P7 to P9. Correlations of the Franklinian floras in the type section of this stage with other areas thus indicate that the Franklinian is of late early Eocene age.

Fultonian Stage

The type section of the Fultonian Stage is stratigraphically above the Franklinian Stage in the Green River Canyon section of the Puget Group (Wolfe, 1968). In the quadrangle north of this area, the upper part of the Fultonian Stage occurs in the lower part of the Tiger Mountain Formation, which is gradational with the underlying marine Raging River Formation. Rau (in Vine, 1969) indicated that the benthic foraminifers from the Raging River Formation belong to Mallory's (1959) upper Ulatisian or possibly lower Narizian. Poore (1976) has shown that in California these two stages overlap in the *Discoaster sublodoensis* (NP14) and *Nannotetrian quadrata* (≈NP15) Zones, which are equivalent to planktonic foraminiferal Zones P9 to part of P12, that is, uppermost lower Eocene through lower middle Eocene.

An upland Fultonian assemblage occurs in the Klondike Mountain Formation near Republic, Washington. A radiometric age of 46.3 + 1.7 m.y. has been obtained on the plant-bearing member of the Klondike Mountain (Engels and others, 1976), which is consistent with a middle Eocene age.

Ravenian Stage

The type section of the Ravenian Stage in Green River Canyon contains two distinct floral assemblages and one assemblage intermediate between these. The Ravenian has thus been informally subdivided (Wolfe, 1977) into three substages. In Alaska, assemblages of early Ravenian age occur in the lower part of the Kushtaka and Kulthieth Formations (Wolfe, 1977). The lower part of the Kushtaka is, in the invertebrate megafossil chronology, of either "Domengine" or early "Tejon" age; relations in California and the Pacific Northwest suggest that this would be Ulatisian or early Narizian (Mallory, 1959). From Poore's (1976) analysis, this age interval would most probably be represented by the *Nannotetrina quadrata* zone (NP15), which is the lower half of the middle Eocene. Although these calibrations suggest that the Fultonian and lower Ravenian could be overlapping, the strong vegetational differences allow a ready differentiation.

By the Ravenian, floral provincialism has become so extreme that no assemblages in the Rocky Mountain region can be correlated with Pacific Coast assemblages on a definitive floral basis. Some significance could, however, be attached to the occurrence of *Populus wyomingensis* in the middle Ravenian. Elsewhere, this species is known in the Green River flora of Wyoming (late Bridgerian in the mammalian sequence) and in the Tallahatta Formation of the Claibornian Stage in the Gulf Coast; that is, this species is elsewhere of middle Eocene age.

A middle Ravenian assemblage in the Kushtaka Formation in Alaska occurs stratigraphically between the assemblages of early Ravenian age and mollusks of "Tejon" (Narizian) age. Similarly, the McIntosh Formation of Washington, which is of "Tejon" (Narizian) age, contains a middle Ravenian assemblage. Thus, the available data indicate that the lower Ravenian is definitely of middle Eocene age, and the middle Ravenian is most probably of late middle or early late Eocene age.

The age of the upper Ravenian in terms of marine-nonmarine relations cannot be readily determined. No assemblage of late Ravenian age has been obtained from beds that have a demonstrable relation to marine beds. At

Figure 1. Map of part of western conterminous United States showing locations of some important floral assemblages.

least four late Ravenian assemblages from Oregon (middle Clarno flora), Nevada (Bull Run, Copper Basin floras), and northeastern California (Lower Cedarville flora) have radiometric ages that range from about 38 to 42 m.y. If the Berggren (1972) time scale is accepted, the upper Ravenian would clearly be of late Eocene (Priabonian) age.

Kummerian Stage

The type section of the Kummerian Stage occurs in the uppermost part of the Puget Group and the lower part of the volcanic rocks that are probably assignable to the Ohanapecosh Formation. Of all the Paleogene floral stages, the Kummerian has the best-known relation to the marine chronology and is largely equivalent to the lower part of the benthic foraminiferal Refugian Stage.

In Alaska, at least 13 localities have produced Kummerian assemblages; in each case, the plant-bearing beds are—on the basis of stratigraphic, molluscan ("Keasey" or "Lincoln" age), or foraminiferal evidence—of Refugian age. In Washington, Kummerian assemblages have been obtained from the lower part of the Twin River Formation and from the Toutle Formation; the latter unit contains the well-known Gries Ranch molluscan fauna of earliest "Lincoln" age. In Oregon, the Comstock flora (Sanborn, 1935) can be assigned to the Kummerian Stage, and the plant-bearing beds are in the basal part of the intertonguing Eugene Formation and the Fisher Formation of Schenck (1927) of Refugian age. Further, the characteristic Kum-

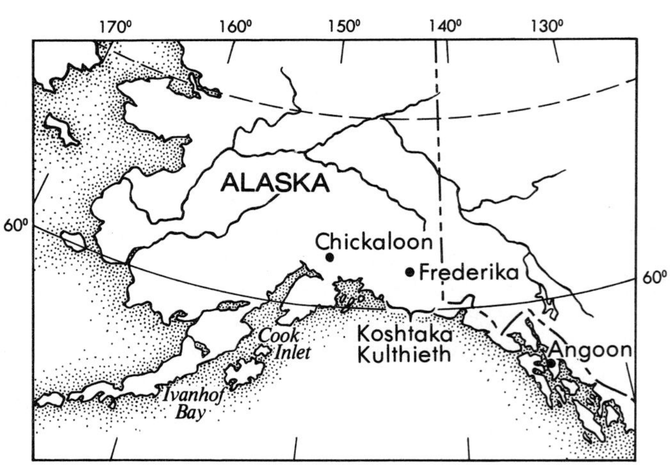

Figure 2. Map of southern Alaska showing locations of some important floral assemblages.

merian *Macclintockia pugetensis* occurs in the lower part of the Keasey Formation. These data clearly indicate that the Kummerian is equivalent to at least part of the Refugian and is most probably of latest Eocene age.

Only one Kummerian flora—that of the Clarno "nut bed" (uppermost part of Clarno Formation) in central Oregon—has been radiometrically dated. The date of 34.0 m.y. B.P. (Evernden and James, 1964) is not consistent with a late Eocene age in the Berggren (1972) time scale, in which the Eocene/Oligocene boundary is placed at 37.5 m.y. B.P. The date of 34 m.y. B.P. is , however, consistent with other radiometric data cited below, and significantly in a review of the radiometric ages on the Eocene/Oligocene boundary, Cavelier and Pomerol (1977) have concluded that this boundary could be anywhere from 34 to 36 m.y. old. In fact, the boundary may be as young as 32 to 33 m.y.

Goshen Type

No formal name has been (or is being) proposed for the unnamed floral assemblage that characterizes the interval between the Kummerian and Angoonian Stages. Goshen-type assemblages occur in a continuous stratigraphic section below the type section of the Angoonian (Wolfe, 1969a, 1977), and Goshen-type assemblages also occur stratigraphically above the type section of the Kummerian and in the Fisher-Eugene sequence (Goshen flora) above a Kummerian assemblage (Comstock flora). I originally thought that the LaPorte flora of the Sierra Nevada was of Kummerian age (Wolfe, 1968), but additional analyses indicate that the LaPorte is of Goshen age.

In terms of the marine chronology, the Goshen type is of late (but not latest) Refugian age. Aside from the late Refugian Goshen assemblage near Eugene, Oregon, Goshen-type assemblages occur in beds along the foothills of the Cascade Range in Oregon near Sweet Home (two localities) and near Scio (three localities). The beds containing these Goshen-type assemblages grade westward into marine beds of the Eugene Formation, which is as young as the *Molopophorus gabbi* zone of Durham (1944) of the "Lincoln stage." Near Goshen, the Fisher-Eugene sequence (Fig. 3) is overlain by nonmarine beds that contain an early Angoonian flora (Willamette flora) which

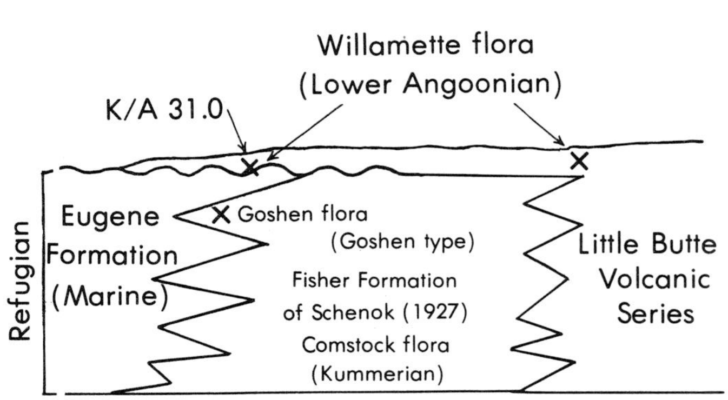

Figure 3. Diagrammatic section south of Eugene, Oregon, showing stratigraphic positions of floral assemblages.

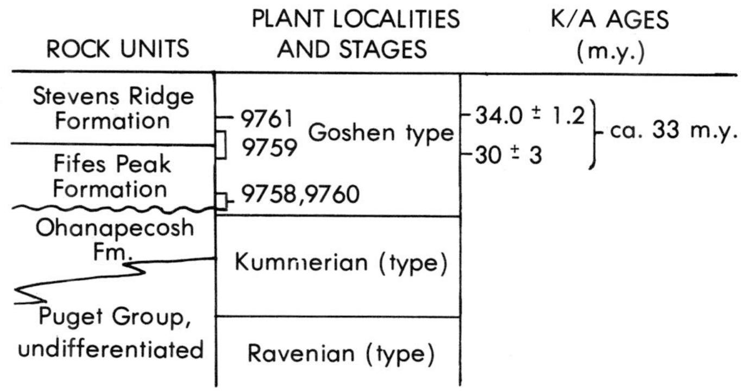

Figure 4. Diagrammatic section southeast of Tacoma, Washington, showing stratigraphic positions of floral assemblages.

has been radiometrically dated at 31.0 m.y. B.P. (Evernden and James, 1964).

Volcanic units in the Cascade Range of Washington (Fig. 4) have produced several Goshen-type assemblages (Wolfe, 1961). These assemblages occur in the uppermost part of the Ohanapecosh Formation and in the overlying Fifes Peak and Stevens Ridge Formations. The Stevens Ridge has an age of 34.0 ± 1.2 m.y. (Fischer, 1976), and the Fifes Peak an age of 30 ± 3 m.y. (Laursen and Hammond, 1974). Considering the stated experimental errors on these two ages, about 33 m.y. would seem a valid approximation of the age of the Goshen-type assemblages.

The Goshen-type LaPorte flora (Potbury, 1935) of California has been dated at 32.3 and 32.5 m.y. B.P. (Evernden and James, 1964). This flora occurs in tuffaceous rocks that are among the oldest known volcanic rocks of Tertiary age in the Sierra Nevada. The only two older Tertiary ages known are 33.2 m.y. from two different rhyolitic tuffs in the Sierra Nevada (Dalrymple, 1964); the significance of these ages centering around 33 m.y. is that they mark the beginning of known Tertiary volcanism in the Sierra Nevada. By inference, the marine Wheatland Formation of Clark and Anderson (1938) in the Sacramento Valley can be no older than 33 m.y. because the Wheatland contains abundant pyroclastic, including rhyolitic, debris.[1] The Wheatland has benthic foraminifers of Refugian age (Kleinpell, 1938), and thus it can be inferred that the Refugian is as young as 33 m.y. If, then, the Refugian is entirely of late Eocene age, the Eocene/Oligocene boundary can be no older than 33 m.y.

An additional indication that the Eocene/Oligocene boundary may be appreciably younger than 37.5 m.y. is the radiometric ages that have been obtained on minerals in strata of Chadronian age in the mammalian chronology. Although the Chadronian has been traditionally considered to be Oligocene, this assignment is based on the correlation of Chadronian faunas with European mam-

[1]A radiometric age of 53.5 m.y. was obtained on a composite of 35 to 40 andesite pebbles from the basal part of the Wheatland Formation, but the radiometric age indicates only that the Wheatland is younger than 53.5 m.y. and that one of the sources of the pebbles is older than 53.5 m.y. (Dalrymple, 1964).

malian faunas such as those contained in the Bembridge beds in England (Thenius, 1959). If Cavelier and Pomerol (1976) are correct and the Bembridge beds are in fact late Eocene (Priabonian), then the Chadronian, at least in part, should also be considered late Eocene. If the range of ages for the Chadronian (31.6 to 35.2 m.y.; Evernden and others, 1964) is considered, by inference the Eocene/Oligocene boundary is no older than about 35 m.y. and could be younger. Also significant is that the Kummerian Clarno "nut bed" flora is associated with Chadronian mammals, an indication that Chadronian = Kummerian = Refugian = late Eocene.

Angoonian Stage

The type section of the Angoonian Stage is in the type locality of the Kootznahoo Formation on Admiralty Island, southeastern Alaska (Wolfe, 1969a, 1977). Two informal subdivisions of the Angoonian can typically be recognized; the early Angoonian assemblages in a given area are typically of cooler aspect than later ones. Angoonian—particularly early—assemblages are widespread in Alaska (Wolfe, 1972). In the conterminous United States, the Bridge Creek and equivalent floras can be assigned to the lower Angoonian. Bridge Creek equivalents are common in the lower part of the Little Butte Volcanic Series of the Cascade Range in Oregon.

As noted above, the Willamette assemblage near Eugene has a radiometric age of 31.0 m.y. (Evernden and James, 1964); the Bridge Creek itself has been dated at 31.1 to 31.5 m.y. B.P. (Evernden and James, 1964). Both these floras are of early Angoonian age, and both presumably fall in the earliest part of the stage. No assemblage of late Angoonian age has been radiometrically dated.

The relation of the Angoonian to the various marine chronologies is not well known. On Sitkinak Island in Alaska, lower Angoonian assemblages occur stratigraphically and conformably below mollusks of late "Blakeley" age. If the equivalence of the "Blakeley" and the benthic foraminiferal Zemorrian Stage (Weaver and others, 1944) is assumed, the Angoonian is equivalent to at least part of the Oligocene. A late Angoonian assemblage occurs in beds near Molalla, Oregon (not, however, to be confused with the Molalla flora of various authors), which are on strike with marine beds apparently assignable to the upper Blakeley *Echinophoria apta* zone (Peck and others, 1964). The Sooke flora of Vancouver Island (LaMotte, 1935) also appears to be of late rather than early Angoonian age; this flora occurs in a marine unit of late Blakeley age (Weaver and others, 1944). Although the Angoonian is not yet satisfactorily calibrated with the marine chronology, this stage appears to be approximately equivalent with the Oligocene.

Seldovian Stage

The type section of the Seldovian Stage is in the middle part of the Kenai Group in the Cook Inlet region, Alaska (Wolfe and others, 1966). Two informal subdivisions of the stage, which is in the Tyonek Formation of the Kenai Group, can be recognized (Wolfe and Tanai, 1980). As with the Angoonian, the early Seldovian assemblages are typically of cooler aspect than the late Seldovian assemblages.

Only one early Seldovian assemblage has been radiometrically dated. The Upper Cedarville flora (LaMotte, 1936; Wolfe, 1969b) has been dated at 19.8 m.y. B.P. (Evernden and James, 1964), which would place this assemblage in the lower Miocene. No early Seldovian assemblage is known in a direct relation to a marine sequence.

The late Seldovian assemblages have been radiometrically dated primarily in the conterminous United States. In Nevada, the Middlegate flora (Axelrod, 1956) is 15.9 m.y. old; in Oregon, the Mascall flora is younger than 15.4 m.y., and the Rockville flora (= Sucker Creek flora of Arnold, 1937; compare to Wolfe, 1969b) is 16.7 m.y. (Evernden and James, 1964). In Washington, the Latah flora at Spokane is older than 15.5 + 0.8 m.y. (Engels and others, 1976), and the Grand Coulee flora is 15.7 to 16.8 m.y. (Gray and Kittleman, 1967). In Alaska, the Frederika Formation in the Wrangell Mountains contains a late Seldovian assemblage (Wolfe, 1972); the overlying Wrangell Lava has radiometric ages as old as 16 m.y. (Eisbacher and Hopkins, 1977). As noted below, Homerian assemblages are as old as 13.1 m.y., and thus the upper limit of the Seldovian is estimated to be about 13 to 14 m.y. These ages indicate that the lower Seldovian may be entirely of early Miocene age and the upper Seldovian of late early and early middle Miocene age (Zones N5 to N12 of the planktonic foraminiferal time scale).

Homerian Stage

The type section of the Homerian Stage is in the Beluga Formation of the Kenai Group in the southern part of the Cook Inlet region, Alaska. Other Homerian assemblages are known in the Columbia Plateau region, the Cascade Range, and Nevada (see Wolfe, 1969b).

Radiometric ages are numerous for the Homerian. In the type section, the upper boundary has been dated at 8.0 m.y. B.P. (Triplehorn and others, 1977), and an additional date in the type section is 11.1 m.y. B.P. Evernden and James (1964) reported the following ages on Homerian floras in the conterminous United States: Ellensburg, 10.1 m.y.; Aldrich Station, 10.7 to 11.2 m.y.; Stinking Water, 12.1 m.y.; and Trout Creek, 13.1 m.y. Additionally, the Molalla flora has been dated at 12.9 + 1.0 m.y. B.P. (J. D. Obradovich, 1966, written commun.). These ages indicate that the Homerian Stage is equivalent to planktonic foraminiferal Zones N13 to part of N16, that is, late middle Miocene to early late Miocene.

Clamgulchian Stage

The type section of the Clamgulchian Stage occurs in the Sterling Formation of the Kenai Group along the east shore of Cook Inlet. Floristic provincialism makes recog-

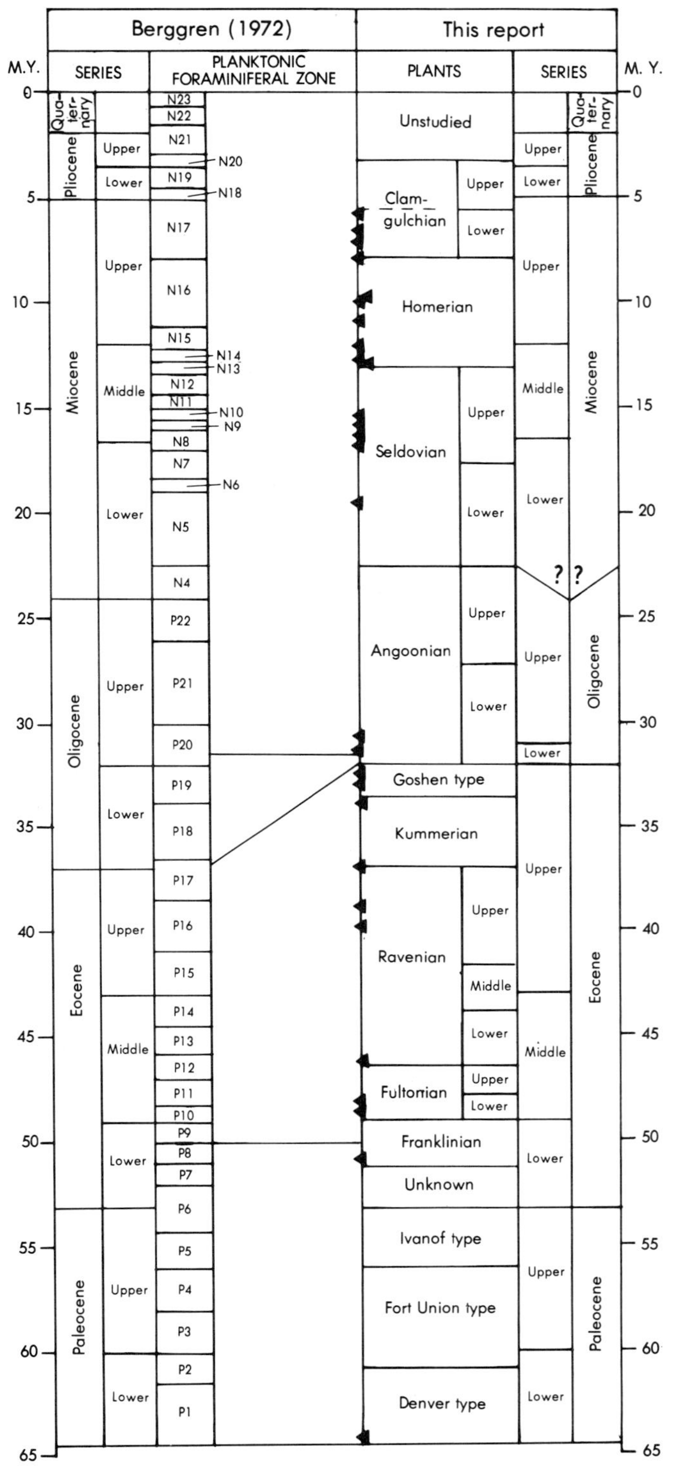

Figure 5. Tentative calibration of plant megafossil sequence in western North America with planktonic foraminiferal zonation. Left-hand column largely after Berggren (1972). Triangles represent radiometric dates.

nition of this stage difficult (if not impossible) at middle latitudes.

The type section of the Clamgulchian has yielded ages as young as 4.2 m.y. (Turner and others, 1980). One age on an apparent late Clamgulchian flora from Seward Peninsula is 5.7 m.y. (Hopkins and others, 1971). Most radiometric ages on Clamgulchian floras in Alaska are indicative of a latest Miocene age, but some ages indicate that the Clamgulchian extends into the Pliocene. Some floras in the Pacific Northwest have radiometric ages (Evernden and James, 1964) indicative of an equivalency to the Clamgulchian, but no biostratigraphic correlations can now be made.

DISCUSSION

A tentative calibration of the plant megafossil stages in relation to the planktonic chronologies is given in Figure 5. Refined calibration in the Paleocene and Oligocene is considerably uncertain, although the broad equivalencies are apparent.

Most of the plant stages and substages do not conform to traditional European stages and subseries, a fact that should not be surprising. The plant framework is, at least in part, closely related to climatic episodes. For example, the late Ravenian and Homerian represent cooler intervals relative to the early Ravenian and late Seldovian, respectively. if such intervals are the result of climatic factors that also affected areas such as Europe, direct correlations should be possible between Europe and western North America based on the climatic implications of their respective floral assemblages. That this is a valid possibility is indicated by the fact that the floras of the upper Bracklesham and Bournemouth beds in England (Zones P8 to P11) represent a cooler interval than the floras of the London Clay (P6 to P7) or the Barton beds (P13 to P14). In western North America, the cool interval during the interval represented by the Franklinian and Fultonian stages appears to correlate with the same planktonic zones as the upper Bracklesham and Bournemouth.

Another intriguing possiblility is that the temperature fluctuations during the Tertiary were cyclic. In a study of late Cretaceous pollen floras from the Atlantic Coastal Plain (Wolfe, 1976), a series of apparently temperature-related fluctuations in the abundance of certain major pollen types was observed. When the radiometric ages based on a number of glauconites were compared to ages predicted according to the assumption that these apparent temperature fluctuations belonged to the same series of cycles as those during the Tertiary, the glauconite ages were remarkably concordant with the predicted ages (Wolfe, 1976, Table 2).

If the temperature fluctuations were cyclic during the Tertiary, and if they are worldwide in scope, they would provide an additional tool for long-distance correlations. In Figure 5, the assumption has been made that such fluctuations are cyclic, and the placement of the boundaries of the plant units is made according to this assumption.

REFERENCES CITED

Arnold, C. A., 1937, Observations on the fossil flora of eastern and southeastern Oregon: Michigan University Museum of Paleontology Contributions, v. 5, no. 8, p. 79-102, 10 pls.

Axelrod, D. I., 1956, Mio-Pliocene floras from west-central Nevada: University of California Publications in Geological Sciences, v. 33, 322 p., 32 pls.

Berggren, W. A., 1972, A Cenozoic time-scale—Some implications for regional geology and paleobiogeography: Lethaia, v. 5, p. 195-215.

Berggren, W. A., McKenna, M. C., and Hardenbol, J., 1978, Revised Cenozoic polarity time scale by D. M. Tarling and J. G. Mitchell: Journal of Geology, v. 86, p. 67-81.

Brown, R. W., 1962, Paleocene flora of the Rocky Mountains and Great Plains: U.S. Geological Survey Professional Paper 375, 119 p., 69 pls.

Bukry, David, 1973, Low-latitude coccolity biostratigraphic zonation, *in* Initial reports of the Deep Sea Drilling Project, Volume 15: Washington, D.C., U.S. Government Printing Office, p. 685-703.

Burk, C. A., 1965, Geology of the Alaskan Peninsula—Island arc and continental margin: Geological Society of America Memoir 99, 250 p.

Butler, R. F., and others, 1977, Magnetostratigraphy of the Cretaceous-Tertiary boundary in the San Juan Basin, New Mexico: Nature, v. 267, no. 5609, p. 318-323.

Cavelier, Claude, and Pomerol, Charles, 1976, Les rapports entre la Bartonien et le Priabonien Incidence sur la position de la limite Eocene moyen-Eocene superieur: Société Géologique de France, Bulletin, Compte Rendu, pt. 2, p. 49-51.

——1977, Proposition d'une echelle stratigraphique standard pour le Paleogene: Newsletter of Stratigraphy, v. 6, no. 1, p. 56-65.

Chaney, R. W., 1967, Miocene forests of the Pacific Basin: Their ancestors and their descendants: Hokkaido University, Sapporo, Jubilee Publication commemorating Professor Sasa's 60th Birthday, p. 209-239.

Clark, B. L., and Anderson, C. A., 1938, Wheatland formation and its relation to early Tertiary andesites in the Sierra Nevada: Geological Society of America Bulletin, v. 49, p. 931-956.

Dalrymple, G. B., 1964, Cenozoic chronology of the Sierra Nevada, California: University of California Publications in Geological Sciences, v. 47, 41 p.

Durham, J. W., 1944, Megafaunal zones of the Oligocene of northwestern Washington: University of California Publications, Department of Geological Sciences Bulletin, v. 27, p. 101-212.

Eisbacher, G. H., and Hopkins, S. L., 1977, Mid-Cenozoic paleogeomorphology and tectonic setting of the St. Elias Mountains, Yukon Territory: Canada Geological Survey Paper 77-1B, p. 319-335.

Engels, J. C., and others, 1976, Summary of K-Ar, Rb-Sr, U-Pb, Pb, and fission-track ages of rocks from Washington State prior to 1975 (exclusive of Columbia Plateau basalts): U.S. Geological Survey Miscellaneous Field Studies Map MF-710.

Evernden, J. F., and James, G. T., 1964, Potassium-argon dates and the Tertiary floras of North America: American Journal of Science, v. 262, p. 945-974.

Evernden, J. F., and others, 1964, Potassium-argon dates and the Cenozoic mammalian chronology of North America: American Journal of Science, v. 262, p. 145-198.

Fischer, J. F., 1976, K-Ar dates from the Stevens Ridge Formation, Cascade Range, central Washington: Isochron/West, no. 16, p. 31.

Ghosh, P., 1972, Use of bentonites and glauconites in potassium-40/argon-40 dating in Gulf Coast stratigraphy [Ph.D. thesis]: Houston, Texas, Rice University, 136 p.

Gray, Jane, and Kittleman, L. R., 1967, Geochronometry of the Columbia River basalt and associated flora of eastern Washington and western Idaho: American Journal of Science, v. 265, p. 257-291.

Hickey, L. J., 1977, Stratigraphy and paleobotany of the Golden Valley Formation (early Tertiary) of western North Dakota: Geological Society of America Memoir 150, 183 p., 55 pls.

Hopkins, D. M., and others, 1971, A Pliocene flora and insect fauna from the Bering Strait region: Palaeogeography, Palaeoclimatology, Palaeoecology, v. 9, p. 211-231, 4 figs., 5 tables.

Kleinpell, R. M., 1938, Miocene stratigraphy of California: Tulsa, Oklahoma, American Association of Petroleum Geologists, 450 p.

Knowlton, F. H., 1917, Fossil floras of the Vermejo and Raton formations of Colorado and New Mexico: U.S. Geological Survey Professional Paper 101, p. 223-455, Pls. 30-113.

——1930, The flora of the Denver and associated formations of Colorado: U.S. Geological Survey Professional Paper 155, 142 p., 59 pls.

LaMotte, R. S., 1935, An upper Oligocene florule from Vancouver Island: Carnegie Institution of Washington Publication 455, p. 49-56.

——1936, The Upper Cedarville flora of northwestern Nevada and adjacent California: Carnegie Institution of Washington Publication 455, p. 57-142, 14 pls.

Laursen, J. M., and Hammond, P. E., 1974, Summary of radiometric ages of Oregon and Washington rocks, through June 1972: Isochron/West, no. 9, p. 1-32.

MacGinitie, H. D., 1941, A middle Eocene flora from the central Sierra Nevada: Carnegie Institution of Washington Publication 543, 178 p., 47 pls.

——1953, Fossil plants of the Florissant beds, Colorado: Carnegie Institution of Washington Publication 599, 198 p., 75 pls.

——1974, An early middle Eocene flora from the Yellowstone-Absaroka volcanic province, northwestern Wind River basin, Wyoming: University of California Publications in Geological Sciences, v. 108, 103 p., 45 pls.

Mallory, V. S., 1959, Lower Tertiary biostratigraphy of the California Coast Ranges: Tulsa, Oklahoma, American Association of Petroleum Geologists, 416 p.

Owens, J. P., and Sohl, N. F., 1973, Glauconites from the New Jersey-Maryland Coastal Plain: Their K/Ar ages and application in stratigraphic studies: Geological Society of America Bulletin, v. 84, p. 2811-2838.

Pabst, M. B., 1968, The flora of the Chuckanut formation of northwestern Washington: University of California Publications in Geological Sciences, v. 76, 85 p.

Peck, D. L., and others, 1964, Geology of the central and northern parts of the Western Cascade Range in Oregon: U.S. Geological Survey Professional Paper 449, 56 p.

Poore, R. Z., 1976, Microfossil correlation of California lower Tertiary sections: A comparison: U.S. Geological Survey Professional Paper 743-F, p. F1-F8.

Potbury, S. S., 1935, The LaPorte flora of Plumas County, California: Carnegie Institution of Washington Publication 465, p. 29-82, 19 pls.

Sanborn, E. I., 1935, The Comstock flora of west central Oregon: Carnegie Institution of Washington Publication 465, p. 1-28, 10 pls.

Schenck, H. G., 1927, Marine Oligocene of Oregon: University of

California Publications, Department of Geological Sciences Bulletin, v. 16, p. 449–460.

Thenius, Erich, 1959, Tertiär, pt, 2, Wirbeltierfaunen: Handbuch Stratigraphie und Geologie, v. 3, pt. 2, Stuttgart, Enke Verlag, 328 p.

Triplehorn, D. M., Turner, D. L., and Naeser, C. W., 1977, K-Ar and fission-track dating of ash partings in coal beds from the Kenai Peninsula, Alaska: A revised age for the Homerian Stage-Clamgulchian Stage boundary: Geological Society of America Bulletin, v. 88, p. 1156–1160.

Turner, D. L., and others, 1980, Radiometric dating of ash partings in Alaskan coal beds and upper Tertiary paleobotanical stages: Geology, v. 8, p. 92–96.

Vine, J. D., 1969, Geology and coal resources of the Cumberland, Hobart, and Maple Valley quadrangles, King County, Washington: U.S. Geological Survey Professional Paper 624, 67 p.

Weaver, C. E., and others, 1944, Correlation of the marine Cenozoic formations of western North America: Geological Society of America Bulletin, v. 55, p. 569–598.

Wolfe, J. A., 1961, Age of the Keechelus andesitic series of the Cascade Range, Washington: U.S. Geological Survey Professional Paper 424C, art, 232, p. C228–C230.

——1968, Paleogene biostratigraphy of nonmarine rocks in King County, Washington: U.S. Geological Survey Professional Paper 571, 33 p., 7 pls.

——1969a, Paleogene floras from the Gulf of Alaska region: U.S. Geological Survey Open-File Report, 114 p., 6 pls.

——1969b, Neogene floristic and vegetational history of the Pacific Northwest: Madrono, v. 20, no. 3, p. 83–110, Figs. 1–3, 2 tables.

——1972, An interpretation of Alaskan Tertiary floras, *in* Graham, Alan, ed., Floristics and paleofloristics of Asia and eastern North America: Amsterdam, Elsevier Publishing Company, p. 201–233, 13 pls., 2 figs., 3 tables.

——1976, Stratigraphic distribution of some pollen types from the Campanian and lower Maestrichtian rocks (Upper Cretaceous) of the Middle Atlantic States: U.S. Geological Survey Professional Paper 977, 18 p., 4 pls.

——1977, Paleogene floras from the Gulf of Alaska region: U.S. Geological Survey Professional Paper 997, 108 p.

Wolfe, J. A., Hopkins, D. M., and Leopold, E. B., 1966, Tertiary stratigraphy and paleobotany of the Cook Inlet region, Alaska: U.S. Geological Survey Professional Paper 398-A, p. A1–A29, 2 pls.

Wolfe, J. A., and Tanai, Toshimasa, 1980, The Miocene Seldovia Point flora of the Kenai Group, Alaska: U.S. Geological Survey Professional Paper 1105, 52 p., 25 pls.

MANUSCRIPT RECEIVED BY THE SOCIETY MARCH 14, 1980
MANUSCRIPT ACCEPTED MAY 7, 1980

Printed in U.S.A.

Geological Society of America
Special Paper 184
1981

Palynologic biostratigraphy of some early Tertiary nonmarine formations in central and western Washington

KARL R. NEWMAN
Colorado School of Mines, Department of Geology, Golden, Colorado 80401

ABSTRACT

Palynologic data have been obtained from 30 localities in the Swauk, Naches, and Roslyn Formations of central Washington and in the Puget Group of western Washington. Although the Swauk Formation previously has been assigned ages as old as Paleocene or Cretaceous, palynologic results show that the type Swauk Formation is Eocene and that the Swauk Formation of the Chiwaukum graben and Wenatchee areas is Eocene and Oligocene. Correlations based on palynomorph biozones show that the Swauk, Naches, Teanaway, and Roslyn Formations of central Washington are time-equivalent to the Puget Group of western Washington and are part of the same depositional sequences of Eocene and Oligocene age.

Palynomorph biozones found useful in this study include the *Pistillipollenites* and *Platycarya* concurrent-range zone, early(?) and middle Eocene; the *Platycarya* and *Bursera* concurrent-range zone, late Eocene; and the *Gothanipollis* and *Elaeagnus* concurrent-range zone, Oligocene. These concurrent-range zones can be used to correlate the central Washington formations and Puget Group with other strata in southwestern Washington and in the Bellingham, Washington, and Vancouver, British Columbia, areas.

INTRODUCTION

This reconnaissance investigation began in 1967 as a study of the relative age of the nonmarine, arkosic Swauk Formation in its type area in Kittitas County, Washington. Reviews of the literature showed that the Swauk Formation as mapped in central Washington (Huntting and others, 1961) has been assigned ages ranging from Cretaceous to Paleocene or Eocene, with a Paleocene age being favored (see reviews by Foster, 1960; Rector, 1962). These age assignments were based on stratigraphic position and fossil leaves, which were correlated with leaves in the Laramie, Denver, and Fort Union Formations of the distant Rocky Mountains and Great Plains regions. The emphasis of much of the paleobotanical interpretation through the years was in terms of the overall Swauk flora; that is, to list all taxons together from various collections and to analyze the results for age interpretations. No invertebrate or vertebrate fossils that would provide other lines of evidence have been reported from the Swauk Formation.

This study is stratigraphic and palynologic; I have used palynomorphs (organic-walled plant microfossils) to determine if biostratigraphic zonation is possible and to see if such zonation would lead to biostratigraphic and time-stratigraphic correlations with other formations in the Pacific Northwest. Initial results from samples collected in the Swauk Creek and Swauk Pass areas, Kittitas County, indicated a probable Eocene age for the rocks in the type Swauk area (Newman, 1968, unpub. paper, Northwest Scientific Association). Sampling then was extended into the Chiwaukum graben and Wenatchee areas, Chelan and Douglas Counties, where there is an extremely thick sedimentary section, and into the Cle Elum syncline area, Kittitas County, to sample the Roslyn and Naches Formations. Results indicated Eocene to Oligocene ages for the entire complex of formations (Newman, 1971). As a preliminary palynologic zonation for the combined Swauk, Naches, and Roslyn strata evolved, samples for comparative study were collected from the Green River canyon section of the Eocene-Oligocene Puget Group in King County, western Washington (Newman, 1975, 1977). A similar zonation found there is the basis for the east-to-west correlation across the Cascade Mountains presented in this paper.

The major general result of this project is that no palynomorphs suggesting Cretaceous or Paleocene ages have been found in central Washington, which are ages suggested for the Swauk Formation by some previous workers. All results indicate Eocene or Eocene to Oligocene ages. Significant elements of the oldest palynomorph assemblages found are *Alnus, Caryapollenites, Polypodiidites, Cicatricosisporites, Proteacidites, Pistillipollenites,*

Platycarya, Tilia, and *Bombacacidites.* My interpretation, based on the published literature and my own work on samples from many parts of western North America, is that this assemblage does not appear to be older than Eocene. Unless some significantly older rocks can be found at the base of the Swauk Formation, the older notions of Cretaceous and Paleocene ages will have to be abandoned.

This is the first publication on stratigraphic palynology of these formations. Previous reports from other parts of the region that aided this study include Rouse (1962) on the Burrard Formation of southwestern British Columbia, Crickmay and Pocock (1963) on the Cretaceous and lower Tertiary rocks of southwestern British Columbia, Hills (1965) on early Tertiary basins in interior British Columbia, Hills and Baadsgaard (1967) on potassium-argon dating of some Eocene strata in interior British Columbia, Hopkins (1966, 1968, 1969) on the Whatcom basin of southwestern British Columbia and northwestern Washington and on the Kitsilano Formation of southwestern British Columbia, Sparks (1967) on Eocene formations in southwestern Washington, Griggs (1970) on the Chuckanut Formation of northwestern Washington, Newman (1970) on some subsurface Tertiary rocks in south-central Washington, Rouse and others (1970) on several lower Tertiary deposits in British Columbia and Alberta, Piel (1971) on Oligocene strata of central British Columbia, and Hopkins and others (1972) on some Oligocene rocks in the Rocky Mountain trench of British Columbia.

SUMMARY OF STRUCTURE AND STRATIGRAPHY

The stratigraphic terminology and correlations and the biostratigraphic interpretations that have been published in the past are, in my opinion, rather confusing. Furthermore, much basic information is contained in unpublished theses. Therefore, a brief review of structure, stratigraphy, and biostratigraphy—based on both published and unpublished sources—is presented in order to establish the framework in which this study was made.

Structural Setting

The Cascade Range trends in a northward direction through the state of Washington. The southern half of the range in Washington is covered by Cenozoic volcanic rocks, whereas the northern half displays rocks of many ages and types (Huntting and others, 1961). The core of the study area in central Washington is the uplifted Mount Stuart batholith that bears the prominent Stuart Range along its southern margin (Fig. 1). The batholith, a Cretaceous intrusive complex, is covered and flanked in places by patches or slices of older metamorphic and intrusive rocks.

Tertiary sedimentary rocks surrounding the Mount Stuart batholith lie in two structural settings, the Chiwaukum graben and the Cle Elum syncline (Figs. 1, 2). To the east lies the Chiwaukum graben, an elongate northwest-trending graben bounded by the Leavenworth fault on the west and by the Entiat fault on the east. The Eagle Creek anticline within the graben exposes older metamorphic rocks overlain by the Swauk Formation. To the south and east of Wenatchee, the Swauk Formation is overlain unconformably by the Columbia River Basalt.

The trend of the Mount Stuart uplift continues southeastward in the Wenatchee Mountains which separate the Chiwaukum graben and Wenatchee areas from the Cle Elum syncline area to the southwest. The Cle Elum syncline area contains a section of Swauk Formation, Teanaway Basalt, and Roslyn Formation. Westward and southward from Cle Elum are found scattered patches of Guye Formation and Naches Formation among older metamorphic and younger volcanic and intrusive rocks. All formations are unconformably overlain by the Columbia River Basalt at the eastern and southern margins of the area. Near Snoqualmie Pass, the Snoqualmie batholith intrudes the Swauk, Naches, and Guye Formations.

The Green River canyon area lies about 80 km west of the central Washington study area and contains some of the geographically closest sedimentary rocks in western Washington that could be sampled for comparative purposes. The area is near Black Diamond—about 48 km southeast of Seattle, along the west side of the Cascade Range—and is underlain by Tertiary and Quaternary sedimentary, volcanic, and intrusive rocks. The Puget Group is exposed in the canyon of the Green River where, although it is folded and faulted, a continuous section can be measured and described.

Mount Stuart Batholith

The Mount Stuart batholith is exposed across an area of about 780 km^2. It intruded the Chiwaukum Schist, Hawkins Formation, and Peshastin Formation, which may range from late Paleozoic to early Mesozoic in age, and the Ingalls Peridotite, which may be Late Jurassic or Early Cretaceous in age (Southwick, 1974). The batholith, generally described as granodiorite in the literature, ranges in composition from gabbro to granite (Erikson, 1976). Engels and Crowder (1971) reported several potassium-argon dates based on biotite and hornblende pairs from which they interpret a time of crystallization of 88 m.y. ago, within the Cretaceous Period.

The Mount Stuart batholith is bordered by faults along its west, south, and east margins (Huntting and others, 1961). During the rise and deroofing of the batholith, the Ingalls Peridotite along the south margin of the horst was deeply weathered to form iron-rich laterite, some of which was reworked in basal Swauk strata (Lupher, 1944). Deposition of parts of the Swauk Formation was concurrent with faulting along margins of the adjacent Chiwaukum graben (Cashman and Whetten, 1976; Whetten, 1976, 1977; Frizzell and Tabor, 1977).

It has been thought that the Mount Stuart batholith reached the surface in Late Cretaceous or Paleocene time (Engels and Crowder, 1971) or in Eocene time (Erikson and

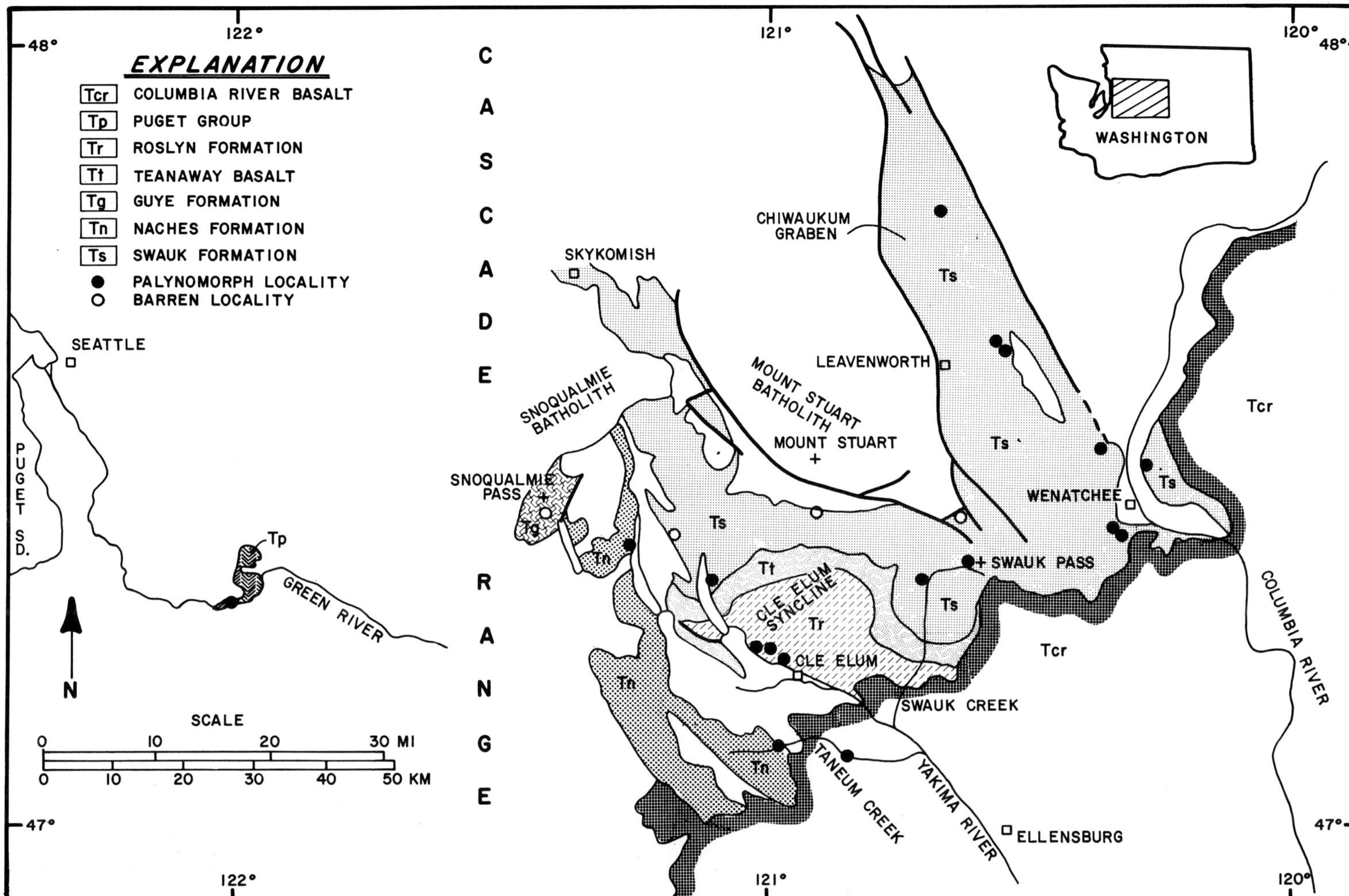

Figure 1. Geologic and sample-locality map of study areas in central and western Washington. Geology is simplified from Huntting and others (1961).

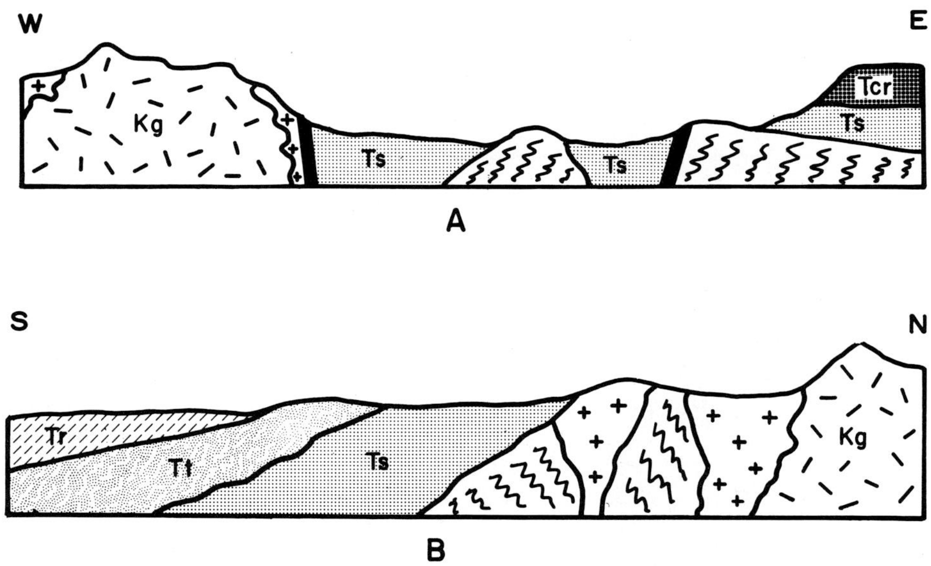

Figure 2. Generalized structure sections: (A) Chiwaukum graben, (B) Cle Elum syncline. Kg, Mount Stuart Granodiorite. Other formation codes as in Figure 1.

Williams, 1976). Evidence presented in this paper shows that arkosic sedimentary units of the Swauk Formation may be no older than Eocene. Therefore, the Mount Stuart batholith probably did not shed sediments until Eocene time.

Swauk Formation

The Swauk Formation is the major sedimentary rock unit in central Washington. As seen on the "Geologic Map of Washington" (Huntting and others, 1961; Weissenborn, 1969), it occupies a northwest-trending U-shaped pattern (TKc) around the Mount Stuart horst in an area of approximately 1,800 km^2.

The formation was named by Russell (1899) for the Swauk mining district in Kittitas County, and although he did not designate a type section, the Swauk Creek area between Swauk Pass and Liberty has been taken historically to be the type area for the unit. In the Swauk Creek valley, the formation is mostly light-colored, poorly indurated arkosic sandstone, with lesser amounts of dark shale and conglomerate, and has a total thickness of about 2,134 m (Rector, 1962). The base of the formation is not exposed in the type area, but the formation is overlain unconformably by the Eocene(?) Teanaway Basalt. In areas west of the type area, reported thicknesses of the Swauk Formation range from 1,219 to 4,572 m (Foster, 1960; Rector, 1962; Galster, 1956).

Greater thicknesses of Swauk Formation are present in the Chiwaukum graben, east of the Mount Stuart horst. Lupe (1971) measured a total thickness of 8,535 m for the Swauk Formation in the northern half of the graben. In the core of the graben near Leavenworth, Whetten (1976) estimated a minimum thickness of 5,800 m.

Structure within the graben is basically anticlinal, with the Eagle Creek anticline trending parallel to the bounding faults and exposing metamorphic basement rocks in its core. Therefore, the oldest Swauk rocks are found along the Eagle Creek anticline, and the youngest are found toward the margins and ends of the graben.

Southward toward Wenatchee the Leavenworth and Entiat faults die out, and in the area south of Wenatchee the nature of the Swauk Formation becomes more complex. Folds, faults, and intrusions are present; an intraformational unconformity has long been suspected (Chappell, 1936; Newman, 1971) and two have been recently mapped (Gresens, 1976b). Mapping in progress within the graben and Wenatchee areas will result in new names for parts of what originally was all mapped as Swauk Formation (Gresens, 1976a; Whetten, 1976; Tabor and others, 1977). The youngest Swauk rocks south and east of Wenatchee (the Wenatchee formation of Gresens, 1976a) reach a thickness of about 300 m and are bounded above and below by angular unconformities.

Correlation of the Swauk Formation from the Chiwaukum graben and Wenatchee areas to the Swauk Creek and Cle Elum syncline areas has been difficult because of structural complexity, lack of continuous beds, and the presence of the Teanaway Basalt only in the Swauk Creek and Cle Elum syncline areas.

The age of the entire Swauk Formation complex has been based in the past on stratigraphic position and fossil leaves. The Swauk Formation is younger than the Mount Stuart batholith, which has been dated as Late Cretaceous (88 m.y. B.P.) by Engels and Crowder (1971). In the Cle Elum syncline the Teanaway Basalt unconformably overlies the Swauk Formation and is in turn conformably overlain by the Roslyn Formation. The latter has been dated as middle and/or late Eocene on the basis of leaves, fish, and turtles (Smith, 1904; Hesse, 1936; Wheeler, 1955). The Teanaway Basalt has been assumed to be Eocene also because of its position directly beneath the Roslyn and because of its similarity to other Eocene volcanic rocks in Washington. Therefore, on the basis of its stratigraphic position, the Swauk Formation in the Cle Elum syncline has an age somewhere between Late Cretaceous and

Eocene. The Swauk Formation in the other areas has been assumed to have a similar age.

Fossil leaves are common in many of the dark shales of the Swauk Formation. Reports or discussions are found in G. O. Smith (1904), W. S. Smith (1916), Waters (1930), Chappell (1936), Willis (1950), LaMotte (1952), Alexander (1956), Yeats (1958), and Rector (1962). Willis and Rector each summarized all previous work, and each concluded that the bulk of the taxons indicated a Paleocene age, with certain reservations. As Willis stated (1950, p. 93):

> The identification and correlation of the fossil leaves of the Swauk Formation by Knowlton, Duror, Brown, and LaMotte suggest that the Swauk Formation is, for the most part, Paleocene in age. However the limitations of paleobotanical correlations and the preliminary status of the work must be recognized. Parts of the Swauk Formation may be uppermost Cretaceous or early Eocene in age.

Rector (1962, Pl. 1) charted 50 leaf taxons and their reported occurrences in the Swauk and other formations in the United States, Canada, and Greenland. According to Rector's chart, 4 taxons from the Swauk indicate Cretaceous to Eocene age, 4 indicate Paleocene age, 7 indicate Paleocene or Eocene age, 4 indicate Eocene age, 6 indicate Paleocene to Oligocene age, 3 indicate Paleocene to Miocene age, 2 indicate Paleocene to Pliocene age, and 1 indicates Paleocene to Pleistocene age. Rector stressed that the stratigraphic positions of collections from the Swauk Formation were not checked and that the age assignments from elsewhere were taken directly as stated in the literature. Since the greatest coincidence of occurrences is Paleocene according to his compilation, Rector concluded that the Swauk Formation is most likely Paleocene.

It is clear from the biostratigrapher's point of view, however, that the range zones (American Commission on Stratigraphic Nomenclature, 1970; International Subcommission on Stratigraphic Classification, 1976) of fossil leaf taxons have not been established in the Swauk Formation or in most other formations. This is true even for the Paleocene Fort Union Formation with which the Swauk has been correlated by several paleobotanists through the years. For example, see the monographic work on the Paleocene flora by Brown (1962) which does not define the range zone of even one taxon in terms of measured stratigraphic sections. The age determinations and correlations based on scattered collections of fossils leaves always will be suspect until local range zones in reference sections are established, region by region. Then only will there be adequate and reliable bases for age determination and correlation by leaves.

Although Brown's work was mainly on the Paleocene flora of the Rocky Mountains and Great Plains, he made the following comment about the Swauk Formation (1962, p. 28):

> Resting upon a basement of igneous and metamorphic rocks, the Swauk is apparently the earliest Tertiary formation of Washington. At some localities it contains leaves that inclined Knowlton to regard the formation as probably of the same age as the Denver and Fort Union formations (Weaver, 1937, p. 47-49). An Eocene age appears to be a more likely assignment.

Unfortunately, however, Brown did not document this statement in any way.

Teanaway Basalt

The Teanaway Basalt is present in the Cle Elum syncline and Swauk Creek areas, unconformably overlying the Swauk Formation and conformably overlain by the Roslyn Formation. The name was proposed by G. O. Smith (1903, 1904) for the river in whose basin the formation is most extensively developed (Smith and Calkins, 1906, p. 5). Foster (1960) stated that the rocks along the Middle Fork of the Teanaway River should be considered the type area, but the complex mixture of lavas and clastic rocks makes a measured and described section of little value in characterizing the sequence.

The Teanaway Basalt ranges from about 305 m to more than 1,524 m in thickness and consists of basaltic lava, tuff, breccia, and some conglomerates, sandstones, and shales. Foster (1958, 1960) interpreted the rocks as being deposited on the slopes of volcanos that perhaps were fed by some of the dikes of the adjacent Teanaway dike swarm.

The only fossils reported previously from the sedimentary rocks of the Teanaway Basalt are a few fragments of mollusks indentified as *Viviparus*, a long-ranging genus (Foster, 1960). Since the Teanaway unconformably overlies the Swauk Formation and is overlain conformably by the Roslyn Formation, dated as middle and/or late Eocene, the Teanaway has been assumed to be Eocene also.

Roslyn Formation

The Roslyn Formation is found mainly in the Cle Elum syncline area, which is the Teanaway River drainage basin. The formation was named by Russell (1899), but a type section was not designated. The best exposures are found along the forks of the Teanaway River where detailed descriptions were made by Bressler (1951, 1957). He found the formation to be about 1,920 m thick and rich in arkosic sandstone and conglomerate. The upper part of the formation contains eight major coal beds in the vicinity of the towns of Roslyn and Cle Elum. This Roslyn coal field is the only one of any significance on the east side of the Cascade Range.

The Roslyn Formation conformably overlies the Teanaway Basalt and is unconformably overlain by the Miocene Columbia River Basalt and possibly by some Oligocene(?) volcanic rocks. The age of the Roslyn Formation has been determined to be middle and/or late Eocene on the basis of leaves by Knowlton (in G. O. Smith, 1904), fish by Hesse (1936), and turtles by Wheeler (1955). Thus it has been the only sedimentary rock unit in the central Washington study area to be dated paleontologically with some confidence.

Naches Formation

The Naches Formation is found along the Naches River valley southwest of the Cle Elum syncline, in scattered patches between the two areas, and in the area between Kachess Lake and Snoqualmie Pass west of the Cle Elum syncline. The formation was named by Smith and Calkins (1906), who described it as interbedded sandstone and basalt most extensively found along the Naches River. It unconformably overlies the Easton Schist and is unconformably overlain by the Keechelus Formation. Foster (1960, 1967) mapped the Naches Formation in the area between Kachess Lake and Snoqualmie Pass, where he described it as a thick, widespread unit of sedimentary rocks and basalt with a few rhyolitic rocks. Basalt generally is more abundant in the lower part of the formation. The total thickness is approximately 1,524 m.

Stout (1964) mapped much of the southern area containing the Naches Formation and described the rock types as arkosic to feldspathic sandstones, basalt, rhyolitic rocks, and andesitic and volcaniclastic rocks. The arkosic to feldspathic sandstones are very similar to those of both the Swauk and Roslyn Formations. Thicknesses of the Naches Formation range from 2,438 to 3,048 m in Stout's area.

Stratigraphic relationships of the Naches Formation with the Swauk and Roslyn Formations have been difficult to determine because of faulting. Most authors have stressed the volcanic content of the Naches as its most significant, and different, characteristic. On the other hand, it is highly arkosic like both the Swauk and Roslyn, and it is coal-bearing like the Roslyn.

Fossil leaves have been reported from several Naches localities by Smith (1904), Smith and Calkins (1906), Beck (1934), Stout (1964), and Foster (1967). Most of the determinations indicate an Eocene age for much of the Naches, and a correlation with the Puget Group has been suggested. Another suggestion—that the Naches Formation and the type Swauk Formation near Liberty are at least in part contemporaneous (Smith and Calkins, 1906; Weaver, 1937)—has not been considered likely because of the longstanding belief that the Swauk is Paleocene (Foster, 1960; Stout, 1964).

Guye Formation

The name Guye Formation was first used by Smith and Calkins (1906) for a group of sedimentary and volcanic rocks in the vicinity of Snoqualmie Pass. Later, Foster (1960) split the original Guye Formation into four formations, one being the Guye Formation (restricted) which is found in two patches at Snoqualmie Pass (TKc on "Geologic Map of Washington," Huntting and others, 1961). The type area is along Coal Creek and its tributaries on the east side of Snoqualmie Pass. The formation consists of at least 1,524 m of black leaf-bearing shales interbedded with dark coarse- and fine-grained arkosic sandstones and conglomerates of subangular fragments of black and gray chert. The Guye Formation appears to be unconformably overlain by the Mt. Catherine Rhyolite which in turn is overlain conformably by the Naches Formation (Foster, 1960). An Eocene or Paleocene age for the Guye Formation was interpreted by Foster (1960) on the basis of leaves identified by R. W. Brown.

The Miocene Snoqualmie batholith intruded the sedimentary rocks in the Snoqualmie Pass area and altered them substantially. Black shale of the type Guye Formation, although still containing abundant, clear leaf impressions, is very hard; since the samples collected for this study were barren of palynomorphs and yielded only carbonized plant debris, the shale must have been heated considerably.

Puget Group

The Puget Group in the Green River canyon area has been described by Vine (1969), who mapped the Cumberland and adjacent quadrangles. Apparently the first use of the name Puget Group was by White (1888, 1889) for the coal-bearing rocks in the Puget Sound lowland. Subsequently the name was applied widely to the nonmarine coal-bearing rocks by Willis (1897, 1898) and Evans (1912), who assigned an Eocene age to these rocks. Later Weaver (1937) stated that "it is probable that the lowest strata in the Puget Group date back into the early Eocene and that the entire sequence was deposited during the middle and late Eocene and possibly to a small extent in the very earliest part of the Oligocene epoch."

Vine (1969) described the Puget Group rocks in the Green River canyon area as nonmarine sedimentary rocks consisting of feldspathic or arkosic sandstone, siltstone, shale, corbonaceous claystone, coal, and minor conglomerate. The Puget Group is exposed in a series of folds; the oldest rocks appear along the axis of the Lawson anticline, and the youngest crop out along the axis of the Kummer syncline, where they are overlain by volcanic rocks of Oligocene age.

The total thickness of the Puget Group in the Green River canyon area is not known because the base is not exposed. However, the exposed rocks have an estimated thickness of 1,890 m (Vine, 1969). Lithologically, the section can be divided into the lowest beds (which contain no significant coal), the Franklin Sandstone, the Franklin coal zone, the Kummer Sandstone, and the Kummer coal zone. These units have yielded no invertebrate fossils that could be used for correlation with marine units in the region. However, many leaf collections have been made through much of the section, and these, combined with physical evidence, have been used for correlation to other sections that contain leaf floras and marine faunas (Vine, 1962, 1969; Wolfe, Gower, and Vine, 1961; Wolfe, 1968). Wolfe (1968) proposed four floral stages based on a succession of leaf assemblages found in the Green River canyon section and vicinity. These stages are the Franklinian, Fultonian, and Ravenian, interpreted by Wolfe to be of early, middle, and late Eocene age, respectively, and Kummerian, interpreted to be of early Oligocene age.

Correlation of the Green River canyon section northward to the Tiger Mountain–Taylor Mountain area by Vine (1962, 1969) suggests that the lower part of the Green River canyon section is equivalent to the Raging River Formation. The upper Raging River Formation has yielded marine invertebrates indicating a middle Eocene age and foraminifera of the *Bulimina* cf. *B. jacksonensis* zone (of Rau, 1958) indicating a late middle Eocene (Ulatisian) to early late Eocene (Narizian) age similar to the McIntosh Formation of southwestern Washington. Recent reviews of these faunas indicate that they are Narizian and that the Narizian Stage is mostly Middle Eocene (J. M. Armentrout, 1978, written commun.). Therefore, it is possible that the lowest Puget Group rocks of the Green River canyon section are no older than middle Eocene. This is an important point when correlations of the central Washington formations are considered.

PALYNOLOGIC INVESTIGATION

Field Methods

Field work for this project was done at various times from 1967 to 1974. Samples were collected from 21 localities in central Washington and 13 localities in the Green River canyon section of western Washington. General locations of collected sections are shown in Figure 1, and Appendix 1 lists the locality names, geographic descriptions, and stratigraphic positions of samples. Central Washington localities include 13 in the Swauk Formation, four in the Roslyn Formation, three in the Naches Formation, and one in the Guye Formation. Three of the Swauk localities and the single Guye locality were barren, however, and provided no data. A total of 13 localities, mostly U.S. Geological Survey paleobotanical localities, was sampled in the Green River canyon section. All of these yielded palynomorphs. More detailed descriptions of all sample localities and lithologies are provided in Appendix 1.

Samples were collected from roadcuts, stream banks, mine dumps or adits, and quarries. In each case an attempt was made to obtain fresh material by digging below weathered surfaces. Spot samples of single rock types were collected for the most part. Gray, brown, and black claystones, siltstones, and carbonaceous shales are the dominant rock types sampled along with a few coals. Multiple samples were collected from some exposures, but for this paper the results from each locality are lumped together. Locations were determined by using published geologic and topographic maps and aerial photographs.

Laboratory Methods

Cleaned rock samples of approximately 2-g size were crushed into small pieces and treated successively with hydrofluoric acid, hydrochloric acid, nitric acid or sodium hypochlorite, and potassium hydroxide or potassium carbonate in order to remove minerals, break down organic matter, and to concentrate the more resistant palynomorphs. Procedures were modified as needed in order to obtain the best recovery from each rock type. Cover glass smears prepared from the residues and Clearcol were cemented to microscope slides with Permount or Histoclad.

Microscope examination utilized magnifications of 100X to 1,000X. Qualitative analyses were made to determine the nature of the assemblages and presence or absence of distinctive taxons, because the aim of the study was to determine if some range zones could be established. Less attention was paid to abundances because quantitative analyses are not as significant in a reconnaissance investigation, especially when preservation is highly variable. A list of identified taxons is given for each productive locality in Appendix 1. The several hundred microscope slides used in this study are stored in my personal collections.

General Results

The quality of preservation and abundance of palynomorphs, almost entirely plant spores and pollen, recovered in this study varied from very poor to good. In general, the older rocks and those most affected by folding, faulting, igneous activity, and weathering yielded the poorest specimens or none in a few cases. The younger and flatter-lying rocks, on the other hand, yielded better and more abundant specimens. The type Swauk Formation, Naches Formation, and parts of the Chiwaukum graben section yielded, in general, specimens of only poor to fair quality and abundance. Much better preserved and more numerous specimens were recovered from the Roslyn Formation, from the youngest Swauk rocks in the Wenatchee area, and from parts of the Puget Group.

Thermal alteration and destruction of palynomorphs are suspected in the areas fringing the Snoqualmie batholith. The Thorp Creek locality in the Swauk Formation northwest of Cle Elum Lake and the Guye Formation locality near Snoqualmie Pass yielded only highly altered, finely divided, carbonized residues indicating extensive thermal alteration caused by temperatures over 300 °F (150 °C) (Wilson, 1964). The Gale Creek locality in the Naches Formation west of Kachess Lake did yield a few palynomorphs, but since they are dark colored and sparse, slightly less thermal alteration probably occurred in this area.

The other barren samples were collected from the Swauk Formation at Iron Peak and Tronsen Creek near the southern margins of the Mount Stuart horst. Both localities yielded some plant debris but no recognizable pollen or spores. Weathering rather than thermal alteration might be the cause of the poor results in these localities. The numerous basalt dikes of the Teanaway dike swarm do not seem to have affected adjacent sedimentary rocks very far from their contacts.

Redeposition of palynomorphs from older rocks into the formations studied is suspected in a few cases. Samples from the type Swauk Formation and the youngest Swauk rocks south of Wenatchee appear to contain some redeposited specimens, judging from their darker color or anomalous assemblages. However, the palynologic biostratigraphy of a region has to be known quite well before patterns of redeposition from older to younger formations can be identified clearly.

Figure 3 illustrates the stratigraphic occurrences of selected palynomorphs in the Cle Elum syncline section, in the Chiwaukum graben section, and in the Green River canyon section.

PALYNOLOGIC BIOSTRATIGRAPHY

Palynomorph Zonation in Central Washington

Results obtained from the stratigraphic succession of type Swauk and Roslyn Formations in the Cle Elum syncline area and from the anticlinal succession of Swauk beds in the Chiwaukum graben and Wenatchee area suggest the presence of three palynomorph concurrent-range zones (American Commission on Stratigraphic Nomenclature, 1970; International Subcommission on Stratigraphic Classification, 1976). The zones are regarded here as informal, operational biostratigraphic units (biozones) whose boundaries are not yet defined because of the reconnaissance nature of this study, the widely scattered samples, and structural complexities. Figure 4 diagrammatically illustrates the relative stratigraphic positions of these biozones and their geologic ages as interpreted in this study. When the many thousands of metres of section involved and the probably long timespan they represent are considered, one concludes that these three zones represent just a few of many that could ultimately be defined in these rocks with detailed sampling.

The oldest palynomorph biozone recognized is the *Pistillipollenites* and *Platycarya* concurrent-range zone, defined by the overlapping range zones of these two pollen genera. Other distinctive pollen and spores commonly present, but not restricted to this zone, include *Proteacidites, Caryapollenites, Alnus, Bombacacidites, Cicatricosisporites, Polypodiidites,* and *Trilites solidus.* This biozone is present in the type Swauk Formation along Swauk Creek, in the upper Swauk Formation at Cle Elum Lake, probably in the Naches Formation along Taneum Creek, and in the lower Swauk Formation of the Chiwaukum graben at the Derby Creek, North Fork Derby Creek, and Easy Street localities. All published and unpublished data from western North America that I have seen indicate that this concurrent-range zone is Eocene, not Paleocene or Cretaceous.

The middle palynomorph biozone is the *Platycarya* and *Bursera* concurrent-range zone, defined by the overlapping range zones of these two pollen genera. *Pistillipollenites* is absent, its source plant having become extinct or having migrated from the region. Other distinctive taxons which may be present, but are not restricted to this zone, include *Liquidambar, Ulmus, Tilia,* cf. *Plicatopollis,* cf. *Quercus*, tricolporate-striate pollen, and triporate-spinate pollen. This biozone is present in the Roslyn Formation at Roslyn and Cle Elum and possibly in the Plain locality of the upper Swauk Formation in the northern Chiwaukum graben. As stated before, the Roslyn Formation has been interpreted as being of middle and/or late Eocene age on the basis of leaves and vertebrates. The upper Roslyn Formation bears palynomorphs of the *Platycarya* and *Bursera* concurrent-range zone, which is interpreted to be late Eocene.

The youngest biozone is the *Gothanipollis* and *Elaeagnus* concurrent-range zone, defined by the overlapping range zones of these two pollen genera. *Platycarya* usually is absent; its rare presence in this biozone might be caused by redeposition. Other distinctive taxons which might be present, but not restricted to the zone, are cf. Acanthaceae, *Symplocoipollenites, Nyssa,* cf. *Engelhardia,* and *Selaginella* cf. *S. sinuites.* This biozone is present in the Lovitt quarry and Clay mine localities south of Wenatchee and in the Blue Grade Road locality east of the Columbia River near Wenatchee (formerly upper Swauk Formation; now Wenatchee formation of Gresens, 1976a). The Blue Grade Road locality is considered to be the best reference section for this biozone because it appears to have little or no redeposition compared to the Lovitt quarry and Clay mine localities. This biozone is considered to be distinctly younger than that represented in the upper Roslyn Formation and is, therefore, of Oligocene age.

Palynomorph Zonation in Puget Group, Western Washington

The palynomorph zonation recognized in the central Washington formations has been found, in part, in the Puget Group at Green River canyon, western Washington. With reference to the Green River canyon section in Figure 3, simplified from Vine (1969, Pl. 2), the *Platycarya* and *Bursera* concurrent-range zone definitely is represented in the lower middle part of the Franklin coal zone. Cf. *Plicatopollis* is present, and *Anacolosidites* is a noticeable taxon in this section.

Pistillipollenites is present in the lowest Puget Group samples below the Franklin Sandstone, the top of its range zone occurring about 152 m above the base of the section. So far, however, *Platycarya* has not been found in this section below the Franklin Sandstone.

The youngest Puget Group beds (the Kummer coal zone) and the overlying volcanic rocks contain some of the accessory taxons of central Washington's upper palynomorph biozone, such as cf. Acanthaceae, cf. *Engelhardia,* and *Selaginella* cf. *S. sinuites.* However, *Gothanipollis* and *Elaeagnus* have not been found. This part of the Green River canyon section may be equivalent to the upper zone of central Washington (Wenatchee formation), or it may be slightly older since it possibly overlaps with the range zone of *Platycarya.* It is difficult to tell until more zonation data become available from Eocene and Oligocene sections in the region.

Palynologic Correlations

The correlations suggested by the palynomorph occurrences in central Washington, in the Green River canyon

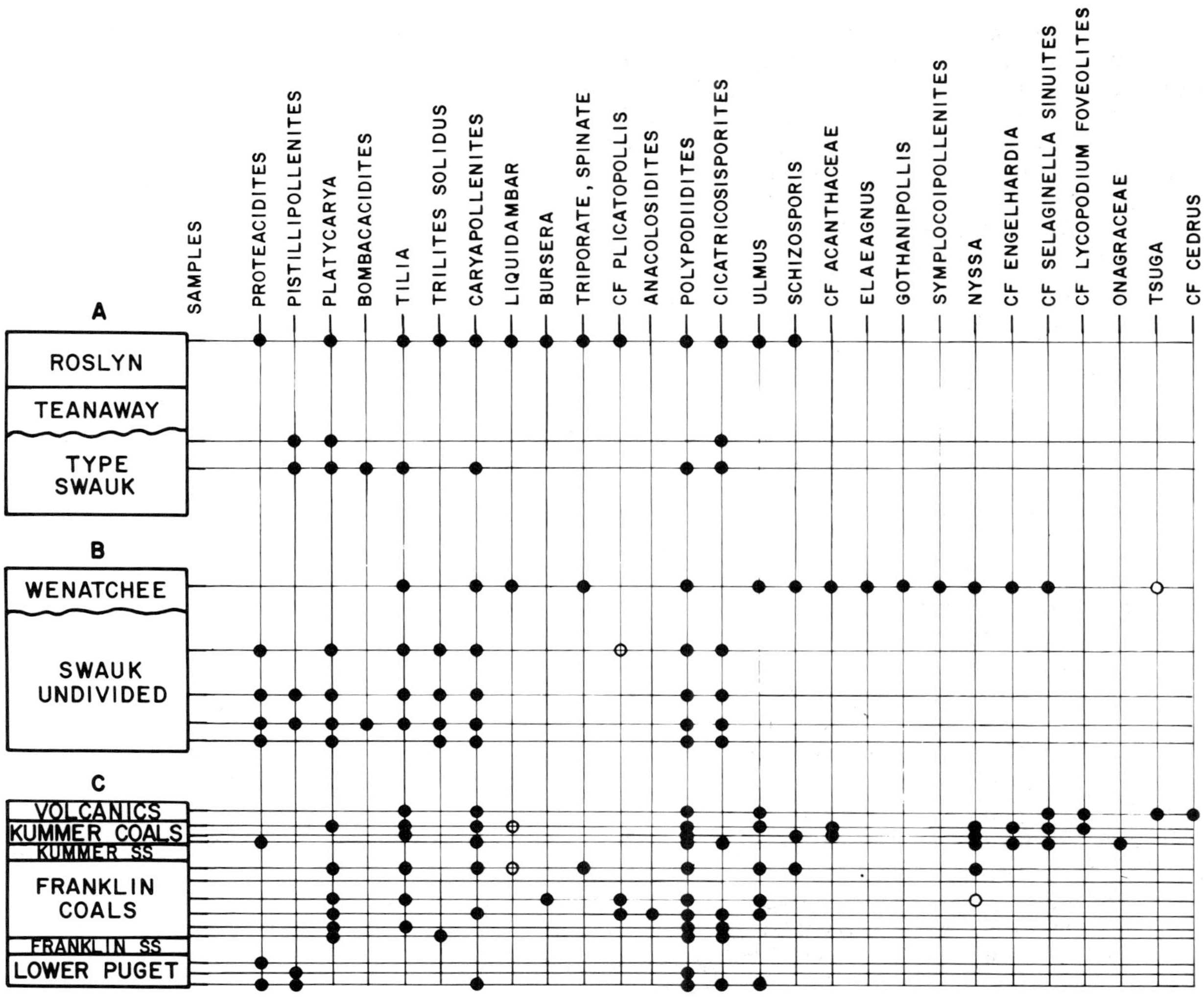

Figure 3. Occurrences of selected palynomorphs in (A) Cle Elum syncline, (B) Chiwaukum graben and Wenatchee area, (C) Green River canyon section of Puget Group. Open circles used where identifications are not certain.

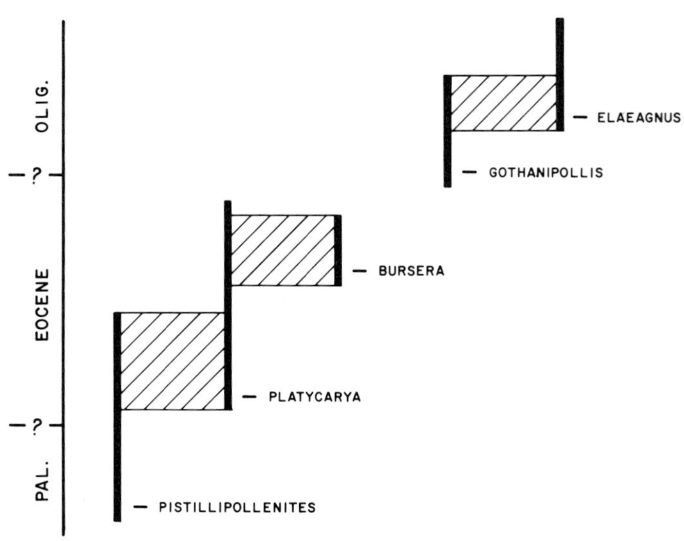

Figure 4. Diagram illustrating relative stratigraphic positions of concurrent-range zones.

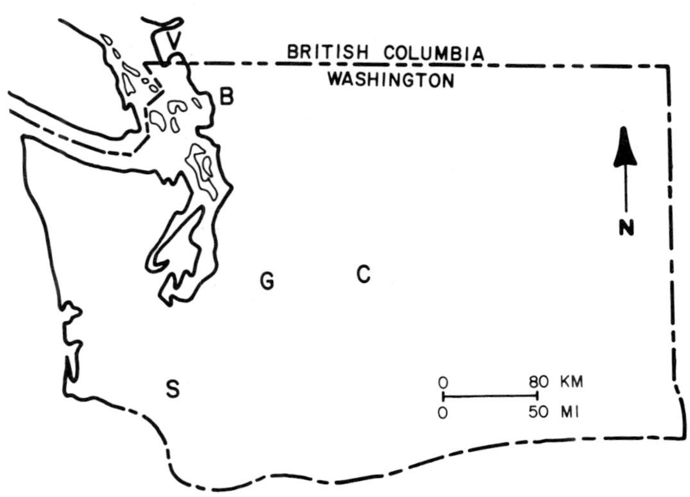

Figure 5. Locations of stratigraphic sections shown in Figure 6: (S) southwestern Washington; (G) Green River canyon; (C) central Washington; (B) Bellingham, Washington; (V) Vancouver, British Columbia.

section of the Puget Group in western Washington, and in other areas in Washington and British Columbia (Fig. 5) are shown in Figure 6. The sequence of type Swauk Formation, Teanaway Basalt, and Roslyn Formation in the Cle Elum syncline is biostratigraphically equivalent and, therefore, probably time-equivalent to most of the Puget Group in the Green River canyon section. The type Swauk Formation probably included rocks older than the lowest exposed rocks of the Puget Group. This conclusion is based on the limited occurrences of *Pistillipollenites* in the lowest exposed Puget Group rocks. The upper, coal-bearing part of the Roslyn Formation appears to be equivalent to the lower middle part of the Franklin coal zone of the Puget Group.

Since the Naches Formation samples from Taneum Creek yielded both *Pistillipollenites* and *Platycarya*, that part of the Naches Formation appears to be time-equivalent to the type Swauk Formation and lowest Puget Group.

The Swauk Formation in the Chiwaukum graben appears to contain rocks equivalent to much or all of the Cle Elum syncline section plus a younger unit (Wenatchee formation of Gresens, 1976a) that unconformably overlies the older sequence. The Wenatchee formation contains some pollen similar to that of the Kummer coal zone and overlying volcanic rocks in the Puget Group section.

Present interpretations of the Puget Group section indicate an Eocene to Oligocene age based on leaves and correlations to sections containing marine invertebrates (Wolfe, 1968; Vine, 1969). As stated earlier in this paper, it is possible that the base of the Puget Group section is no older than middle Eocene. If this is correct, then the central Washington sequence likewise is Eocene to Oligocene in age, and most likely no older than early Eocene at its base (type Swauk Formation and equivalents). Previous age assignments of Cretaceous and Paleocene for the Swauk formation, based largely on fossil leaves, are incorrect. Radiometric ages cited by Gresens (1976a), Whetten (1976), Frizzell and Tabor (1977), and Vance and Naeser (1977) are tending to support the conclusions based on stratigraphic palynology.

Palynomorph assemblages from the lower Chuckanut Formation at Bellingham, Washington, reported by Griggs (1970), lack *Pistillipollenites, Platycarya*, and *Polypodiidites,* but do contain *Alnus* and *Caryapollenites.* Therefore, the lower half of the Chuckanut, or perhaps all of it, is of Tertiary age and older than Eocene; hence it is of Paleocene age, older than all of the formations shown in Figure 6.

The Burrard and Kitsilano Formations of the Vancouver, British Columbia, area contain, for the most part, palynomorphs of the *Pistillipollenites* and *Platycarya* concurrent-range zone as interpreted from the data of Rouse (1962), Hopkins (1969), and Rouse and others (1970). However, the upper part of the Kitsilano Formation lacks *Pistillipollenites* and contains palynomorphs of the *Platycarya* and *Bursera* concurrent-range zone. Therefore, the Burrard and Kitsilano Formations are equivalent to the Swauk, Teanaway, and Roslyn Formations of central Washington and to the lower and middle parts of the Puget Group of the Green River canyon section.

Sparks (1967) reported on palynomorphs from the McIntosh, Skookumchuck, and Lincoln Creek Formations of southwestern Washington and northwestern Oregon. Although it is difficult to determine range zones from his data, the overall aspect of the assemblages through those formations suggests that the *Pistillipollenites* and *Platycarya* concurrent-range zone and the *Platycarya* and *Bursera* concurrent-range zone are represented in the McIntosh and Skookumchuck Formations. These formations have been dated as middle and late Eocene on the basis of marine invertebrates.

SUMMARY AND CONCLUSIONS

In summary, it must be emphasized that this has been a reconnaissance study with quite widely scattered samples that give only a partial biostratigraphic picture for the formations involved. Palynomorphs were recovered from most samples, but the quality of preservation and abundance of specimens are highly variable. The poor or barren samples appear to have been affected by weathering, folding and faulting, or thermal effects of igneous activity.

Three palynomorph biozones that appear to be useful for correlation have been detected, but their boundaries are not yet defined in reference sections. It seems likely that more palynomorph biozones remain to be found, especially in very thick sections such as those in the Chiwaukum graben.

The overall results indicate that the central Washington formations generally are equivalent to the Puget Group and related western Washington formations, and all are part of the same complex of Eocene coastal plain fluvial and swamp sediments and volcanic rocks. The Swauk Formation (undivided), often dated as Cretaceous and Paleocene in the past, is part of this Eocene complex except for the uppermost part in the Wenatchee area which is Oligocene and part of a separate depositional sequence.

The palynologic dating of the type Swauk Formation and basal Swauk Formation in the Chiwaukum graben indicates that the Mount Stuart batholith reached the surface and began to shed sediments in Eocene time and that the Chiwaukum graben subsided and received arkosic sediments during Eocene time.

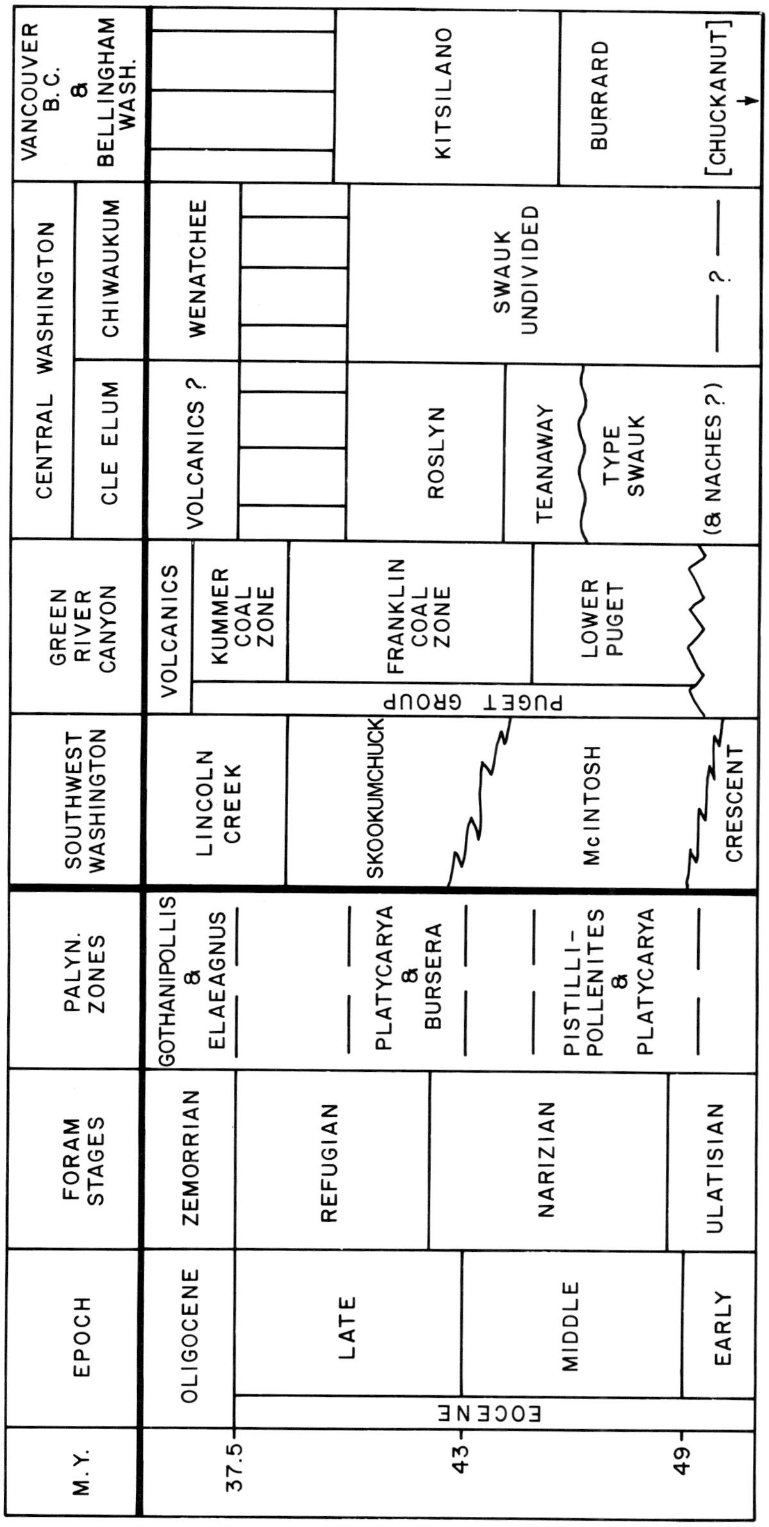

Figure 6. Stratigraphic correlations suggested by palynomorph occurrences in Washington and British Columbia.

APPENDIX 1. SAMPLE LOCALITIES, LITHOLOGIES, AND PALYNOMORPHS

SWAUK FORMATION

Swauk Creek: Center N½ sec. 15, T. 21 N., R. 17 E., Kittitas County, in type area of Swauk Formation. Low roadcut along west side of Hwy. 97 exposes dark shales among lenticular sandstones. Gray shale collected from about 10 ft of section in middle portion of roadcut.

Pollen: Pistillipollenites mcgregorii
Platycarya sp.
cf. Bombacacidites
Caryapollenites sp.
cf. Ulmoideipites
Tricolpites spp.
Triporate, smooth
Tricolporate, microreticulate, small
Tricolpate or tricolporate, striate
Syncolpate, verrucate
Tricolpate, smooth
Taxodiaceaepollenites sp.
Bisaccate pollen
cf. Tilia

Spores: Laevigatosporites spp.
Polypodiidites spp.

Acritarchs: cf. Sigmopollis psilatus

Swauk Pass: Center SW¼ sec. 4, T. 21 N., R. 18 E., Kittitas County, in type area of Swauk Formation. Large roadcut on north side of Hwy. 97 exposes dark shales among lenticular sandstones. Gray claystone and siltstone collected from lower, middle, and upper portions of cut.

Pollen: Pistillipollenites mcgregorii
Platycarya sp.
Alnus sp.
Monoporate?, smooth
Taxodiaceaepollenites sp.

Spores: Cicatricosisporites spp.
cf. Deltoidospora
Laevigatosporites sp.
Fungal spores
Several large spores, unidentifiable, redeposited?
Osmunda sp.

Algae: Pediastrum??

Tronsen Creek: SE¼ NE¼ sec. 18, T. 22 N., R. 18 E., Chelan County, near southeast corner of Mount Stuart horst, near Leavenworth fault zone. Roadcut along north side of Hwy. 97 exposes dark claystones among sandstones. One sample of dark-gray claystone collected; hard, brittle, fractured.

Palynomorphs: None identifiable. Residue contains fuzzy and flaky organic material plus a few black, carbonized spores(?).

Iron Peak: NW¼ SE¼ sec. 12, T. 22 N., R. 15 E., Kittitas County, in basal Swauk Formation south of summit of Iron Peak. One sample of black shale, containing plant impressions, collected from among basalt dikes of Teanaway swarm.

Palynomorphs: None. Residue contains sparse plant debris in fair condition, but no palynomorphs could be found.

Thorp Creek: Center NE¼ sec. 26, T. 22 N., R. 13 E., Kittitas County, in basal(?) Swauk Formation about 100 to 200 ft above contact with Easton Schist. Blocks of black shale collected from first gully east of contact that is visible in trail. Shale contains plant megafossils.

Palynomorphs: None. Residue consists of black, fuzzy organic material only.

Morgan Creek: SW¼ NW¼ sec. 16, T. 21 N., R. 14 E., Kittitas County, in upper Swauk Formation among basalt dikes. Ledges are exposed along shore of Cle Elum Lake at low water stage, south of Morgan Creek. With reference to Foster (1960, p. 108 and Plate 1) and to the Geologic Map of Washington (Huntting and others, 1961), this site originally was thought to be in the lower Teanaway Basalt. However, it probably is upper Swauk Formation (oral comm., R. W. Tabor, 1977). One sample of dark shale collected from bed below a conglomerate containing large dark shale clasts.

Pollen: Pistillipollenites mcgregorii
Platycarya sp.
Alnus sp.
Triporate, smooth
Taxodiaceaepollenites sp.

Spores: Cicatricosisporites sp.
Osmunda sp.
Laevigatosporites sp.
Deltoidospora sp.
Fungal spores

Derby Creek: NE¼ sec. 2, T. 24 N., R. 18 E., Chelan County, in basal Swauk Formation, west side of Eagle Creek anticline, along Road #256. Lowest exposures of Swauk are west of the contact with metamorphics, in a cut on north side of road, estimated to be a few hundred feet above the base of the formation. One sample of tan, carbonaceous siltstone collected from among sandstone, conglomerate, claystone, and tuff; sample a few feet below tuff.

Pollen: Platycarya sp.
Caryapollenites sp.
cf. Ulmoideipites
cf. Pterocarya
cf. Bombacacidites
cf. Tilia
Proteacidites?
Alnus sp.
Tricolpites sp.
Tricolpate spp. (foveolate; striate)
Tricolporate spp. (smooth, small; microreticulate, small)
Bisaccate pollen

Spores: Osmunda sp.
Laevigatosporites sp.
Trilites solidus
Cicatricosisporites sp.
Polypodiidites sp.

North Fork Derby Creek: SW¼ SE¼ sec. 34, T. 25 N., R. 18 E., Chelan County, in lower Swauk Formation, west side of Eagle Creek anticline, estimated 1,000 to 2,000 ft above base of formation. Exposures are along Road #256A, about 1 mile north of junction with Road #256. Samples of gray claystone and siltstone collected from intervals between sandstones.

Pollen: Pistillipollenites mcgregorii
Platycarya sp.
Ilex?
Caryapollenites sp.
Alnus sp.
Proteacidites sp.
cf. Bombacacidites
cf. Tilia
Pterocarya sp.
Erdtmanipollis sp.
Triporate, smooth, spp.
Tricolporate spp.
cf. Liliacidites
Taxodiaceaepollenites sp.
Bisaccate pollen

Spores: Laevigatosporites sp.
Cicatricosisporites sp.
Gleicheniidites sp.
Deltoidospora spp.
Polypodiidites spp.
Trilites solidus
cf. Lycopodiumsporites

Plain: NW¼ sec. 1, T. 26 N., R. 17 E., Chelan County, in upper Swauk Formation of northern Chiwaukum graben. Three samples of dark shale collected from roadcut, west side of road, about 2 miles NNW of Plain, just north of junction with Road #2736.

Pollen: Platycarya sp.
cf. Plicatopollis(?)
Ilex sp.
Proteacidites sp.
Alnus sp.
cf. Tilia
Caryapollenites sp.
Monosulcites sp.
Tricolpate spp.
Tricolporate spp.
Taxodiaceaepollenites sp.
Bisaccate pollen

Spores: Laevigatosporites sp.
Polypodiidites spp.
Trilites solidus
Cicatricosisporites spp.
Deltoidospora sp.
cf. Selaginella?

Easy Street: NE¼ sec. 19, T. 23 N., R. 20 E., Chelan County, on Easy Street west of Sunnyslope and just north of Highway 2. Samples of gray, carbonaceous siltstone collected from middle part of roadcut on north side of road, in lower(?) Swauk Formation of southeast end of Chiwaukum graben.

Pollen: Pistillipollenites mcgregorii
Platycarya sp.
cf. Tilia
cf. Ulmus
Alnus sp.
Caryapollenites sp.
Pterocarya sp.
Proteacidites sp.
Ilex sp.
Triporate, smooth
Tricolpate and tricolporate spp.
Taxodiaceaepollenites sp.
Bisaccate pollen

Spores: Laevigatosporites sp.
Polypodiidites spp.
Deltoidospora sp.
Cicatricosisporites sp.
Osmunda sp.
Trilites solidus
Cingulatisporites?

Lovitt Quarry: SW¼ NE¼ sec. 21, T. 22 N., R. 20 E., Chelan County, in upper Swauk Formation south of Wenatchee (Wenatchee formation of Gresens, 1976a) Sample of gray siltstone collected by V. E. Livingston -- his locality W-22, a fossil leaf locality. This is an active silica sand quarry, known now as quarry of Wenatchee Silica Products, Inc.

Pollen: Elaeagnus sp.
Gothanipollis sp.
Symplocoipollenites spp.
Liquidambar sp.
Tilia sp.
Ilex sp.
cf. Engelhardia (15-20 microns)
Ulmus sp.
Caryapollenites sp.
Pterocarya sp.
Alnus sp.
Juglans?
Acer?
Tricolporate, smooth, small (abundant)
Tricolpate and tricolporate spp.
Triporate, smooth spp.
Schizosporis sp., smooth
cf. Sterculiaceae
Monosulcites?
Taxodiaceaepollenites sp.
Bisaccate pollen
Platycarya sp. (redeposited??)
Proteacidites sp. "
cf. Plicatopollis? "

Spores: Cicatricosisporites sp. (redeposited??)
Laevigatosporites sp.
Polypodiidites spp.
Osmunda sp.
Fungal spores
Trilete?, spinate, 20-25 microns

Clay Mine: NW corner of sec. 27, T. 22 N., R. 20 E., Chelan County, in upper Swauk Formation (Wenatchee formation of Gresens, 1976a). Sample of dark gray to black carbonaceous claystone collected by V. E. Livingston from adit of an old clay mine -- his locality W-20.

Pollen: Alnus sp.
Caryapollenites sp.
cf. Engelhardia (16-19 microns)
cf. Quercus
cf. Sterculiaceae
Liquidambar?
Tricolpate and tricolporate spp.
Platycarya sp. (redeposited??)
cf. Plicatopollis? "

Spores: Polypodiidites sp.
Cicatricosisporites sp. (redeposited??)

Blue Grade Road: Center sec. 23, T. 23 N., R. 20 E., Douglas County, in upper Swauk Formation east of Columbia River (Wenatchee formation of Gresens, 1976a). Samples collected from two pits along north side of road, from approximately 42 ft of section; 8 samples of gray siltstone collected from western pit, and 5 samples of brown, gray, and black claystone, siltstone, and carbonaceous shale collected from eastern pit.

Pollen: Elaeagnus sp.
Gothanipollis spp.
cf. Acanthaceae
cf. Nyssa
Symplocoipollenites sp.
Tricolpate and tricolporate spp.
Tetracolporate spp.
Tilia sp.
Liquidambar sp.
cf. Sterculiaceae
Juglans sp.
Pterocarya sp.
Caryapollenites sp.
cf. Engelhardia
Ulmus sp.
Alnus sp.
Triporate, smooth spp.
Triporate, spinate, small
cf. Salix
Schizosporis sp., smooth
Taxodiaceaepollenites sp.
Bisaccate pollen
cf. Tsuga

Spores: Polypodiidites spp.
Laevigatosporites sp.
Selaginella sinuites
Deltoidospora sp.
Fungal spores
cf. Lycopodium?

ROSLYN FORMATION

No. 3 Mine: SE¼ sec. 7, T. 20 N., R. 15 E., Kittitas County, in upper Roslyn Formation. Sample of coal collected from mine dump, just east of town of Ronald.

Pollen: Platycarya sp.
Bursera sp.
Alnus sp.
cf. Ulmoideipites
Caryapollenites sp.
cf. Momipites
Proteacidites?
Tilia sp.
Schizosporis sp., smooth
Monosulcites sp.
Tricolpate and tricolporate spp.
Triporate, smooth spp.
Taxodiaceaepollenites sp.

Spores: Polypodiidites spp.
Laevigatosporites spp.
Cicatricosisporites sp.
Osmunda sp.
Deltoidospora sp.

No. 5 Mine: SE¼ sec. 21, T. 20 N., R. 15 E., Kittitas County, in upper Roslyn Formation. Sample of coal collected from mine dump, half-way between the towns of Roslyn and Cle Elum.

Pollen: Platycarya sp.
Bursera sp.
Plicatopollis? sp.
Caryapollenites sp.
cf. Ulmus
cf. Nyssa
Ilex sp.
Tilia sp.
cf. Juglans
cf. Quercus
Liquidambar sp.
cf. Momipites
Triporate, spinate, small
Tricolpate and tricolporate spp.
Tricolporate, striate
Bisaccate pollen
Alnus sp.
Monosulcites sp.

Spores: Cicatricosisporites spp.
Polypodiidites spp.
Osmunda sp.
Laevigatosporites sp.

No. 6 Mine: NW¼ sec. 16, T. 20 N., R. 15 E., Kittitas County, in upper Roslyn Formation. Sample of coal collected from mine dump on northeast edge of town of Roslyn.

Pollen: Platycarya sp.
cf. Ulmoideipites
Monosulcites sp.
Caryapollenites sp.
Alnus sp.
cf. Momipites
cf. Ulmus

Spores: Laevigatosporites sp.

No. 6 Canyon: NW¼ sec. 16, T. 20 N., R. 15 E., Kittitas County, in upper Roslyn Formation. Four samples of claystone and siltstone collected from outcrops along No. 6 Canyon, northeast of the town of Roslyn.

Pollen:
Platycarya sp.
Bursera sp.
Proteacidites sp.
cf. Ulmus
Alnus sp.
Tilia sp.
Caryapollenites sp.
Pterocarya sp.
cf. Quercus
Ilex sp.
Taxodiaceaepollenites sp.
Tiliaepollenites sp.
Triporate, smooth spp.
cf. Sterculiaceae
cf. Chenopodium?
Tricolpate and tricolporate spp.
Triporate, scabrate to granulate sp.

Spores:
Cicatricosisporites spp.
Trilites solidus
Osmunda sp.
Polypodiidites sp.
Laevigatosporites sp.

NACHES FORMATION

Lower Taneum Creek (Wilson Mine): SW¼ NE¼ sec. 33, T. 19 N., R. 16 E., Kittitas County, in Naches Formation (Stout, 1964) but position in formation unknown. Mine adit is on south side of Taneum Creek at elevation of 2,355 ft, approximately.

Sample LTC-1: Coal from dump at caved-in mine adit.

Pollen:
Pistillipollenites mcgregorii (dominant)
Tilia?
Caryapollenites sp.
Alnus sp.
Triporate, smooth spp.
Taxodiaceaepollenites sp.
Tricolpate and tricolporate spp.

Spores:
Polypodiidites sp.
Deltoidospora sp.
Cicatricosisporites sp.
Osmunda sp.
Fungal spores
Laevigatosporites sp.

Sample LTC-2: Gray shale from outcrop a few feet downslope and slightly west of adit.

Pollen:
Pistillipollenites mcgregorii
Proteacidites sp.
Tilia sp.
cf. Bombacacidites
Alnus sp.
Caryapollenites sp.
Triporate, smooth spp.
Taxodiaceaepollenites sp.
Bisaccate pollen
Syncolporate? pollen

Spores:
Laevigatosporites sp.
Trilites solidus
Osmunda sp.
Gleicheniidites sp.
Cicatricosisporites sp.
cf. Lycopodium
Azolla sp.
Fungal spores
Deltoidospora sp.

South Fork Taneum Creek: NE¼ sec. 33, T. 19 N., R. 15 E., Kittitas County, in Naches Formation (Stout, 1964) but position in formation unknown. Samples of gray shale and black carbonaceous shale collected from cuts along north edge of road.

Pollen:
Platycarya? sp. (probable)
Caryapollenites sp.
Alnus sp.
cf. Duplopollis?
cf. Arecipites
Monosulcites?
cf. Ulmus
Schizosporis, smooth
Taxodiaceaepollenites sp.
Tricolpate and tricolporate spp.
Triporate, smooth

Spores:
Trilites solidus
Deltoidospora sp.
Laevigatosporites sp.
cf. Gleicheniidites
Polypodiidites spp.

Gale Creek: SW¼ NW¼ sec. 32, T. 22 N., R. 13 E., Kittitas County, in Naches Formation (Foster, 1960, 1967) but position in formation unknown. Samples of black carbonaceous shale collected from outcrops on north side of Gale Creek, above bridge on road that goes to campground on Kachess Lake.

Palynomorphs: Sparse, dark brown to black, carbonized, and fragmented. The dominant recognizable taxon is the spore Osmunda sp.

GUYE FORMATION

Hyak: SW¼ NW¼ sec. 15, T. 22 N., R. 11 E., Kittitas County, in type area of Guye Formation (Foster, 1960) but position in formation unknown. Samples of black shale and mudstone collected just west of entrance road road to Hyak ski area, from rock bulldozed up during road construction. Rock is very hard and sheared, but has abundant leaf impressions.

Palynomorphs: None; rock apparently has been heated beyond critical temperature for palynomorphs by Snoqualmie batholith; residue consists of black, fuzzy, carbonized material.

PUGET GROUP: Samples were collected from 13 stratigraphic levels in the Green River canyon section, T. 21 N., R. 6 and 7 E., King County, using the maps, measured section, and locality descriptions of Vine (1969) and Wolfe (1968). Palynomorph sample localities are expressed in feet above the base of the Puget Group as exposed in Green River canyon.

PG 6550: Center sec. 36, T. 21 N., R. 6 E., in volcanic rocks overlying Puget Group, approximately 6,550 ft above base of Puget Group, USGS Loc. 9690. One sample of red and brown claystone containing leaf fossils collected from small roadcut.

Pollen:
Tilia sp.
Caryapollenites sp.
Ulmus sp.
Tsuga sp.
cf. Cedrus
Alnus sp.
Taxodiaceaepollenites sp.
Pterocarya sp.
cf. Gramineae?
cf. Sterculiaceae
Ilex?
Triporate, smooth sp.
Tricolporate and tricolpate spp.
cf. Fagus?

Spores:
Laevigatosporites sp.
Polypodiidites spp.
Fungal spores
Selaginella sinuites
cf. Lycopodium foveolites

PG 6000: SW¼ sec. 25, T. 21 N., R. 6 E., in upper Puget Group, Kummer coal zone, in vicinity of Kummer No. 5 coal, about 6,000 ft above base of Puget Group. Five samples of black shale and black carbonaceous shale collected from outcrops along south side of river, below bridge.

Pollen:
Platycarya sp. (several specimens seen - in place? If not redeposited, this level may be near the top of the range zone of Platycarya).
Tilia sp.
Caryapollenites sp.
Liquidambar?
Ulmus sp.
Acanthaceae?
Nyssa sp.
cf. Engelhardia, small
Alnus sp.
Ilex sp.
Pterocarya sp.
Arecipites?
Taxodiaceaepollenites sp.
Tsuga sp. (?)
Tricolpate and tricolporate spp.
Triporate, smooth spp.
cf. Sterculiaceae?
Bisaccate pollen

Spores:
Polypodiidites spp.
Osmunda sp.
Laevigatosporites spp.
Fungal spores
Deltoidospora sp.
cf. Lycopodium
cf. Lycopodium foveolites
Selaginella sinuites

PG 5600: NE¼ sec. 26, T. 21 N., R. 6 E., in upper Puget Group, Kummer coal zone, near USGS Loc. 9679, about 5,600 ft above base of Puget Group. One sample of coal and one sample of black carbonaceous shale collected from cuts along road leading down to river.

Pollen: Tilia sp.
Schizosporis sp., smooth
cf. Acanthaceae
Nyssa sp.
Alnus sp.
Taxodiaceaepollenites sp.
Onagraceae??
Pterocarya sp.
Triporate, smooth sp.
Bisaccate pollen
Tricolpate and tricolporate pollen

Spores: Polypodiidites spp.
Osmunda sp.
Fungal spores
Deltoidospora sp.
Laevigatosporites spp.

PG 5300: NE¼ sec. 25, T. 21 N., R. 6 E., in upper Puget Group, lower Kummer coal zone, above Kummer No. 0 coal, about 5,300 ft above base of Puget Group. Two samples of coal and one of gray-brown mudstone collected from outcrops along river, south side, between two sandstone cliffs.

Pollen: Caryapollenites sp.
Nyssa sp.
cf. Engelhardia
cf. Onagraceae - Jussiaea?
Ilex spp.
Alnus sp.
Bisaccate pollen
Taxodiaceaepollenites sp.
Triporate, tricolporate, and tricolpate spp.
Proteacidites sp. (1 specimen - redeposited?)

Spores: Polypodiidites spp.
Cicatricosisporites sp. (1 specimen - redeposited?)
Selaginella sinuites
Laevigatosporites sp.
Osmunda sp.

PG 4600: NW¼ SW¼ sec. 19, T. 21 N., R. 7 E., in upper Puget Group, upper Franklin coal zone, near USGS Loc. 9680, about 4,600 ft above base of Puget Group. One sample of gray shale containing leaf fossils collected from outcrops along north side of river.

Pollen: Platycarya sp.
Tilia sp.
Caryapollenites sp.
Liquidambar?
Triporate, spinate, 20 microns
Ulmus sp.
Schizosporis sp., smooth
Nyssa sp.
Alnus sp.
Taxodiaceaepollenites sp.
Triporate, tricolpate, and tricolporate spp.
Bisaccate pollen

Spores: Fungal spores
Laevigatosporites sp.
Polypodiidites spp.
Osmunda sp.
Deltoidospora sp.

PG 3750: SW¼ sec. 19, T. 21 N., R. 7 E., in middle Puget Group, upper Franklin coal zone, near USGS Loc. 9691, near Gem coal, about 3,750 ft above base of Puget Group. Two samples of brown and black carbonaceous shale collected from outcrops near river. Palynomorph recovery poor from these samples.

Pollen: Alnus sp.
Ilex sp.
Ulmus?
Tricolporate sp.

Spores: Osmunda sp.

PG 2930: Sec. 11, T. 21 N., R. 6 E., in middle Puget Group, lower Franklin coal zone, about 2,930 ft above base of Puget Group, northwest side of town of Black Diamond. Two samples of coal and one of carbonaceous black shale collected from Palmer Coking Coal Co. mine dump; this coal is considered to be equivalent to the McKay bed.

Pollen: Platycarya sp.
Tilia sp.
Bursera sp.
cf. Plicatopollis
Ulmus sp.
Pterocarya sp.
cf. Nyssa
Taxodiaceaepollenites sp.
Ilex sp.
Triporate, tricolpate, and tricolporate spp.
Alnus sp.
Caryapollenites?
Monosulcites sp.

Spores: Polypodiidites spp.
Fungal spores
Deltoidospora sp.
Laevigatosporites sp.

PG 2500: SE¼ sec. 18, T. 21 N., R. 7 E., in middle Puget Group, near Franklin No. 11 or No. 12 coal, about 2,500 ft above base of Puget Group. One sample of coal and one sample of black carbonaceous shale collected from strip mine.

Pollen: Platycarya sp.
Caryapollenites sp.
cf. Plicatopollis
Ulmus sp.
Tricolporate or tricolpate, striate sp.
Anacolosidites sp.
Alnus sp.
Ilex sp.
Bisaccate pollen
Monosulcites sp.
Tricolpate and tricolporate spp.
Triporate, smooth sp.
cf. Tilia?
cf. Quercus
Duplopollis?

Spores: Fungal spores
Osmunda sp.
Laevigatosporites sp.
Polypodiidites sp.
Cicatricosisporites sp.
Deltoidospora sp.

PG 2050: NE¼ NE¼ sec. 19, T. 21 N., R. 7 E., in lower middle Puget Group, in lower Franklin coal zone, near USGS Loc. 9833, about 2,050 ft above base of Puget Group. One sample of black carbonaceous shale collected from bulldozer cut near old mine adit.

Pollen: Platycarya sp.
Pterocarya sp.
cf. Ulmoideipites
Bisaccate pollen
cf. Tilia
Tricolpate or tricolporate, striate sp.
Alnus sp.

Spores: Polypodiidites spp.
Osmunda sp.
Monolete, finely reticulate
Trilete?, spinate
Cicatricosisporites sp.

PG 1400: NE¼ SE¼ sec. 8, T. 21 N., R. 7 E., in lower middle Puget Group, below Franklin coal zone, near USGS Loc. 9840, about 1,400 ft above base of Puget Group. Two samples of gray mudstone collected from bulldozer cut on rim of south-facing point above river, at or above 700 ft elevation.

Pollen: Platycarya sp.
Proteacidites?
Bisaccate pollen
Triporate, smooth sp.
Tricolporate sp.

Spores: Polypodiidites sp.
Taurocusporites sp. (redeposited?)
Cicatricosisporites sp.
Deltoidospora sp.

PG 800: SE¼ sec. 8, T. 21 N., R. 7 E., in lower Puget Group, below Franklin Sandstone, near USGS Loc. 9675, about 800 ft above base of Puget Group. One sample of gray mudstone containing leaf fossils collected from outcrops along north side of river.

Pollen: Proteacidites sp.
Alnus sp.
Triporate, smooth sp.
Tricolporate, smooth sp.

Spores: Polypodiidites sp.
Deltoidospora sp.

PG 500: SW¼ sec. 9, T. 21 N., R. 7 E., in lower Puget Group, about 500 ft above base of Puget Group. Five samples of coal and gray mudstone collected from outcrops along river at south-facing point. Palynomorph recoveries poor.

Pollen: Pistillipollenites mcgregorii
Alnus sp.
Tricolpate and tricolporate spp.
Taxodiaceaepollenites sp.
Bisaccate pollen

Spores: Laevigatosporites sp.
Polypodiidites sp.

PG 200: SE¼ sec. 8, T. 21 N., R. 7 E., in lower Puget Group, about 200 ft above base of Puget Group section exposed in Lawson anticline. Two samples of coal and one sample of black shale collected from outcrops along north side of river, in vertical beds on west limb of anticline.

Pollen: Pistillipollenites mcgregorii
Proteacidites sp.
Platycarya??
Caryapollenites sp.
Pterocarya sp.
Ilex sp. (small)
Alnus sp.
Taxodiaceaepollenites sp.
Tricolpate and tricolporate spp.
Triporate, smooth spp.
cf. Ulmoideipites
Ulmus sp.

Spores: Cicatricosisporites sp.
Polypodiidites sp.
Osmunda sp.
Laevigatosporites sp.
Fungal spores

ACKNOWLEDGMENTS

Field work and early stages of laboratory work for this project were aided by grants from the Central Washington State College Faculty Research Fund, Battelle Northwest Laboratories, the Geological Society of America, and Union Oil Company Foundation. Final laboratory work and manuscript preparation were completed at, and with the aid of, the Colorado School of Mines and at my laboratory in Evergreen, Colorado. Many colleagues have contributed to the study in various ways. Their interest, encouragement, and occasional disbelief are appreciated.

The manuscript was reviewed by and improved, thanks to the comments from, J. M. Armentrout, W. C. Elsik, D. W. Engelhardt, R. L. Pierce, J. F. Stone, R. Thrombley, and J. T. Whetten.

REFERENCES CITED

Alexander, F., 1956, Stratigraphic and structural geology of the Blewett-Swauk area, Washington [M.S. thesis]: Seattle, University of Washington, 64 p.

American Commission on Stratigraphic Nomenclature, 1970, Code of stratigraphic nomenclature: Tulsa, Oklahoma, American Association of Petroleum Geologists, 22 p.

Beck, G. F., 1934, Tertiary floras of central Washington [abs.]: Northwest Science, v. 8, no. 3, p. 3.

Bressler, C. T., 1951, The petrology of the Roslyn arkose, central Washington [Ph.D. thesis]: State College, Pennsylvania State College, 147 p.

____1957, The petrology of the Roslyn arkose—A study in tectonic control of sedimentation in the Cascade Range, central Washington: International Geological Congress, 20th Mexico City, sec. 5, p. 439–453.

Brown, R. W., 1962, Paleocene flora of the Rocky Mountains and Great Plains: U.S. Geological Survey Professional Paper 375, 119 p.

Cashman, S. M., and Whetten, J. T., 1976, Low-temperature serpentinization of peridotite fanglomerate on the west margin of the Chiwaukum graben, Washington: Geological Society of America Bulletin, v. 87, p. 1773–1776.

Chappell, W. M., 1936, Geology of the Wenatchee quadrangle, Washington [Ph.D. thesis]: Seattle, University of Washington, 249 p.

Crickmay, C. H., and Pocock, S.A.J., 1963, Cretaceous of Vancouver, British Columbia, Canada: American Association of Petroleum Geologists Bulletin, v. 47, p. 1928–1942.

Engels, J. C., and Crowder, D. F., 1971, Late Cretaceous fission-track and potassium-argon ages of the Mount Stuart Granodiorite and Beckler Peak stock, North Cascades, Washington: U.S. Geological Survey Professional Paper 750-D, p. D39–D43.

Erikson, E. H., Jr., 1976, Structure and petrology of the late Cretaceous Mount Stuart batholith, Cascade Mountains, Washington [abs.]: Geological Society of America Abstracts with Programs, v. 8, p. 371.

Erikson, E. H., Jr., and Williams, A. E., 1976, Implications of apatite fission track ages in the Mount Stuart batholith, Cascade Mountains, Washington [abs.]: Geological Society of America Abstracts with Programs, v. 8, p. 372.

Evans, G. W., 1912, The coal fields of King County: Washington Geological Survey Bulletin 3, 247 p.

Foster, R. J., 1958, The Teanaway dike swarm of central Washington: American Journal of Science, v. 256, p. 644–653.

____1960, Tertiary geology of a portion of the central Cascade Mountains, Washington: Geological Society of America Bulletin, v. 71, p. 99–125.

____1967, Geology of the Rampart Ridge-Keechelus Ridge area, central Cascade Mountains, Washington: Northwest Science, v. 41, p. 32–41.

Frizzell, V. A., Jr., and Tabor, R. W., 1977, Stratigraphy of Tertiary arkoses and their included monolithologic fanglomerates and breccias in the Leavenworth fault zone, central Cascades, Washington [abs.]: Geological Society of America Abstracts with Programs, v. 9, p. 421.

Galster, R. W., 1956, Geology of the Miller-Foss River area, King County [M.S. thesis]: Seattle, University of Washington, 96 p.

Gresens, R. L., 1976a, A new Tertiary formation near Wenatchee, Washington [abs.]: Geological Society of America Abstracts with Programs, v. 8, p. 376–377.

____1976b, Unusual structural features associated with mid-Tertiary deformation near Wenatchee, Washington [abs.]: Geological Society of America Abstracts with Programs, v. 8, p. 892–893.

Griggs, P. H., 1970, Palynological interpretation of the type section, Chuckanut Formation, northwestern Washington, *in* Kosanke, R. M., and Cross, A. T., eds., Symposium on palynology of the late Cretaceous and early Tertiary: Geological Society of America Special Paper 127, p. 169–212.

Hesse, C. J., 1936, A new species of the genus *Priscarara* from the Eocene of Washington: Journal of Geology, v. 44, p. 745–750.

Hills, L. V., 1965, Palynology and age of early Tertiary basins, interior British Columbia [Ph.D. thesis]: Edmonton, University of Alberta, 188 p.

Hills, L. V., and Baadsgaard, H., 1967, Potassium-argon dating of some lower Tertiary strata in British Columbia: Canadian Petroleum Geology Bulletin, v. 15, no. 2, p. 138–149.

Hopkins, W. S., Jr., 1966, Palynology of Tertiary rocks of the Whatcom basin, southwestern British Columbia and northwestern Washington [Ph.D. thesis]: Vancouver, University of British Columbia, 184 p.

____1968, Subsurface Miocene rocks, British Columbia-Washington, a palynological investigation: Geological Society of America Bulletin, v. 79, p. 763–768.

____1969, Palynology of the Eocene Kitsilano Formation, southwest British Columbia: Canadian Journal of Botany, v. 47, no. 7, p. 1101–1131.

Hopkins, W. S., Jr., Rutter, N. W., and Rouse, G. E., 1972, Geology, paleoecology, and palynology of some Oligocene rocks in the Rocky Mountain trench of British Columbia: Canadian Journal of Earth Sciences, v. 9, p. 460–470.

Huntting, M. T., and others, 1961, Geologic map of Washington: Washington Division of Mines and Geology, 1:500,000.

International Subcommission on Stratigraphic Classification of IUGS Commission on Stratigraphy, Hedberg, H. D., ed., 1976, International stratigraphic guide: New York, Wiley, 200 p.
LaMotte, R. S., 1952, Catalogue of the Cenozoic plants of North America through 1950: Geological Society of America Memoir 51, 381 p.
Lupe, R. D., 1971, Stratigraphy and petrology of the Swauk Formation in the Wenatchee Lake area, Washington [M.S. thesis]: Seattle, University of Washington, 27 p.
Lupher, R. L., 1944, Stratigraphic aspects of the Blewett-Cle Elum iron ore zone, Chelan and Kittitas Counties, Washington: Washington Division of Geology Report of Investigations 11, 63 p.
Newman, K. R., 1970, Palynology of interflow sediments from the Standard Oil Co. of California Rattlesnake Hills No. 1 well, Benton County, Washington, *in* Gilmour, E. H., and Stradling, D., eds., Proceedings, Second Columbia River Basalt symposium: Cheney, Eastern Washington State College Press, p. 201-207.
——1971, Age ranges of strata mapped as Swauk Formation, central Washington [abs.]: Geological Society of America Abstracts with Programs, v. 3, p. 397-398.
——1975, Palynomorph sequence in the early Tertiary Swauk and Roslyn Formations, central Washington [abs.]: Geoscience and Man, v. 11, p. 158.
——1977, Palynologic biostratigraphy of some early Tertiary nonmarine formations in central and western Washington [abs.]: Geological Society of America Abstracts with Programs, v. 9, p. 1113.
Piel, K. M., 1971, Palynology of Oligocene sediments from central British Columbia: Canadian Journal of Botany, v. 49, p. 1885-1920.
Rau, W. W., 1958, Stratigraphy and foraminiferal zonation in some of the Tertiary rocks of southwestern Washington: U.S. Geological Survey Oil and Gas Investigations Chart OC-57. 2 sheets.
Rector, R. J., 1962, Geology of the east half of the Swauk Creek mining district [M.S. thesis]: Seattle, University of Washington, 73 p.
Rouse, G. E., 1962, Plant microfossils from the Burrard Formation of western British Columbia: Micropaleontology, v. 8, p. 187-218.
Rouse, G. E., Hopkins, W. S., Jr., and Piel, K. M., 1970, Palynology of some Late Cretaceous and early Tertiary deposits in British Columbia and adjacent Alberta, *in* Kosanke, R. M., and Cross, A. T., eds., Symposium on palynology of the Late Cretaceous and early Tertiary: Geological Society of America Special Paper 127, p. 213-246.
Russell, I. C., 1899, A preliminary paper on the geology of the Cascade Mountains in northern Washington: U.S. Geological Survey 20th Annual Report, Pt. 2, p. 83-210.
Smith, G. O., 1903, Contributions to the geology of Washington: The geology and physiography of central Washington: U.S. Geological Survey Professional Paper 19, p. 9-39.
——1904, Description of the Mount Stuart quadrangle: U.S. Geological Survey Geologic Atlas, Folio 106, 10 p.
Smith, G. O., and Calkins, F. C., 1906, Description of the Snoqualmie quadrangle: U.S. Geological Survey Geologic Atlas, Folio 139, 14 p.
Smith, W. S., 1916, Stratigraphy of the Skykomish basin, Washington: Journal of Geology, v. 24, p. 559-582.
Southwick, D. L., 1974, Geology of the alpine-type ultramafic complex near Mount Stuart, Washington: Geological Society of America Bulletin, v. 85, p. 391-402.
Sparks, D. M., 1967, Microfloral zonation and correlation of some lower Tertiary rocks in southwest Washington and some conclusions concerning the paleoecology of the flora [Ph.D. thesis]: East Lansing, Michigan State University, 168 p.
Stout, M. L., 1964, Geology of a part of the south-central Cascade Mountains, Washington: Geological Society of America Bulletin, v. 75, p. 317-334.
Tabor, R. W., and others 1977, Preliminary map of the Wenatchee 1:100,000 quadrangle, Washington: U.S. Geological Survey Open-File Map 77-531.
Vance, J. A., and Naeser, C. W., 1977, Fission track geochronology of the Tertiary volcanic rocks of the central Cascade Mountains, Washington [abs.]: Geological Society of America Abstracts with Programs, v. 9, p. 520.
Vine, J. D., 1962, Stratigraphy of Eocene rocks in a part of King County, Washington: Washington Division of Mines and Geology Report of Investigations 21, 20 p.
——1969, Geology and coal resources of the Cumberland, Hobart, and Maple Valley quadrangles, King County, Washington: U.S. Geological Survey Professional Paper 624, 67 p.
Waters, A. C., 1930, Geology of the southern half of the Chelan quadrangle, Washington [Ph.D. thesis]: New Haven, Yale University 256 p.
Weaver, C. E., 1937, Tertiary stratigraphy of western Washington and northwestern Oregon: University of Washington Publications in Geology, v. 4, 266 p.
Weissenborn, A. E., 1969, Geologic map of Washington: U.S. Geological Survey Miscellaneous Geologic Investigations Map I-583, scale 1:2,000,000.
Wheeler, H. E., 1955, Dermatemidid turtle from the Roslyn Formation of Washington [abs.]: Geological Society of America Bulletin, v. 66, p. 1668.
Whetten, J. T., 1976, Tertiary sedimentary rocks in the central part of the Chiwaukum graben, Washington [abs.]: Geological Society of America Abstracts with Programs, v. 8, p. 420.
——1977, Sedimentology and structure of part of the Chiwaukum graben, Washington [abs.]: Geological Society of America Abstracts with Programs, v. 9, p. 527.
White, C. A., 1888, On the Puget Group of Washington territory: American Journal of Science, 3rd ser., v. 36, p. 443-450.
——1889, On invertebrate fossils from the Pacific coast: U.S. Geological Survey Bulletin 51, 102 p.
Willis, B., 1897, Stratigraphy and structure of the Puget group, Washington: Geological Society of America Bulletin, v. 9, p. 2-6.
——1898, Some coal fields of Puget Sound: U.S. Geological Survey Annual Report 18, pt. 3, p. 393-436.
Willis, C. L., 1950, Geology of the northeastern quarter of the Chiwaukum quadrangle, Washington [Ph.D. thesis]: Seattle, University of Washington, 158 p.
Wilson, L. R., 1964, Recycling, stratigraphic leakage, and faulty techniques in palynology: Grana Palynologica, v. 5, no. 3, p. 425-436.
Wolfe, J. A., 1968, Paleogene biostratigraphy of nonmarine rocks in King County, Washington: U.S. Geological Survey Professional Paper 571, 33 p.
Wolfe, J. A., Gower, H. P., and Vine, J. D., 1961, Age and correlation of the Puget Group, King County, Washington: U.S. Geological Survey Professional Paper 424-C, p. 230-232.
Yeats, R. S., 1958, Geology of the Skykomish area in the Cascade Mountains, Washington [Ph.D. thesis]: Seattle, University of Washington, 243 p.

MANUSCRIPT RECEIVED BY THE SOCIETY MARCH 14, 1980
MANUSCRIPT ACCEPTED MAY 7, 1980

Printed in U.S.A.

Geological Society of America
Special Paper 184
1981

Pacific Northwest Tertiary benthic foraminiferal biostratigraphic framework—An overview

WELDON W. RAU
Washington Division of Geology and Earth Resources, Department of Natural Resources and *U.S. Geological Survey, Olympia, Washington 98504*

ABSTRACT

The benthic foraminiferal biostratigraphic framework for marine Tertiary strata of the Pacific Northwest, as it has developed over the past 30 years, has been fundamental to geologic mapping by the U.S. Geological Survey and the state of Washington. However, the refinement of foraminiferal zonation in this area varies considerably because foraminiferal faunas sometimes do not differ greatly throughout thick sequences or the available data from certain areas have been limited. Furthermore, correlations are most reliable within local areas because differences in benthic foraminiferal faunas are largely the result of paleoecologic variations through time, as well as over geographic area, rather than evolutionary change.

Studies, beginning in southwest Washington in the late 1940s and continuing into the 1950s, resulted in the initial correlation of Washington's Tertiary sequence with California foraminiferal stages and the establishment of provincial zones. This framework was later extended to the early and middle Tertiary rocks of northwest Washington. Refinements and additions were made in the 1960s in connection with Washington State and U.S. Geological Survey geologic mapping in the Grays Harbor basin. In western Oregon during much of the past 30 years, foraminiferal studies have also been an integral part of U.S. Geological Survey geologic mapping. California stages have been recognized and the local faunal distribution has been informally established. In the 1970s, foraminiferal studies and geologic mapping in several coastal areas of the Olympic Peninsula by both the U.S. Geological Survey and the state of Washington made it possible to present biostratigraphic and paleoecologic interpretations for the Tertiary rocks of that area.

As a result of these studies, eight foraminiferal zones, two informal subdivisions of a stage, and four distinctive assemblages from additional parts of the Pacific Northwest Tertiary sequence have been described. The characteristics of these early Eocene to Pliocene faunal divisions are summarized in this report.

INTRODUCTION

Over the past 30 years, foraminiferal studies have been conducted in conjunction with both U.S. Geological Survey and Washington State geologic mapping and other surveys in the Tertiary province of the Pacific Northwest. During this time geologic mapping has become fundamentally dependent upon foraminiferal biostratigraphy. Few and poorly exposed outcrops, relatively few continuously exposed sections, and substantial thicknesses of massive relatively uniform lithology have all contributed to the need for biostratigraphic control. Much of the foraminiferal data and biostratigraphic interpretations developed from these studies has been presented in various publications through the years. Thus, foraminiferal biostratigraphic concepts evolved from, and were built upon, a number of geologic investigations in different parts of the Pacific Northwest. It therefore seems appropriate to review and synthesize these studies for a better understanding of their contribution to Pacific Northwest foraminiferal biostratigraphy.

Limitations and Nature of Biostratigraphic Units

Because a foraminiferal biostratigraphic framework had already been established in California, early studies in western Washington and Oregon necessarily relied heavily on correlations with California biostratigraphy. However, it was soon recognized that although broad divisions, such as stages of Kleinpell (1938), could be generally recognized in western Washington and Oregon, the faunal characteristics of smaller California units, such as zones, were frequently unrecognizable (Rau, 1958). It therefore became necessary, when practical, to develop local subdivisions or zones. The extent to which local subdivisions could be defined depended on several considerations. First, the stratigraphic variability of assemblages within a local area had to be considered. In some areas, for example, the Zemorrian Stage of Kleinpell (1938) is readily recognized, but its faunal character varies little throughout hundreds of metres of section. Thus, although it was desirable, in reality it was not practical to define local subdivisions of the stage, even though a substantial amount of data was available. In other areas and/or in parts of the Tertiary sequence, faunal characteristics were found to vary sufficiently so that local subdivisions of stages could be defined. In still other areas only a lower and upper part of a stage could be generally recognized with no finite boundary definable between. Some geologic studies were limited and did not provide extensive collections of foraminiferal material; in those cases only a broad framework, mainly of stages, was locally developed and utilized. For a number of reasons, then, and as is apparent in the following historical review, the refinement of foraminiferal biostratigraphy varies considerably from area to area in western Washington and Oregon.

Benthic foraminifers are relatively abundant and diverse in Tertiary rocks of the Pacific Northwest, whereas planktonic foraminifers are of rare or scattered occurrence; therefore, early uses of foraminifers for stratigraphic purposes were based largely on benthic rather than on planktonic taxa. The successful use of benthic faunas for stratigraphic purposes requires assessing the reality of ecologic control on faunal occurrence. Furthermore, the effects of ecological differences can be and are reflected stratigraphically as well as geographically. Thus, the stratigraphic distribution of benthic foraminifers in local sequences is perhaps far more the result of changing environment through time than it is the result of evolutionary change. The detailed characteristics of a local biostratigraphic unit can be recognized only as widespread as was the paleoenvironment of the geologic time that it represents. If a relatively uniform environment existed over a wide area during a given geologic time, that geologic time may be represented by a characteristic fauna throughout that same geographic area. As geologic time progressed, changes took place in ecologic conditions. Such changes reflect stratigraphic variations in faunas. These differences become the characteristics of local biostratigraphic zones. This concept of "dual environmental variablility" has been recognized in foraminiferal biostratigraphic studies of the Pacific Northwest (Rau, 1964, 1966, 1967, 1970) and is particularly apparent, for example, in the lower and middle Miocene strata of southwest Washington as is discussed in a following section of this report.

REVIEW OF EARLY PUBLICATIONS

Prior to the late 1940s, a few taxonomic reports were published on isolated collections from various rock units of the Northwest Tertiary sequence. Among these are reports by Hanna and Hanna (1924) and by Beck (1943) on the Eocene Cowlitz Formation of Washington Cushman and Frizzell (1943) on the Lincoln Creek Formation of Washington; and in Oregon, Cushman and Schenck (1928) described foraminifera from the Keasey and Bastendorff beds. In the late 1940s Detling (1946) described foraminifera from the lower Tertiary of Coos Bay; Cushman and others (1947, 1949) published on foraminiferal faunas from the Miocene Astoria Formation, Eocene Toledo and Coaledo Formations and Helmick Hills beds of Oregon, and the Pliocene Quinault Formation of Washington. Rau (1948a, 1948b, 1951) described an additional fauna from the Lincoln Creek Formation, a small Miocene assemblage from southwest Washington, and faunas ranging in age from Eocene to Miocene from a Tertiary section exposed in the Willapa Valley of southwest Washington.

FORAMINIFERAL BIOSTRATIGRAPHY, 1950-1979

The Willapa Valley study (Rau, 1951) was an early attempt to describe the foraminifera and to point out faunal differences in several stratigraphic units that occur in a major Pacific Northwest Tertiary sequence. Furthermore, suggested correlations to California stages—the Refugian of Schenck and Kleinpell (1936) and the Zemorrian and Saucesian of Kleinpell (1938)—was an early recognition of these units in the Pacific Northwest.

The first attempt to formally divide a Pacific Northwest Tertiary sequence into biostratigraphic units on the basis of foraminifers developed from U.S. Geological Survey foraminiferal studies and mapping in the Centralia-Chehalis area of southwest Washington (Fig. 1). In the resulting report Rau (in Snavely and others, 1958) defined 10 zones of Eocene and Oligocene age, the details of which are summarized in Figure 2. These divisions were not necessarily intended to be "Oppelian" zones in the strict sense, but rather they were designed to point out significant groups of species that occur in certain segments of local sections. With the exception of the *Cibicides hodgei* Zone, each faunal zone was assigned to a relatively distinct stratigraphic position within the local Tertiary section. The *C. hodgei* Zone is represented by a provincial shallow-water-facies fauna that locally occurs in the basal part of what is known today as the Lincoln Creek Formation (Beikman and others, 1967). Although this zone is in part equivalent to the *Sigmomorphina schencki* Zone, in places it extends higher in the section.

A continuation of geologic mapping by the U.S. Geological Survey both west and southwest from the Centralia-Chehalis area (Pease and Hoover, 1957; Roberts, 1958) prompted additional foraminiferal studies of a much larger area in southwest Washington. Some 13 surface sections and wells served as a base from which eight foraminiferal zones were proposed for the Tertiary strata of southwest Washington (Rau, 1958). Although these zones are fewer in number than those of the local Centralia-Chehalis area, their characteristics were presented in greater detail and correlations were more comprehen-

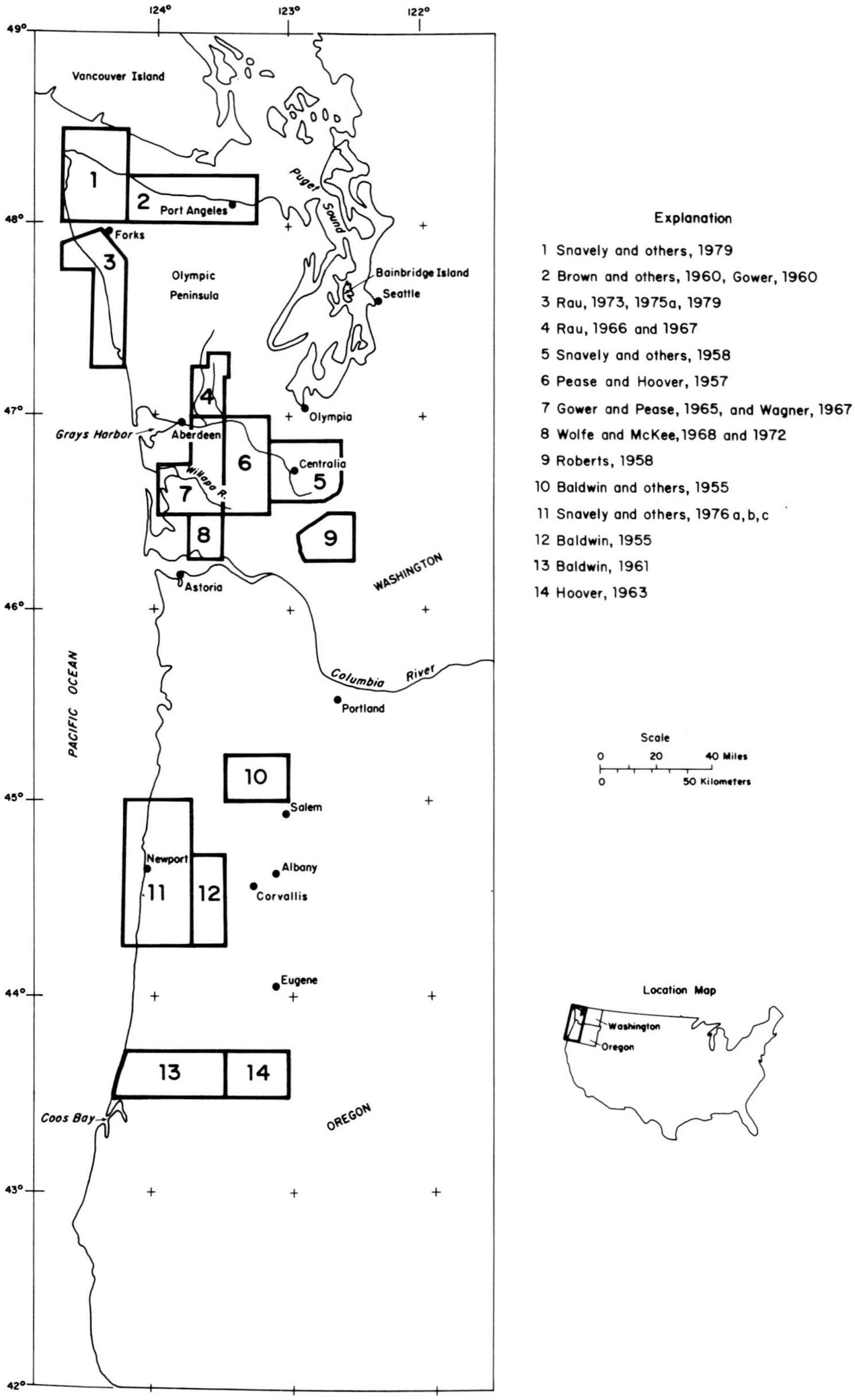

Figure 1. Index map showing areas discussed in this report.

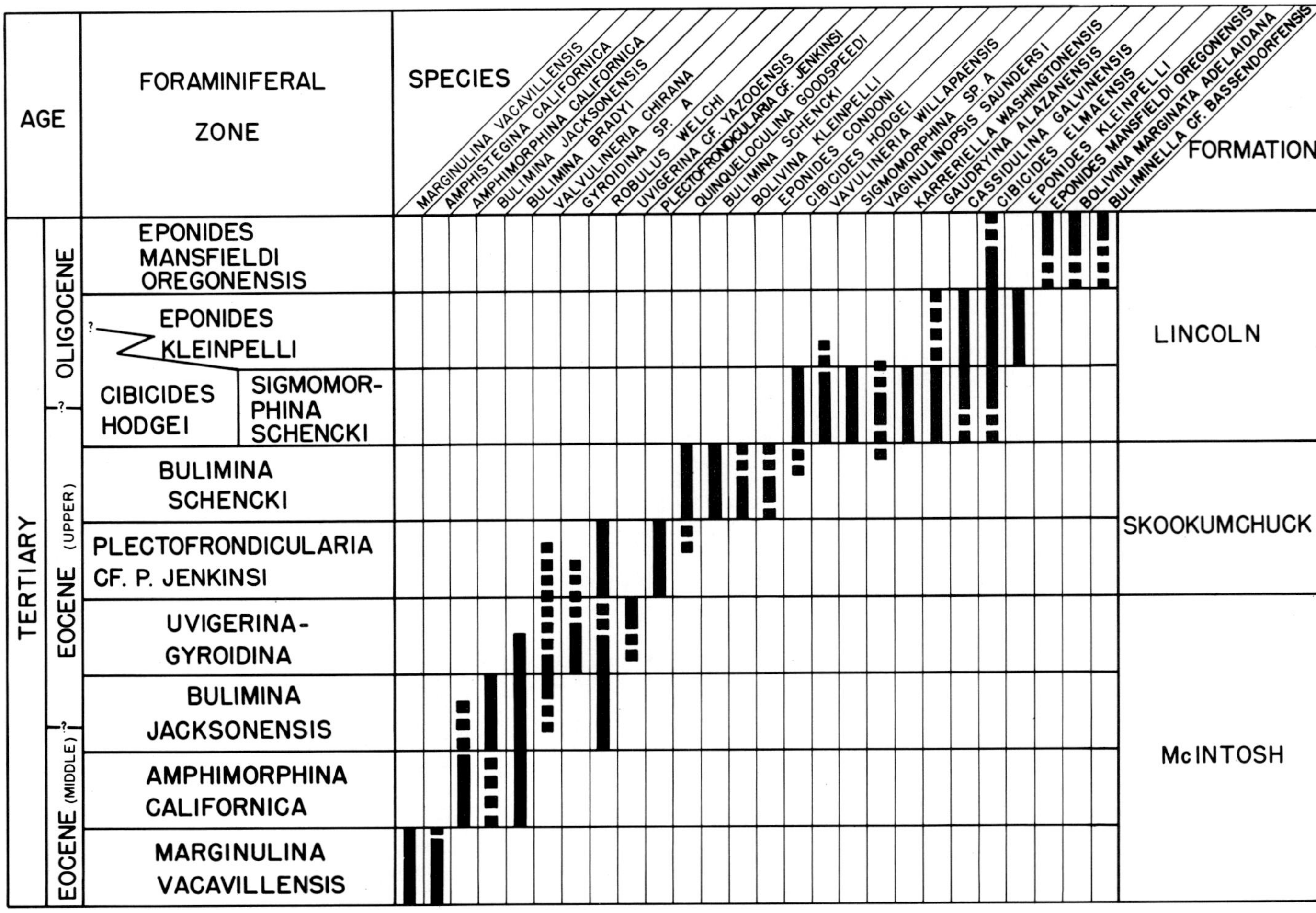

Figure 2. Foraminiferal zones and stratigraphic distribution of significant species in the Tertiary sequence of the Centralia-Chehalis area, southwest Washington, as originally applied in the 1950s. Changes have since been made in zonal concepts, zone names, and taxonomy. See text for discussion.

sive (Fig. 3). Furthermore, they are more nearly "Oppelian" in design because they are of chronologic as well as biostratigraphic significance and were constructed from substantial data from a number of Tertiary sequences over a relatively large area of the Pacific Northwest. Many of the zonal names were perpetuated or modified from usage in the Centralia-Chehalis area. The *Epistominella parva* Zone, broadly including the faunas of Miocene age and primarily representing the Astoria(?) Formation of southwest Washington, was added to the framework. Eocene zones were generally related to Laiming's (1940) Eocene zones of California on the basis of similarities in the stratigraphic occurrence of species. Furthermore, the California Refugian Stage of Schenck and Kleinpell (1936) and the Zemorrian and Saucesian Stages of Kleinpell (1938) were more formally recognized. It was not intended to imply a precise correlation to these stages but rather to point out that faunal sequences occurring in the Tertiary of California, at least to the stage magnitude, could also be recognized in the same chronologic succession in Tertiary strata of western Washington. The 1958 report of Rau constitutes the fundamental framework for benthic foraminiferal biostratigraphy in a major part of the Pacific Northwest Tertiary sequence. Additions and refinements have since been made, particularly in the Neogene, but most of the original zonal names remain the same and only minor conceptual changes have been made.

Geologic mapping in the late 1950s by the U.S. Geological Survey in the northern Olympic Peninsula area of Washington (Brown and others, 1960; Gower, 1960) required further expansion of Pacific Northwest foraminiferal studies to the Tertiary sequence of that area (Fig. 1). Over 40 species significant to correlation were illustrated and described, and correlations were made to the established biostratigraphic framework of southwest Washington, as well as other West Coast areas (Rau, 1964). References to Eocene stages of Mallory (1959) were made for the first time by the U.S. Geological Survey in Pacific Northwest Tertiary strata and were applied not only to the Eocene sequence of the northern Olympic Peninsula but to the Eocene zones of southwest Washington (Fig. 4). Foraminiferal assemblages were referred to stages ranging from the early Eocene Ulatisian Stage of Mallory through the early Miocene Saucesian Stage of Kleinpell. Because the observed stratigraphic range of many taxa varied appreciably from section to section, reliable correlations within

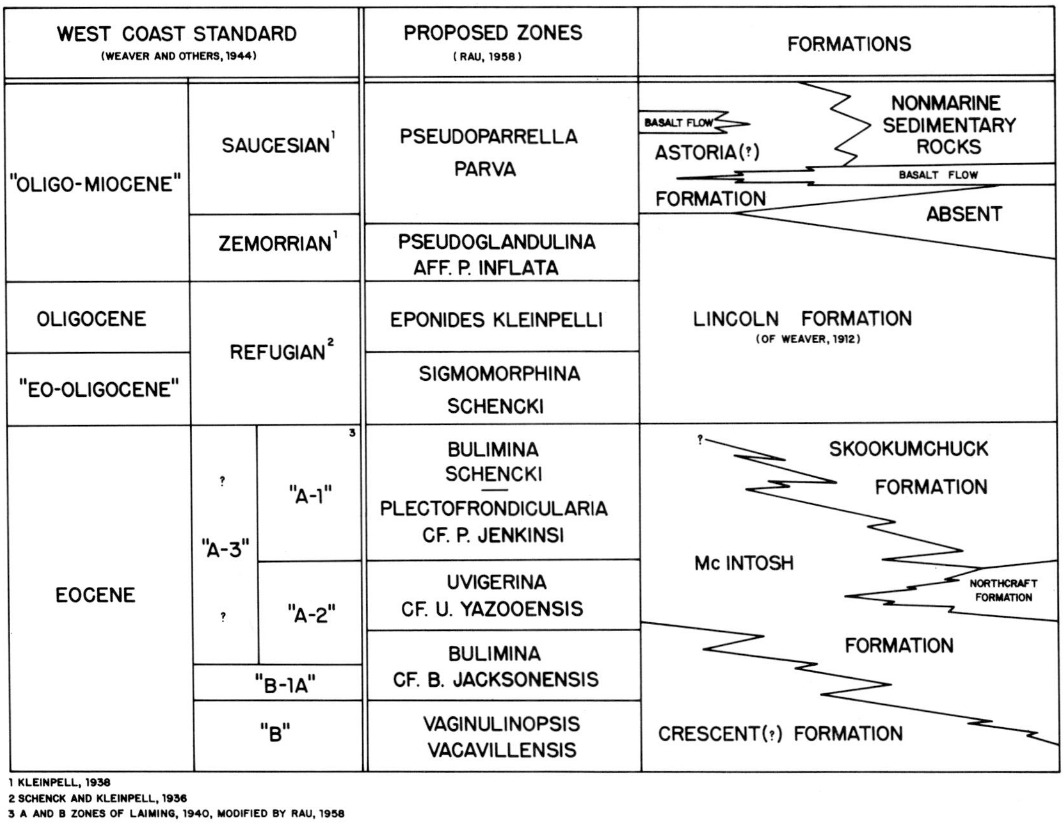

Figure 3. Foraminiferal zones and stratigraphy of southwest Washington as proposed by Rau in 1958. Subsequent changes in zonal nomenclature are discussed in the text.

the study area, as well as to other areas, were limited largely to the stage level of magnitude. The thick Twin River Formation was found to contain assemblages representative of the Narizian, Refugian, and Zemorrian Stages. A twofold division of the latter stage was recognized and informally referred to as "upper" and "lower" Zemorrian. Key taxa for each unit were noted as shown in Figure 5. Due to the growing need for paleoenvironmental analysis, a greater emphasis has been placed on paleoecologic interpretation in this study than in previous works.

Recent studies in the northern Olympic Peninsula have demonstrated that the Saucesian Stage is in part represented in the Twin River Formation (Rau, in Addicott, 1976b). The Zemorrian/Saucesian boundary is within the upper part of the formation and at least 45 m below the top of the molluscan "Blakeley" Stage of Weaver and others (1944) (= Matlockian Stage of Armentrout, 1975).

Geologic mapping and related research in western Oregon by the U.S. Geological Survey during the past 30 years have yielded extensive foraminiferal data, particularly from the Newport embayment (Snavely and others, 1976a, 1976b, 1976c), where foraminiferal studies were an integral part of the mapping project (Fig. 1). Extensive sampling from measured sections along the Yaquina and Siletz Rivers, as well as from sections along the shores of Alsea Bay, provided most of the basic biostratigraphic information concerning foraminiferal occurrences throughout this major Tertiary basin. Several reports were presented dealing with foraminiferal data (Rau, in Snavely and others, 1964, 1969), but the major use of these studies was as a direct aid to geologic mapping. Faunal characteristics of each formation of the local sequence were stressed rather than defining and describing new biostratigraphic zones. Suggested correlations were made to California foraminiferal stages and to some zones (Fig. 6). Rau (1975b) located and defined the Refugian/Zemorrian boundary in the type section of the Alsea Formation at Alsea Bay (Fig. 7) and suggested that faunas of this formation can be correlated to those of his *Cassidulina galvinensis* and *Pseudoglandulina* cf. *P. inflata* Zones of the southwest Washington sequence. Additional foraminiferal studies were made in support of geologic mapping in the Oregon Coast Range and the west side of the Willamette Valley, particularly for projects conducted by Baldwin (1955), Baldwin and others (1955), Baldwin (1961), and Hoover (1963) (see Fig. 1).

In the 1960s Washington State began stratigraphic studies on the northern flank of the Grays Harbor basin, mainly in tributaries of the Wynoochee and Satsop River valleys (Rau, 1966; see Fig. 1). The project was initiated for the purpose of conducting detailed biostratigraphic studies where the Refugian and Zemorrian Stages are well represented in the thick and widespread Lincoln Creek Formation. Modifications and refinements were made

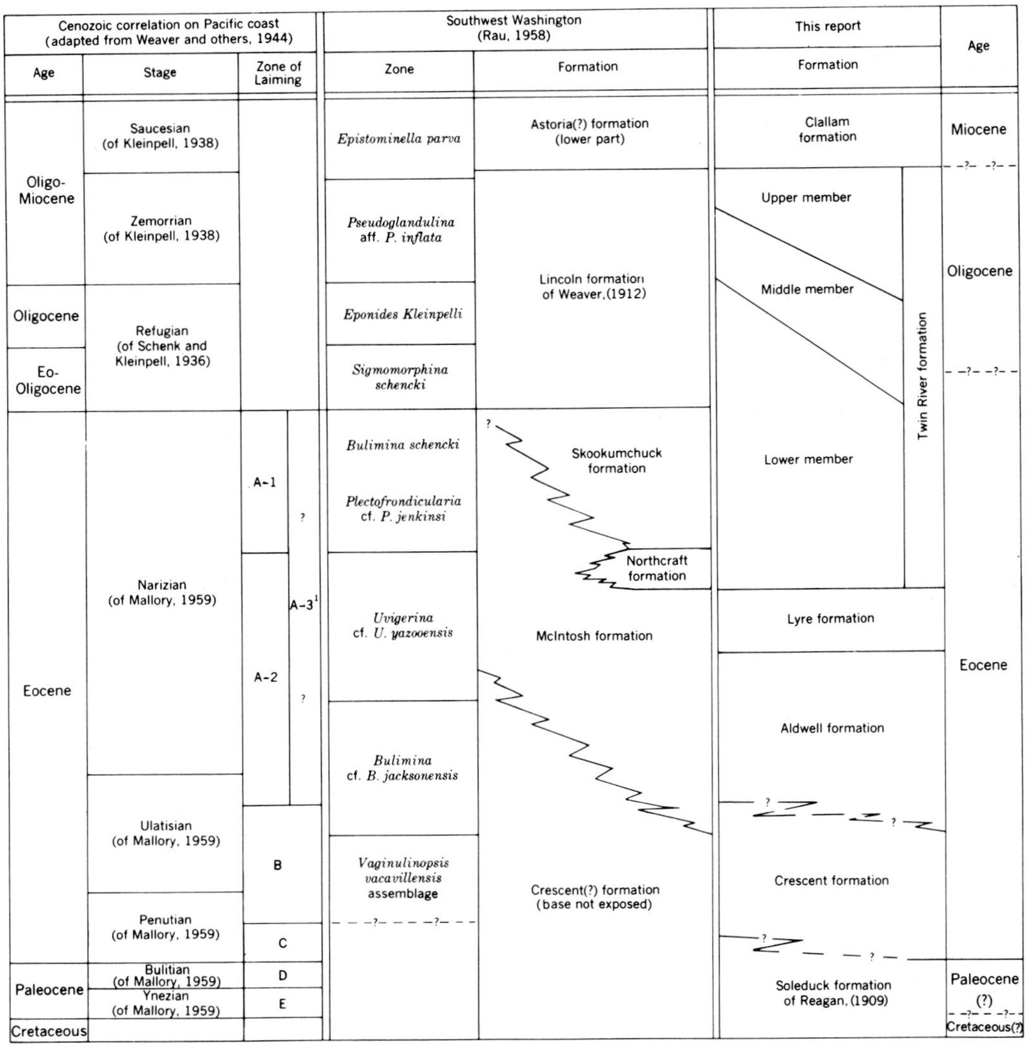

[1] Position of A-3 zone modified by Rau (1958)

[2] B zones not differentiated

Figure 4. Stratigraphic relations—northern Olympic Peninsula, southwest Washington, and a West Coast standard as proposed by Rau in 1964. Subsequently, the zone name *Eponides kleinpelli* has been changed to *Cassidulina galvinensis*, and the single *Epistominella parva* Zone has been divided into three newly named zones (see text).

especially to faunal definitions for provincial subdivisions of the Refugian and Zemorrian Stages. The previously named *Eponides kleinpelli* Zone (Rau, 1958) of the Refugian Stage was renamed the *Cassidulina galvinensis* Zone because the latter species was found to be more widely distributed (Fig. 8). In addition, a twofold division of the Zemorrian Stage was recognized. These subdivisions were not given formal names, mainly because faunal characteristics are gradational between them without a precise boundary. Concurrent ranges of several long-ranging species not restricted to the Zemorrian Stage characterize these subdivisions. The two parts of the Zemorrian Stage were regarded as general correlatives of two previously recognized subdivisions of the stage in the northern Olympic Peninsula area.

Subsequent geologic mapping southward into the main valleys of the Wynoochee and Satsop Rivers, where large areas of the Miocene Astoria(?) Formation and Montesano Formation are exposed, provided the opportunity to determine the stratigraphic occurrence of Miocene foraminifers of the Grays Harbor basin (Rau, 1967). Three zones, all occurring within rocks of the Astoria(?) Formation and broadly referred to as the *Pseudoparella parva* Zone of Rau (1958) were named and described from bottom to top as the *Siphogenerina kleinpelli* Zone, *Baggina washingtonensis* Zone, and *Rotalia becki* Zone (Fig. 8). It was emplasized that the faunas of the three zones actually reflect differences in depositional environments through time within a given geographic area. The local basin of deposition apparently was deepest during the early part of the Miocene and is represented by the relatively deep-water, perhaps bathyal, faunas of the *Siphogenerina*

Figure 5. Frequency of occurrence of selected species of foraminifera in the Twin River Formation as presented by Rau in 1964.

kleinpelli Zone. As the basin became shallower, faunal characteristics changed to those of the *Baggina washingtonensis* Zone and eventually to those of the relatively shallow-water shelf faunas of the *Rotalia becki* Zone. Thus, within the basin of deposition for the Astoria(?) Formation, deep-water faunas are essentially restricted to the lowest part, whereas the upper part of the Astoria(?) Formation contains only shallow-water taxa. Even though shallow-water faunas are more widespread in the upper part of the sequence, they are not necessarily confined to the upper part because they presumably existed as basin margin features throughout the depositional cycle.

The *Siphogenerina kleinpelli* Zone was correlated to the lower part of the Saucesian Stage, the *Baggina washingtonensis* Zone to the upper part of the Saucesian Stage and probably the lower part of the Relizian Stage, and the *Rotalia becki* Zone, largely on the basis of superposition, to an upper part of the Relizian Stage and possibly a part of the Luisian Stage.

Foraminifera of the Montesano Formation were collected only from key areas as a means of identifying the formation. No attempt was made to conduct detailed biostratigraphic or paleoecologic studies with foraminifers of these rocks. Foraminifers of the Montesano Formation were broadly referred to the upper Miocene undifferentiated Mohnian and Delmontian Stages of Kleinpell

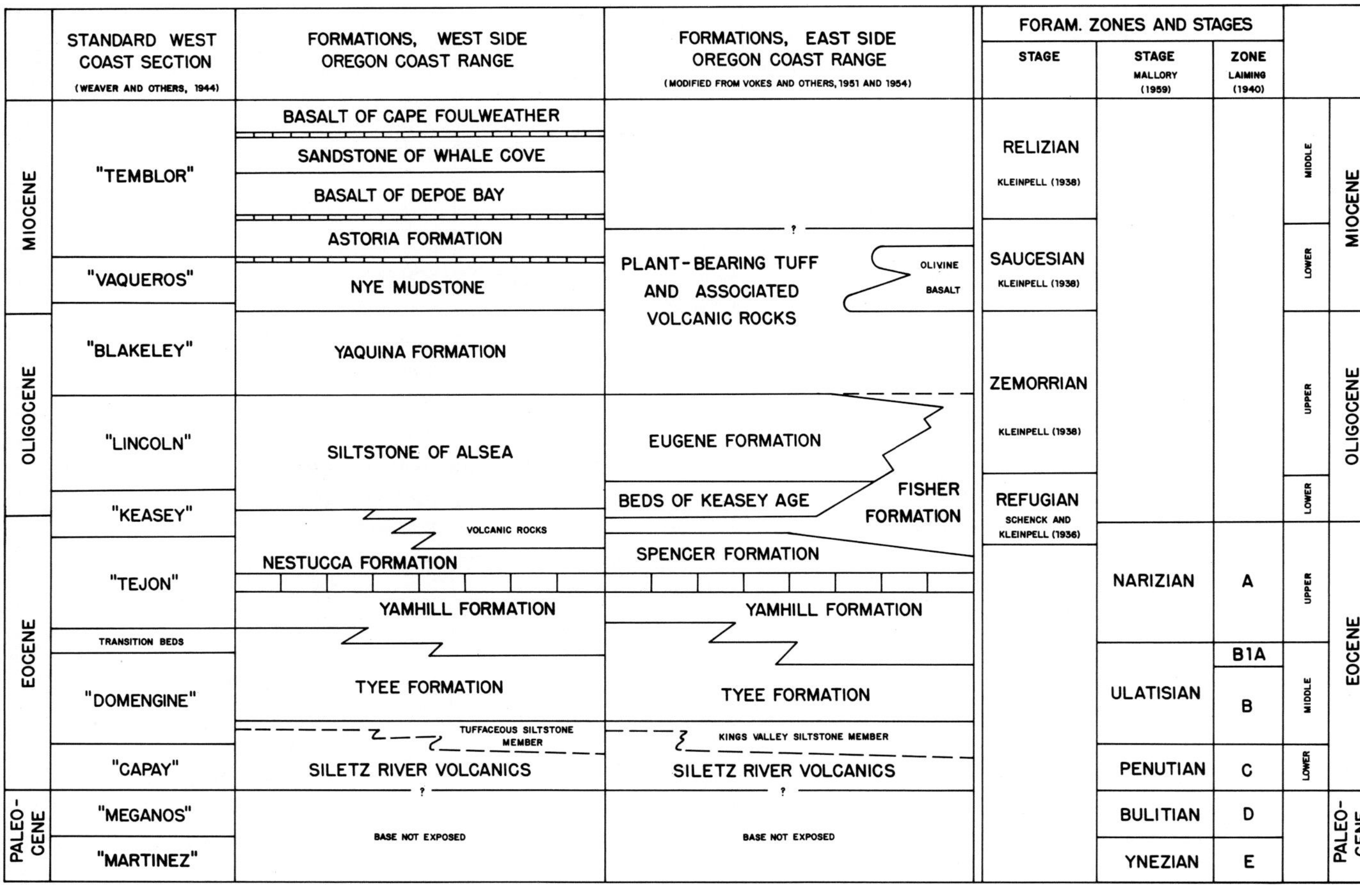

Figure 6. Correlations—west and east sides of Oregon Coast Range (Snavely and others, 1969). The Eocene/Oligocene boundary has been placed variously on the diagrams of the present report with respect to the base of the Refugian Stage of Schenck and Kleinpell (1936). The authors of this diagram considered the Eocene/Oligocene boundary to be with the Refugian Stage at that time.

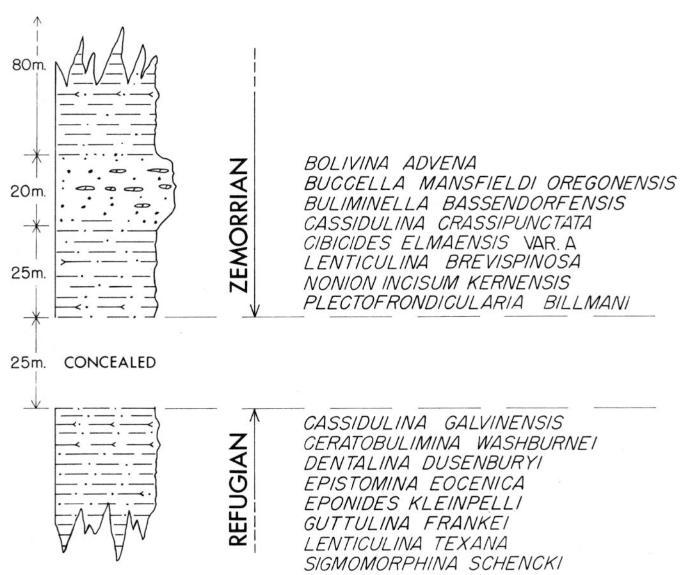

Figure 7. Foraminifera used in recognizing the Refugian/Zemorrian boundary, type section of Alsea Formation, north shore of Alsea Bay, Oregon (Rau, 1975b).

(1938). Fowler (1965) presented considerable detail on the stratigraphic and paleoecologic significance of foraminifers of the Montesano Formation.

Also in the 1960s, cooperative state and federal geologic mapping was conducted in the Grays River area of the extreme southwestern part of Washington (Wolfe and McKee, 1968). This project provided additional foraminiferal information and allowed an extended application of the foraminiferal zonation of southwest Washington (Rau, in Wolfe and McKee, 1972).

Support of U.S. Geological Survey geologic mapping in the western part of the Grays Harbor basin (Gower and Pease, 1965; Wagner, 1967) further extended the foraminiferal biostratigraphy of Washington to this area (see Fig. 1).

Increasing interest in offshore exploration in the late 1960s prompted Washington State to begin stratigraphic studies of the onshore Olympic coastal area. The project began with detailed biostratigraphic studies of the Pliocene Quinault Formation (Rau, 1970) in which four major surface sections were carefully measured, sampled, and described (Fig. 9). Foraminifers from these sections were described and illustrated, and their stratigraphic and paleoecologic significance discussed. Correlations were suggested to other West Coast Pliocene sequences. Formal

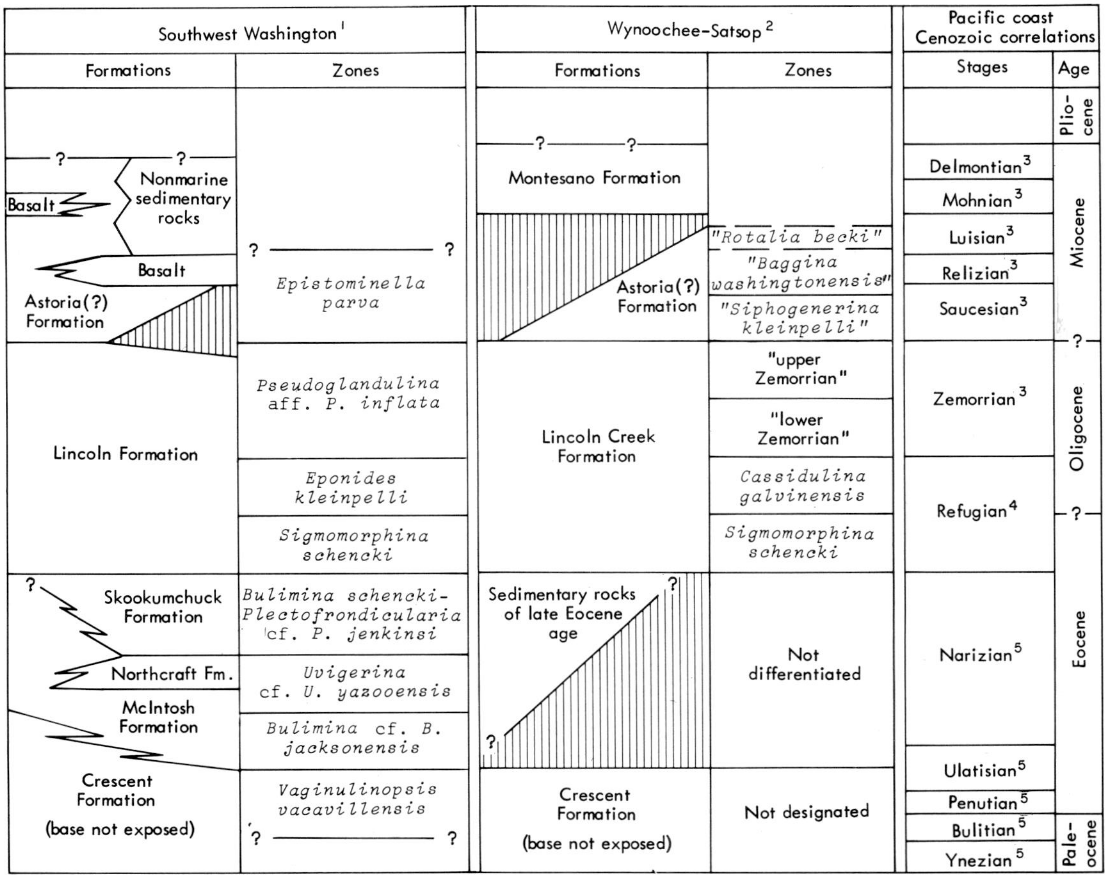

1 Adapted from report by Pease and Hoover, 1957; Snavely and others, 1958; Rau, 1958.
2 Includes area of report by Rau, 1966.
3 Of Kleinpell, 1938.
4 Of Schenck and Kleinpell, 1936.
5 Of Mallory, 1959.

Figure 8. Correlations—southwest Washington and the Wynoochee-Satsop area as presented by Rau in 1967. See text regarding differences in zonal nomenclature, southwest Washington, 1958, versus Wynoochee-Satsop area, 1966 and 1967.

biostratigraphic subdivisions were not presented for the Quinault sequence because stratigraphic relations between each section were uncertain (Fig. 10). However, paleoecologic studies suggested that the Quinault Formation may have been deposited under shoaling conditions beginning with upper bathyal depths and terminating at littoral depths. Thus, the deepest water foraminifers may be confined largely to the oldest Quinault strata and the youngest beds should contain only relatively shallow-water faunas, a depositional model similar to that previously suggested for the early and middle Miocene Astoria(?) Formation of the Grays Harbor basin.

Continuing stratigraphic studies along the western flank of the Olympic Mountains in pre-Pliocene coastal rocks (Rau, 1973, 1975a, 1979) have shown that it is difficult to conduct detailed biostratigraphy in these rocks because faunas are sparse and structures are complex (Fig. 1). Stage-level assignments have been made to faunas from isolated outcrops mostly of Oligocene and Miocene age. A few older assemblages of both middle and late Eocene age have been recorded. However, available sections are not sufficiently complete to adequately verify that the stratigraphic order of occurrence of these faunas coincides with that of the West Coast stages.

Perhaps equally significant to relative age and correlation of these coastal strata have been the paleoecologic data that the contained faunas have produced. Almost all foraminiferal assemblages indicate substantial water depths in an oceanic environment—an environment distinct from that indicated by the shelf faunas that occur in most of the rocks of the Grays Harbor basin and northern Olympic Peninsula area of Washington. This evidence in itself supports the currently employed concept that most of the Tertiary rocks of the Olympic Mountains and west coastal area have been accreted to the North American plate from the oceanic plate and are therefore of a different ecological province than rocks of roughly the same age that flank the Olympic Mountains to the north, east, and south (Rau, 1973; Cady, 1975; Snavely and others, 1977; Tabor and Cady, 1978).

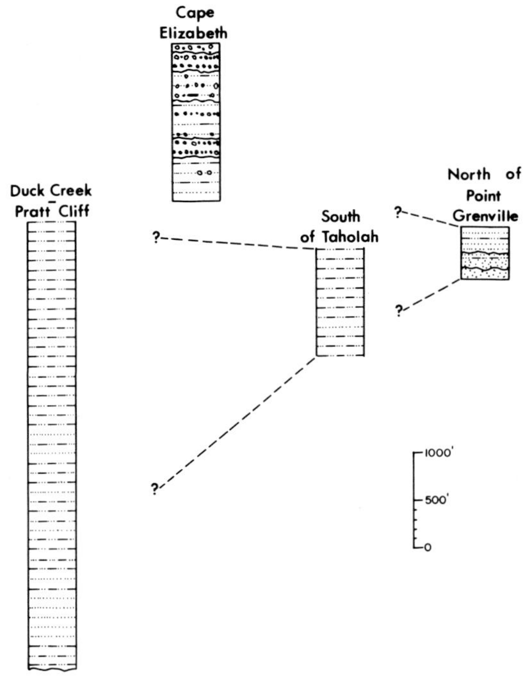

Figure 9. Suggested stratigraphic relations of four major coastal sections of the Pliocene Quinault Formation. Details of these sections are discussed in the original report (Rau, 1970).

Current geologic mapping by the U.S. Geological Survey in the northwest part of the Olympic Peninsula has provided foraminiferal data from a sequence of a deep, marginal basin facies (Rau, in Snavely and others, 1979). Although faunas are noticeably distinct from those of the adjacent northern Olympic Peninsula area to the east, West Coast foraminiferal stages have been recognized and local subdivisions are being developed as mapping progresses.

SUMMARY

Faunal differences in foraminiferal assemblages of the Pacific Northwest Tertiary sequence largely reflect variations in ecological conditions not only over a geographic area but through geologic time as well. The details of faunal characteristics for each zone are, therefore, most reliable for stratigraphic correlation within the provincial Tertiary sequence of southwest Washington where the zones were originally defined. Nevertheless, the general faunal characteristics of most of these biostratigraphic divisions are also recognizable over a larger area and are broadly applicable to correlations throughout western Washington and Oregon. Furthermore, a geographically widespread consistency of ecological evolution through time is obvious in foraminiferal assemblages of the entire Tertiary sequence of the West Coast, thus allowing broad stratigraphic correlations throughout this region at the stage level of magnitude.

The following is a brief synthesis of faunal characteristics for each zone or assemblage described and defined over the years and reviewed in the preceding section of this report (Fig. 11). Greater details of these faunal characteristics are presented in the original reports.

Pre-Ulatisian Assemblages

A few collections from scattered areas in western Washington that usually occur in association with the Crescent Formation have been referred to either the Penutian Stage of Mallory (1959) or a general early Eocene age. Such faunas are perhaps even more widespread in western Oregon. Thoms (1965) described some of these faunas from the Umpqua Group. Planktonic foraminifers are usually more prevalent in early Tertiary strata than in other parts of the Tertiary. Studies by Miles (1977), McKeel (1972), and McKeel and Lipps (1975) are important contributions to information on these foraminifers. Taxa listed below typically occur in, although are not necessarily restricted to, faunas that are assigned to a pre-Ulatisian age:

Anomalina dorri aragonensis Nuttall
Cibicides martinezensis Cushman and Barksdale
Globorotalia aragonensis Nuttall
Glomospira charoides (Jones and Parker)
Loxostomum applinae (Plummer)
Nodosaria velascoensis Cushman
Nonion micrum Cole
Silicosigmoilina californica Cushman and Church
Siphona cf. *S. wilcoxensis* Cushman
Spiroplectammina directa (Cushman and Siegfus)
Spiroplectammina richardi Martin
Tritaxilina colei Cushman and Siegfus
Vaginulinopsis asperuliformis Nuttall

Vaginulinopsis vacavillensis Assemblage

Although this group of species is not treated as characterizing a formal zone, its stratigraphic occurrence is consistent in Pacific Northwest Tertiary sequences. The faunal composition of the assemblage may be compared generally to that of the combined "B" zones of Laiming (1940), and its relative stratigraphic position in the Pacific Northwest sections is also comparable to that of the "B" zones in California, as well as a lower part of the Ulatisian Stage of Mallory (1959). The following listed species are among the most commonly occurring, and they typify the assemblage:

Amphistegina californica Cushman and M. A. Hanna
Asterigerina crassaformis Cushman and Siegfus
Bulimina lirata Cushman and Parker
Baggina teninoensis Rau
Cibicides coalingensis Cushman and G. D. Hanna
Cibicides martinezensis Cushman and Barksdale
Cibicides spiropunctatus Galloway and Morrey
Gaudryina coalingensis (Cushman and G. D. Hanna)
Marginulina subbullata Hantken
Nodosaria latejugata Gumbel
Pseudoglandulina nallpeensis Rau
Spiroplectammina mississippiensis (Cushman)
Tritaxilina colei Cushman and Siegfus
Vaginulinopsis asperuliformis (Nuttall)

Vaginulinopsis vacavillensis (G. D. Hanna) [*mexacana* var. A-D of Laiming]

Bulimina cf. *B. jacksonensis* Zone

This zone was defined from surface sections of southwest Washington (Rau, 1958) and has been broadly recognized throughout western Washington and parts of western Oregon. It is correlated with the upper part of the Ulatisian Stage of Mallory (1959). The following species occur typically in the zone:

Amphimorphina californica Cushman and McMasters
Asterigerina crassaformis Cushman and Siegfus
Bulimina cf. *B. jacksonensis* Cushman
Bulimina corrugata Cushman and Siegfus
Baggina teninoensis Rau
Cibicides warreni Cushman, Stewart, and Stewart
Gaudryina coalingensis (Cushman and G. D. Hanna)
Lenticulina sp. [robust, raised limbate sutures]
Nodosaria latejugata Gumbel

Uvigerina cf. *U. yazooensis* Zone

This zone was defined from surface sections and wells of southwest Washington (Rau, 1958), and it generally represents the lower part of the Narizian Stage of Mallory in the Pacific Northwest. Its faunal characteristics vary somewhat from area to area. However, the following taxa are common in the zone throughout western Washington and Oregon:

Asterigerina crassaformis Cushman and Siegfus
Bolivina basisenta Cushman and Stone
Bulimina corrugata Cushman and Siegfus
Cibicides warreni Cushman, Stewart, and Stewart
Eponides yeguaensis Weinzierl and Applin
Gyroidina simiensis Cushman and McMasters
Gyroidina soldanii octocamerata Cushman and G.D. Hanna
Lenticulina welchi (Church)
Plectofrondicularia searsi Cushman, Stewart, and Stewart
Pleurostomella acuta Hantken
Uvigerina cf. *U. yazooensis* Cushman
Uvigerina garzaensis Cushman and Siegfus
Valvulineria tumeyensis Cushman and Simonson
Valvulineria jacksonensis welcomensis Mallory
Vulvulina cf. *V. curta* Cushman and Siegfus

Bulimina schencki-Plectofrondicularia cf. *P. jenkinsi* Zone

Although the boundary between this zone and the subjacent *Uvigerina* cf. *U. yazooensis* Zone is gradational, the upper limit of the *Bulimina schencki-Plectofrondicularia* cf. *P. jenkinsi* Zone is relatively distinct, particularly in southwest Washington where the zone was originally defined (Rau, 1958). Many species ranging from below are not normally found above the top of this zone. The *Bulimina schencki-Plectofrondicularia* cf. *P. jenkinsi* Zone represents an upper part of the Narizian Stage of Mallory (1959) in the Pacific Northwest. A large number of species representing a wide range of depositional environments occur in this zone. The following are among the more commonly occurring species in the zone:

Alabamina kernensis Smith
Angulogerina hannai Beck
Bolivina basisenta Cushman and Stone
Bolivina kleinpelli Beck
Bulimina alazanensis Cushman
Bulimina schencki Beck
Cibicides haydoni (Cushman and Schenck)
Cibicides mcmastersi Beck
Cibicides natlandi Beck
Cibicides warreni Cushman, Stewart, and Stewart
Globocassidulina globosa (Hantken)
Gyroidina condoni (Cushman and Schenck)
Lenticulina welchi (Church)
Planularia cf. *P. markleyana* Church
Plectofrondicularia cf. *P. jenkinsi* Church
Plectofrondicularia cf. *P. searsi* Cushman, Stewart, and Stewart
Pseudoglandulina conica (Neugeboren)
Quinqueloculina goodspeedi Hanna and Hanna
Spiroloculina wilcoxensis Cushman and Garrett
Uvigerina garzaensis Cushman and Siegfus
Valvulineria aff. *V. willapaensis* Rau [a small form]
Valvulineria tumeyensis Cushman and Simonson
Virgulina hobsoni Beck

Sigmomorphina schencki Zone

This zone was originally defined from surface sections and wells of southwest Washington where it occurs in the lower part of the Lincoln Creek Formation (Rau, 1958). Its base is well defined by the lowest occurrence of many of the taxa listed below. The upper limits of the zone are less well defined and are based largely on the absence of species that are known to occur in the superjacent *Cassidulina galvinensis* Zone. Locally, at least, as indicated on the checklist, a few species make their highest appearance at or near the top of the zone. Many of the taxa that first appear at the base of the *Sigmomorphina schencki* Zone continue to occur throughout this and the overlying *Cassidulina galvinensis* Zone, thus suggesting very little or at least gradational change in environmental conditions through time. Although some variations in faunal character occur with increased distance away from its type area, generally the *Sigmomorphina schencki* Zone is recognizable throughout western Washington and at least as far south as the Alsea Bay area of western Oregon.

The following species are significant to the recognition of the *Sigmomorphina schencki* Zone (- indicates species is particularly common and first to appear at base; $^+$ indicates highest appearance at or near top of zone):

\+ *Angulogerina* cf. *A. occidentalis* (Cushman)
Cancris joaquinensis Smith
\+ *Cassidulina* cf. *C. kernensis* Smith
\+ *Ceratobulimina washburnei* Cushman and Schenck

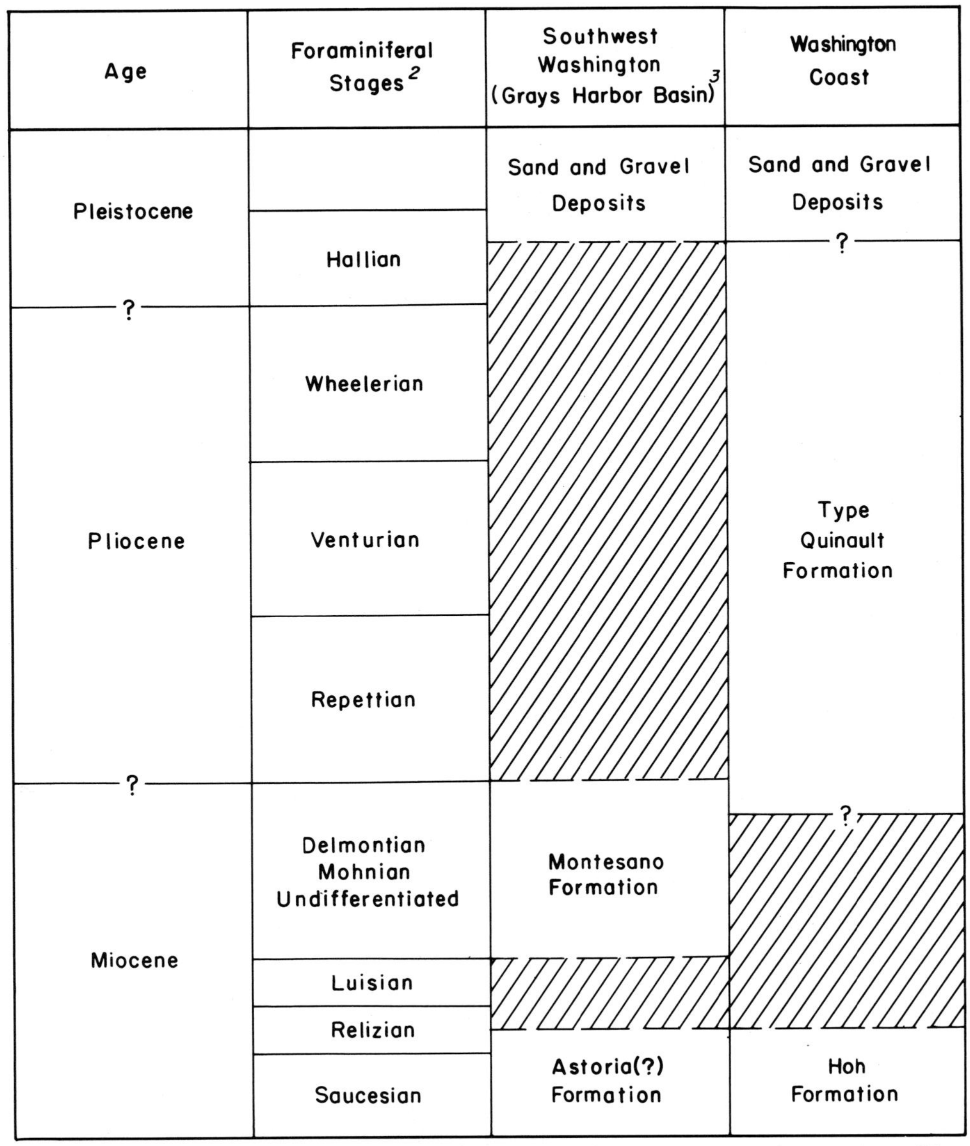

1 Berggren (1969), based on radiometric data, suggests that the Mio-Pliocene boundary may fall somewhere within the Wheelerian stage. Various workers place the Plio-Pleistocene boundary either in or at the base of the Hallian stage.

2 Miocene stages of Kleinpell (1938); Plio-Pleistocene stages of Natland (1952).

3 After Rau (1967).

Figure 10. Suggested stratigraphic relations of the Washington coast Quinault Formation, formations of the Grays Harbor basin, and West Coast foraminiferal stages (Rau, 1970).

- *Cibicides elemaensis* Rau
  Cibicides haydoni (Cushman and Schenck)
  Cibicides hodgei Cushman and Schenck
  Dentalina dusenburyi Beck
  Eponides yeguaensis Weinzierl and Applin
- *Gaudryina alazanensis* Cushman
- *Karreriella washingtonensis* Rau
+ *Lenticulina becki* (Rau)
+ *Lenticulina texanus* (Cushman and Applin)
  Listerella nodulosa (Cushman)
+ *Marginulina alazanensis* Nuttall
+ *Melonis halkyardi* (Cushman)
  Plectofrondicularia packardi Cushman and Schenck *sensu lato*
  Pseudoglandulina inflata (Bornemann)
  Quinqueloculina imperialis Hanna and Hanna
  Quinqueloculina weaveri Rau
- *Sigmomorphina* sp. A [similar to *S. schencki* but with edges less acute]
+ *Sigmomorphina schencki* Cushman and Ozawa
  Spiroloculina texanus Cushman and Ellisor
- *Uvigerina atwilli* Cushman and Simonson
+ *Uvigerina cocoaensis* Cushman
  Uvigerina cf. *U. garzaensis* Cushman and Siegfus

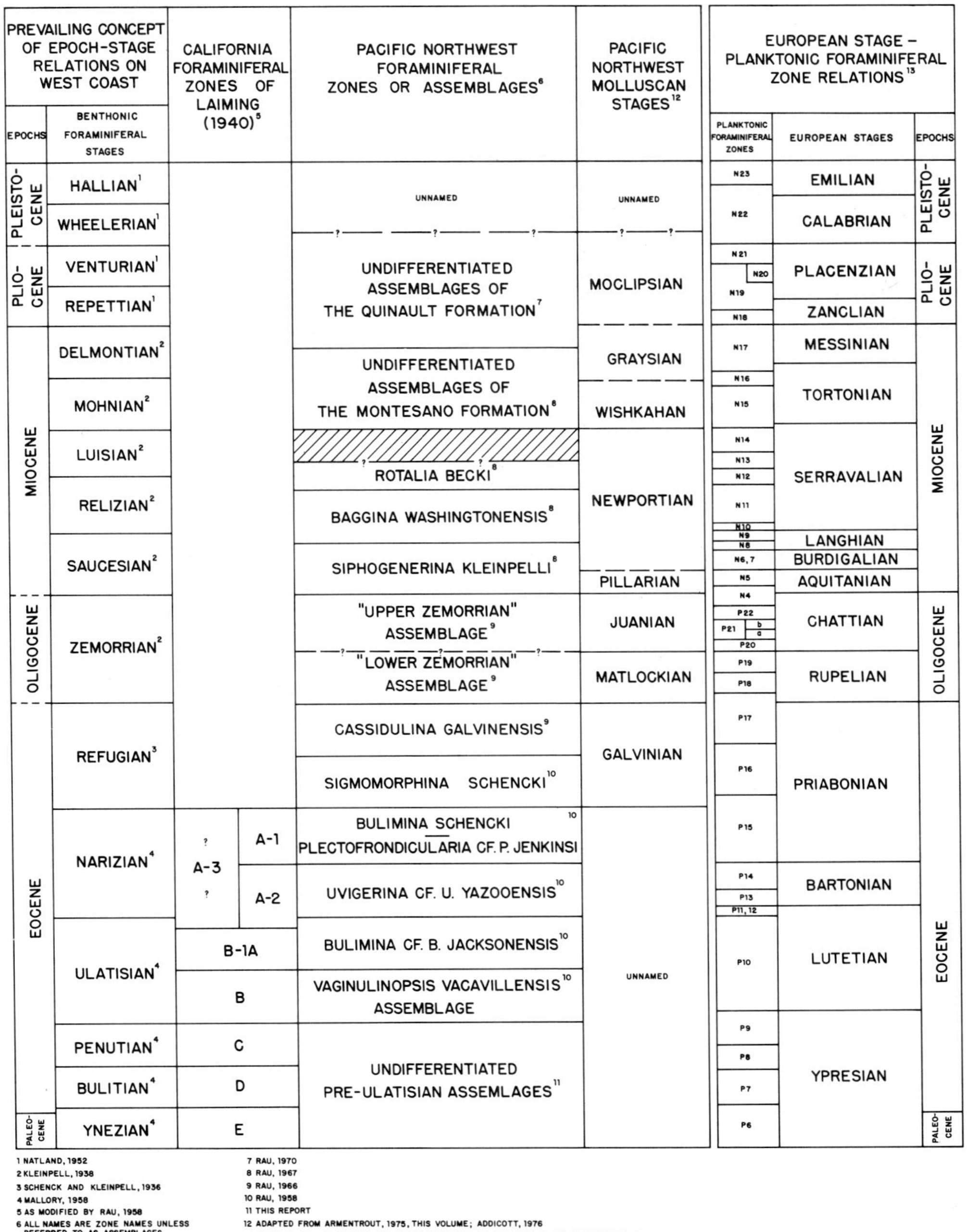

Figure 11. Pacific Northwest foraminiferal biostratigraphic units with correlations to other West Coast standards and a suggested correlation of European stages with planktonic foraminiferal zones.

\+ *Vaginulinopsis saundersi* (Hanna and Hanna)
\- *Valvulineria willapaensis* Rau

Cassidulina galvinensis Zone

This zone locally represents an upper part of the Refugian Stage. It was first defined from surface sections and wells of southwest Washington and named the *Eponides kleinpelli* Zone (Rau, 1958). It was later redefined from additional surface sections of southwest Washington and renamed the *Cassidulina galvinensis* Zone (Rau, 1966). It is represented in southwest Washington in a middle part of the Lincoln Creek Formation. The *Cassidulina galvinensis* Zone lacks clear definition both at the base and at the top because faunal changes through both Refugian and the superjacent Zemorrian time were not only few but gradational. Thus, locally, the upper Refugian foraminiferal faunas are transitional in character between that of the lower Refugian *Sigmomorphina schencki* Zone and the overlying faunas of the Zemorrian Stage. The faunal character of the upper Refugian *Cassidulina galvinensis* Zone is broadly recognizable throughout western Washington and at least as far south as the Alsea Bay area of Oregon. The following species are locally common in the *Cassidulina galvinensis* Zone; the occurrence together of those making either their highest or lowest appearance characterizes this zone (+ indicates first appearance at base of zone; + indicates highest appearance at top of zone):

+ *Angulogerina* cf. *A. occidentalis* (Cushman)
+ *Cassidulina galvinensis* Cushman and Frizzell
+ *Cassidulina* cf. *C. kernensis* Smith
- *Cassidulina* cf. *C. oblonga* Reuss
Cibicides elmaensis Rau
Epistomina eocenica (Cushman and Hanna)
Eponides kleinpelli Cushman and Frizzell
+ *Guttulina hantkeni* Cushman and Ozawa
Gyroidina orbicularis planata Cushman
+ *Karreriella chilostoma* (Reuss)
- *Plectofrondicularia* cf. *P. packardi multilineata* Cushman and Simonson
Pseudoglandulina inflata (Bornemann)
Quinqueloculina imperialis Hanna and Hanna
- *Virgulina bramlettei* Galloway and Morrey

Faunas of the Zemorrian Stage

The Zemorrian Stage is well represented in rocks of western Washington and Oregon. In southwest Washington the stage is contained in some 2,000 m of section within the Lincoln Creek Formation. It therefore would be desirable to be able to recognize local subdivisions of the stage. Unfortunately, as previously pointed out, depositional conditions were relatively uniform and faunal changes are minimal throughout the thick section in which the stage is represented. Although formal names have not been assigned, the terms "lower" and "upper" Zemorrian have been employed in both southwest Washington and the northern Olympic Peninsula (Rau, 1964, 1966, 1967). The recognition of these informal subdivisions is based largely on combinations of species ranging from above and below, and the reader is referred to the original reports for a more complete discussion. The boundary between the lower and upper parts of the Zemorrian Stage is not defined, even though in a general way the lower and upper parts of the stage can be distinguished faunally. The Zemorrian Stage is also well represented in the Newport-Alsea Bay area of west-central Oregon where the lower and upper parts of the stage are also generally distinguishable on many of the same faunal criteria as are presented for western Washington. In addition, the stage is well represented in a thick sequence exposed on the southern part of Bainbridge Island and nearby areas of Puget Sound. Fulmer (1975) has made detailed studies of the foraminifers of these strata.

A typical congregation of foraminifers occurring in the lower part of Zemorrian Stage in the Pacific Northwest is listed below (- indicates range from below the Zemorrian Stage; + indicates range from above the "lower" Zemorrian):

+ *Anomalina californiensis* Cushman and Hobson
+ *Buccella mansfieldi oregonensis* (Cushman, Stewart, and Stewart)
+ *Bulimina alsatica* Cushman and Parker
+ *Cassidulina crassipunctata* Cushman and Hobson
+ *Cassidulinoides* sp.
- *Cassidulina* cf. *C. oblonga* Reuss
- *Cornuspira byramensis* Cushman
- *Eggerella bradyi* (Cushman)
+ *Entosolenia* sp.
- *Guttulina problema* d'Orbigny
Lagena hexagona (Williamson)
Marginulina cf. *M. alazanensis* Nuttall
+ *Plectofrondicularia* cf. *P. californica* Cushman and Stewart
- *Plectofrondicularia packardi multilineata* Cushman and Simonson
Pseudopolymorphina cf. *P. ligua* (Roemer)
- *Quinqueloculina weaveri* Rau
Sigmoilina tenuis (Czjzek)
+ *Sphaeroidina variabilis* Reuss

Although many species range throughout the Zemorrian Stage, others are far more common to, or are not found at all below, the upper part of the stage. Furthermore, species ranging from below the Zemorrian Stage are fewer in the upper part than they are in the lower part of the stage. The following congregation of species is typical of upper Zemorrian strata (- indicates range from below the "upper" Zemorrian; + indicates range from above the Zemorrian Stage; 0 indicates occurrence is most common in the "upper" Zemorrian):

+ *Anomalina californiensis* Cushman and Hobson
+ *Buccella mansfieldi oregonensis* (Cushman, Stewart, and Stewart)
+ *Bolivina marginata adelaidana* Cushman and Kleinpell
0+ *Buliminella subfusiformis* Cushman
+ *Bulimina alsatica* Cushman and Parker
+ *Cassidulina crassipunctata* Cushman and Hobson
+ *Cassidulinoides* sp.
0- *Cibicides elmaensis* Rau var. A [highly convexed ventrally]
0 *Dentalina quadrulata* Cushman and Laiming
0+ *Elphidium* cf. *E. minutum* (Reuss)
+ *Entosolenia* sp.
0+ *Florilus incisum* (Cushman)
0 *Lenticulina* cf. *L.* calcar (Linne)
- *Plectofrondicularia packardi multilineata* Cushman and Simonson
- *Quinqueloculina weaveri* Rau
0 *Siphogenerina nodifera* Cushman and Kleinpell
0 *Sphaeroidina variabilis* Reuss
+ *Uvigerinella obesa impolita* Cushman and Laiming

Siphogenerina kleinpelli Zone

This is the lowest of three zones that were defined from the early and middle Miocene Astoria(?) Formation of southwest Washington (Rau, 1967). The faunal composition of each of the three zones reflects changing ecological conditions through time in a gradually filling depositional basin. Deepest conditions apparently existed during earliest Miocene time and are represented by the faunas of the *Siphogenerina kleinpelli* Zone. The following species are commonly found in this zone:

Bolivina advena Cushman
Buccella mansfieldi oregonensis (Cushman, Stewart, and Stewart)
Cassidulina pulchella d'Orbigny
Epistominella parva (Cushman and Laiming)
Florilus costiferum (Cushman)
Planulina astoriensis Cushman, Stewart, and Stewart
Plectofrondicularia californica Cushman and Stewart
Siphogenerina kleinpelli Cushman
Sphaeroidina variabilis Reuss
Uvigerinella obesa impolita Cushman and Laiming
Valvulineria araucana (d'Orbigny)

Baggina washingtonensis Zone

Intermediate depths are believed to have dominated during the deposition of the overlying strata of the *Baggina washingtonensis* Zone. The following species are common in this zone:

Baggina washingtonensis Rau
Bolivina chehalisensis Rau
Buccella mansfieldi oregonensis (Cushman, Stewart, and Stewart)
Cassidulina pulchella d'Orbigny
Cibicides elmaensis Rau var. A
Epistominella parva (Cushman and Laiming)
Florilus costiferum Cushman
Uvigerinella californica Cushman

Rotalia becki Zone

Shallow-water conditions are believed to have prevailed during the deposition of the uppermost part of the lower to middle Miocene Astoria(?) Formation. During this time faunas of the *Rotalia becki* Zone dominated. Listed below is a typical congregation of commonly occurring species for this zone:

Bolivina chehalisensis Rau
Buccella cf. *B. frigida* (Cushman)
Buliminella elegantissima (d'Orbigny)
Cassidulina pulchella d'Orbigny
Elphidium cf. *E. minutum* Cushman
Rotalia becki Bandy and Arnal
Uvigerina auberiana d'Orbigny
Valvulineria menloensis Rau

The three preceding zones, although applicable in southwest Washington, have not been recognized in the Miocene Clallam Formation of the northern Olympic Peninsula of Washington, nor the Astoria Formation and Nye Mudstone of Oregon, particularly the Newport area of west-central Oregon. In the latter area the faunas of the Nye Mudstone and the Astoria Formation have been listed by Rau (in Snavely and others, 1964, 1969); their characteristics have been discussed, and they have been referred to the Saucesian Stage. Species characteristically common to the Nye Mudstone of Newport embayment are listed below:

Bolivina advena Cushman
Bolivina marginata adelaidana Cushman and Kleinpell
Buccella mansfieldi oregonensis (Cushman, Stewart, and Stewart)
Buliminella subfusiformis Cushman
Bulimina inflata alligata Cushman and Laiming
Cassidulina laevigata carinata Cushman
Epistominella parva (Cushman and Laiming)
Gyroidina soldanii d'Orbigny
Florilus costiferum (Cushman)
Florilus incisum (Cushman)
Plectofrondicularia californica Cushman and Stewart
Uvigerina auberiana d'Orbigny
Uvigerinella obesa impolita Cushman and Laiming
Virgulina californiensis Cushman

Species characteristically common to the Astoria Formation of Newport embayment are listed below:

Bolivina advena Cushman
Buccella mansfieldi oregonensis (Cushman, Stewart, and Stewart)
Buliminella elegantissima (d'Orbigny)
Buliminella subfusiformis Cushman
Epistominella parva (Cushman and Laiming)
Florilus costiferum (Cushman)
Nonionella miocenica Cushman
Lenticulina mayi Cushman and Parker
Uvigerinella californica ornata Cushman

Although many of the same species occur in both the Astoria Formation and the Nye Mudstone, the assemblage of the Nye Mudstone is more diverse and therefore easier to recognize. A few Astoria species, particularly *Buliminella elegantissima* and *Lenticulina mayi*, are not known to occur in the Nye Mudstone and are useful for distinguishing the faunas of the Astoria Formation from those of the Nye Mudstone.

The subjacent Yaquina Formation, on the basis of a limited number of significant species, has been generally regarded as representing an upper part of the Zemorrian Stage. However, the reported occurrence of *Florilus costiferum* (R. G. Gonsalves, 1963, personal communs.) in the upper part of the formation suggests that at least a part of the formation may be as young as Saucesian.

Assemblages of the Upper Miocene Montesano Formation

Onshore outcrops of upper Miocene strata in Washington are largely confined to the Montesano Formation of southwest Washington. Although our studies of foraminifers were limited to use as an aid to geologic mapping, other workers have presented more comprehensive observations of these foraminifers. Fowler (1965) discussed foraminiferal biostratigraphy of the Montesano Formation in considerable detail and presented extensive lists of foraminifers. Bergen and Bird (1972), in their discussion of the biostratigraphy of the Ocean City area, presented additional data on late Miocene foraminiferal biostratigraphy. The following species are common to or characteristic of the Montesano Formation:

Bolivina brevior Cushman
Bolivina rankini Kleinpell
Buccella frigida (Cushman)
Buliminella curta Cushman
Cibicides mckannai Galloway and Wissler
Epistominella pacifica (Cushman)
Epistominella smithi (Stewart and Stewart)
Nonionella cf. *N. basispinata* (Cushman and Moyer)
Nonionella miocenica Cushman
Pullenia salisburyi Stewart and Stewart
Rotalia cf. *R. garveyensis* Natland
Uvigerina hootsi Rankin
Uvigerina modeloensis Cushman and Kleinpell
Uvigerina subperegrina Cushman and Kleinpell
Virgulina californiensis ticensis (Cushman and Kleinpell)
Virgulina subplana Barbat and Johnson

Assemblages of the Pliocene Quinault Formation

Marine rocks of Pliocene age in western Washington are confined to the Olympic coastal area and constitute a major part of the Quinault Formation. Outcrops of the lowermost part of the formation are generally regarded as late Miocene in age, largely on the basis of the rare occurrence of *Rotalia garveyensis*. Furthermore, offshore data indicate that strata of late Miocene age are extensive beneath rocks of Pliocene age (Rau, 1970).

Biostratigraphic subdivisions of the Quinault Pliocene have not been defined largely because stratigraphic control between isolated surface sections is insufficient. The following listed foraminifers are either among those characteristically common in the Quinault Formation or are particularly useful for general correlation of the formation:

Angulogerina semitrigona (Galloway and Wissler)
Anomalinoides quinaultensis Rau
Bolivina acuminata Natland
Bolivina advena Cushman
Buccella inusitata Andersen
Bulimina subacuminata Cushman and Todd
Buliminella elegantissima (d'Orbigny)
Cassidulina islandica Norvang
Cassidulina limbata Cushman and Hughes
Cassidulina reflexa Galloway and Wissler
Cassidulina translucens Cushman and Hughes
Cibicides conoideus Galloway and Wissler
Cibicides fletcheri Galloway and Wissler
Cibicides mckannai Galloway and Wissler
Elphidiella hannai (Cushman and Grant)
Elphidium hughesi foraminosum Cushman
Elphidium hughesi hughesi Cushman and Grant
Elphidium microgranulosum (Galloway and Wissler)
Epistominella pacifica (Cushman)
Florilus basispinatum (Cushman and Moyer)
Globigerina bulloides d'Orbigny
Globigerina pachyderma (Ehrenberg)
Globobulimina auriculata (Bailey)
Globorotalia crassaformis (Galloway and Wissler)
Lagena pliocenica timmsana Cushman and Gray
Nonionella miocenica Cushman
Oolina borealis Loeblich and Tappan
Pullenia cf. *P. salisburyi* Stewart and Stewart
Rotalia cf. *R. garveyensis* Natland
Uvigerina juncea Cushman and Todd
Uvigerina cf. *U. Hootsi* Rankin
Uvigerina senticosa Cushman
Uvigerina subperegrina Cushman and Kleinpell
Virgulina californiensis ticensis Cushman and Kleinpell

ACKNOWLEDGMENTS

In addition to the studies of governmental agencies, numerous workers of various other affiliations have contributed much to the present knowledge of foraminiferal biostratigraphy of the Pacific Northwest. Significant studies were made by petroleum stratigraphers and paleontologists, but because of the competitive nature of these positions, much of their work has not been published. Among these researchers are H. G. Billman, formerly of the Union Oil Company of California; T. W. Rothwell, Jr., formerly of the Richfield Oil Corporation; and Charles V. Fulmer, formerly of the Standard Oil Company of California. Although some studies of the late R. E. Stewart and K. C. Stewart were published, they conducted many additional unpublished studies as well. Several university professors and their students have also contributed much to the foraminiferal biostratigraphy of the Pacific Northwest. Among these in particular are V. S. Mallory of the University of Washington; R. E. Thoms of Portland State University; G. A. Fowler, formerly of Oregon State University; J. C. Ingle, Jr., of Stanford University; R. M. Kleinpell, professor emeritus of the University of California; and the late O. L. Bandy of the University of Southern California. In recent years additional important studies have been, or are being, conducted by Ann Tipton Donnelly, D. R. McKeel, Kristin McDougall, G. A. Miles, J. M. Armentrout, and Annalisa Berta.

Technical reviews by W. O. Addicott, J. M. Armentrout, Kristin McDougall, and Ron Echols have added substantially to this report and are gratefully acknowledged.

REFERENCES CITED

Addicott, W. O., 1976a, Neogene molluscan stages of Oregon and Washington, *in* Fritsche, A. E., and others, eds., The Neogene symposium—Selected papers on paleontology, sedimentology, petrology, tectonics, and geologic history of the Pacific coast of North America: Society of Economic Paleontologists and Mineralogists, Pacific Section, Annual Meeting, San Francisco, p. 95-115.

——1976b, New molluscan assemblages from the upper member of the Twin River Formation, western Washington: Significance in Neogene chronostratigraphy: U.S. Geological Survey Journal of Research, v. 4, p. 437-447.

Armentrout, J. M., 1975, Molluscan biostratigraphy of the Lincoln Creek Formation, southwest Washington, *in* Paleogene symposium: Society of Economic Paleontologists and Mineralogists, Pacific Section, Annual Meeting, Long Beach, California, p. 14-48.

Baldwin, E. M., 1955, Geology of the Alsea and Marys Peak quadrangles, Oregon: U.S. Geological Survey Oil and Gas Investigations Map OM-162, scale 1:62,500.

——1961, Geologic map of the lower Umpqua River area, Oregon: U.S. Geological Survey Oil and Gas Investigations Map OM-204, scale 1:62,000.

Baldwin, E. M., Brown, R. D., Jr., Gair, J. E., and Pease, M. H., Jr., 1955, Geology of the Sheridan and McMinnville quadrangles, Oregon: U.S. Geological Survey Oil and Gas Investigations Map OM-155, scale 1:62,500.

Beck, R. S., 1943, Eocene Foraminifera from the Cowlitz River, Lewis County, Washington: Journal of Paleontology, v. 17, p. 584-614, Pls. 98-109.

Beikman, H. M., Rau, W. W., and Wagner, H. C., 1967, The Lincoln Creek Formation, Grays Harbor basin, southwestern Washington: U.S. Geological Survey Bulletin 1244-I, 14 p.

Bergen, F. W., and Bird, K. J., 1972, The biostratigraphy of the Ocean City area, *in* Pacific Coast Miocene Biostratigraphic symposium, Proceedings: Society of Economic Paleontologists and Mineralogists, Pacific Section, Bakersfield, California, p. 173-198.

Berggren, W. A., 1969, Cenozoic chronostratigraphy, planktonic foraminiferal zonation, and the radiometric time scale: Nature, v. 224, no. 5224, p. 1072-1075.

Berggren, W. A., and Van Couvering, J. A., 1974, Biostratigraphy, geochronology, and paleoclimatology of the last 15 million years in marine and continental sequences: Palaeogeography, Palaeoclimatology, Palaeoecology, v. 16, no. 1/2, p. 216.

Brown, R. D., Jr., Gower, H. D., and Snavely, P. D., Jr., 1960, Geology of the Port Angeles-Lake Crescent area, Clallam County, Washington: U.S. Geological Survey Oil and Gas Investigations Map OM-203, scale 1:62,500.

Cady, W. M., 1975, Tectonic setting of the Tertiary volcanic rocks of the Olympic Peninsula, Washington: U.S. Geological Survey Journal of Research, v. 3, p. 573-582.

Cushman, J. A., and Frizzell, D. L., 1943, Foraminifera from the type area of the Lincoln Formation (Oligocene) of Washington State: Cushman Laboratory for Foraminiferal Research Contributions, v. 19, pt. 4, p. 79-89, Pls. 14, 15.

Cushman, J. A., and Schenck, H. G., 1928, Two foraminiferal faunules from the Oregon Tertiary: University of California Publications, Department of Geological Sciences Bulletin, v. 17, no. 9, p. 305-324, Pls. 42-45.

Cushman, J. A., Stewart, R. E., and Stewart, K. C., 1947, Five papers on Foraminifera from the Tertiary of western Oregon: Oregon Department of Geology and Mineral Industries Bulletin 36, pts. 1-5, 111 p., 13 pls.

——1949, Three papers on Foraminifera from the Tertiary of western Oregon and Washington: Oregon Department of Geology and Mineral Industries Bulletin 36, pts. 6-8, p. 113-208, Pls. 14-18.

Detling, M. R., 1946, Foraminifera from the Coos Bay lower Tertiary, Coos County, Oregon: Journal of Paleontology, v. 20, p. 348-361, Pls. 46-51.

Fowler, G. A., 1965, The stratigraphy, Foraminifera, and paleoecology of the Montesano Formation (Miocene-Pliocene), Grays Harbor County, Washington [Ph.D. dissert.]: Los Angeles, University of Southern California, 355 p.

Fulmer, C. V., 1975, Stratigraphy and paleontology of the type Blakeley and Blakeley Harbor Formations, *in* Paleogene symposium: Society of Economic Paleontologists and Mineralogists, Pacific Section, Annual Meeting, Long Beach, California, p. 210-271.

Gower, H. D., 1960, Geology of the Pysht quadrangle, Washington: U.S. Geological Survey Geologic Quadrangle Map GQ-129, scale 1:62,500.

Gower, H. D., and Pease, M. H., Jr., 1965, Geology of the Montesano quadrangle, Washington: U.S. Geological Survey Geologic Quadrangle Map GQ-374, scale 1:62,500.

Hanna, G. D., and Hanna, M. A., 1924, Foraminifera from the Eocene of Cowlitz River, Lewis County, Washington: University of Washington Publications in Geology, v. 1, no. 4, p. 57-64, Pl. 13.

Hardenbol, J., and Berggren, W. A., 1978, A new Paleogene numerical time scale, *in* Cohee, G. V., and others, The geologic time scale: American Association of Petroleum Geologists Studies in Geology VI, p. 213-234.

Hoover, Linn, 1963, Geology of the Anlauf and Drain quadrangles, Douglas and Lane Counties, Oregon: U.S. Geological Survey Bulletin 1122-D, 62 p., 1 map.

Kleinpell, R. M., 1938, Miocene stratigraphy of California: Tulsa, Oklahoma, American Association of Petroleum Geologists, 450 p., 14 figs., 22 pls.

Laiming, B. G., 1940, Some foraminiferal correlations in the Eocene of San Joaquin Valley, California: Pacific Congress, 6th, 1939, Proceedings, v. 2, p. 535-568.

Mallory, V. S., 1959, Lower Tertiary biostratigraphy of the California Coast Ranges: Tulsa, Oklahoma, American Association of Petroleum Geologists, 416 p., 42 pls.

McKeel, D. R., 1972, Systematics, biostratigraphy, and paleoceanography of Tertiary planktonic foraminiferal assemblages, Oregon Coast Range [M.S. thesis]: Davis, University of California, 115 p.

McKeel, D. R., and Lipps, J. H., 1975, Eocene and Oligocene planktonic Foraminifera from the central and southern Oregon Coast Range: Journal of Foraminiferal Research, v. 5, p. 249-269.

Miles, G. A., 1977, Planktonic Foraminifera of the lower Tertiary Roseburg, Lookingglass, and Flournoy Formations, southern Oregon [Ph.D. dissert.]: Eugene, University of Oregon, 360 p.

Natland, M. L., 1952, Pleistocene and Pliocene stratigraphy of southern California [Ph.D. dissert.]: Los Angeles, University of California, 165 p., 20 pls.

Pease, M. H., and Hoover, Linn, 1957, Geology of the Doty-Minot Peak area, Washington: U.S. Geological Survey Oil and Gas Investigations Map OM-188, scale 1:62,500.

Rau, W. W., 1948a, Foraminifera from the Porter shale (Lincoln Formation), Grays Harbor County, Washington: Journal of Paleontology, v. 22, p. 152-174, Pls. 27-31.

——1948b, Foraminifera from the Miocene Astoria Formation in southwestern Washington: Journal of Paleontology, v. 22, p. 774-782, Pl. 119.

——1951, Tertiary Foraminifera from the Willapa River valley of southwest Washington: Journal of Paleontology, v. 25, p. 417-453, Pls. 63-67.

——1958, Stratigraphy and foraminiferal zonation in some of the Tertiary rocks of southwestern Washington: U.S. Geological Survey Oil and Gas Investigations Chart OC-57, 2 sheets, scale 1 inch to 300 feet.

——1964, Foraminifera from the northern Olympic Peninsula, Washington: U.S. Geological Survey Professional Paper 374-G, p. 1-33, 7 pls.

——1966, Stratigraphy and Foraminifera of the Satsop River

area, southern Olympic Peninsula, Washington: Washington Division of Mines and Geology Bulletin 53, 66 p.
____1967, Geology of the Wynoochee Valley quadrangle, Grays Harbor County, Washington: Washington Division of Mines and Geology Bulletin 56, 51 p., 1 map.
____1970, Foraminifera, stratigraphy, and paleontology of the Quinault Formation, Point Grenville-Raft River coastal area, Washington: Washington Division of Mines and Geology Bulletin 62, 40 p. 3 pls.
____1973, Geology of the Washington coast between Point Grenville and the Hoh River: Washington Division of Geology and Earth Resources Bulletin 66, 58 p.
____1975a, Geology of the Destruction Island and Taholah quadrangles, Washington: Washington Division of Geology and Earth Resources Geologic Map GM-13.
____1975b, Foraminifera and biostratigraphy of the Alsea Formation of western Oregon, *in* Paleogene symposium: Society of Economic Paleontologists and Mineralogists, Pacific Section, Annual Meeting, Long Beach, California, p. 409-416.
____1979, Geologic map in the vicinity of the lower Bogachiel and Hoh River valleys, and the Washington coast: Washington Division of Geology and Earth Resources Geologic Map GM-24.

Reagan, A. B., 1909, Some notes on the Olympic Peninsula, Washington: Kansas Academy of Science Transactions, v. 22, p. 131-238.

Roberts, A. E., 1958, Geology and coal resources of the Toledo-Castle Rock district, Cowlitz and Lewis Counties, Washington: U.S. Geological Survey Bulletin 1062, 71 p.

Schenck, H. G., and Kleinpell, R. M., 1936, Refugian Stage of Pacific Coast Tertiary: American Association of Petroleum Geologists Bulletin, v. 20, p. 215-225.

Snavely, P. D., Jr., Brown, R. D., Jr., Roberts, A. E., and Rau W. W., 1958, Geology and coal resources of the Centralia-Chehalis district, Washington: U.S. Geological Survey Bulletin 1053, 159 p., 25 figs., 1 chart, 13 pls.

Snavely, P. D., Jr., MacLeod, N. S., and Rau, W. W., 1969, Geology of the Newport area, Oregon: The Ore Bin, v. 31, nos. 2 and 3, p. 25-71.

Snavely, P. D., Jr., MacLeod, N. S., Wagner, H. C., and Rau, W. W., 1976a, Geologic map of the Waldport and Tidewater quadrangles, Oregon: U.S. Geological Survey Miscellaneous Investigations Map I-866, scale 1:62,500.
____1976b, Geologic map of the Yaquina and Toledo quadrangles, Lincoln County, Oregon: U.S. Geological Survey Miscellaneous Investigations Map I-867, scale 1:62,500.
____1976c, Geologic map of the Cape Foulweather and Ucher Mountain quadrangles, Oregon: U.S. Geological Survey Miscellaneous Investigations Map I-868, scale 1:62,500.

Snavely, P. D., Jr., Niem, A. R., MacLeod, N. S., Pearl, J. E., and Rau, W. W., 1979, Makah Formation—An upper Eocene and Oligocene deep marginal basin sedimentary sequence in the northwestern Olympic Peninsula, Washington: U.S. Geological Survey Open-File Report 79-581, 75 p.

Snavely, P. D., Jr., Pearl, J. E., and Landers, D. L., 1977, Interim report on petroleum resources potential and geologic hazards in the outer continental shelf—Oregon and Washington Tertiary province: U.S. Geological Survey Open-File Report 77-282, 64 p.

Snavely, P. D., Jr., Rau, W. W., and Wagner, H. C., 1964, Miocene stratigraphy of the Yaquina Bay area, Newport, Oregon: The Ore Bin, v. 26, no. 8, p. 133-151, 3 figs., 1 pl.

Tabor, R. W., and Cady, W. M., 1978, The structure of the Olympic Mountains, Washington—Analysis of a subduction zone: U.S. Geological Survey Professional Paper 1033, 38 p.

Thoms, R. E., 1965, Biostratigraphy of the Umpqua Formation, southwest Oregon [Ph.D. dissert.]: Los Angeles, University of California, 219 p.

Vokes, H. E., Myers, D. A., and Hoover, Linn, 1954, Geology of the west-central border area of the Willamette Valley, Oregon: U.S. Geological Survey Oil and Gas Investigations Map OM-150, scale 1:62,500.

Vokes, H. E., Snavely, P. D., Jr., and Myers, D. A., 1951, Geology of the southern and southwestern border areas of the Willamette Valley, Oregon; U.S. Geological Survey Oil and Gas Investigations Map OM-110, scale 1 inch to 1 mile.

Wagner, H. C., 1967, Preliminary geologic maps of the Raymond and South Bend quadrangles, Pacific County, Washington: U.S. Geological Survey Open-File Report, 2 maps, scale 1:62,500.

Weaver, C. E., 1912, A preliminary report on the Tertiary paleontology of western Washington: Washington Geological Survey Bulletin 15, 80 p.

Weaver, C. E., and others, 1944, Correlation of the marine Cenozoic formations of western North America: Geological Society of American Bulletin, v. 55, p. 569-598.

Wolfe, E. W., and McKee, E. H., 1968, Geology of the Grays River quadrangle, Wahkiakum and Pacific Counties, Washington: Washington Division of Mines and Geology Geologic Map GM-4, scale 1:62,500.
____1972, Sedimentary and igneous rocks of the Grays River quadrangle, Washington: U.S. Geological Survey Bulletin 1335, 70 p.

MANUSCRIPT RECEIVED BY THE SOCIETY MARCH 14, 1980
MANUSCRIPT ACCEPTED MAY 7, 1980

Printed in U.S.A.

Geological Society of America
Special Paper 184
1981

Planktonic foraminifers of the lower Tertiary Roseburg, Lookingglass, and Flournoy Formations (Umpqua Group), southwest Oregon

GREGORY A. MILES
Exxon Company, U.S.A., Exploration Department, P.O. Box 4279, Houston, Texas 77001

ABSTRACT

Microfossils recovered from the Roseburg, Lookingglass, and Flournoy Formations (Umpqua Group) of southwest Oregon include 33 species and subspecies of planktonic foraminifers. Some of the faunas from these units include as many as 14 species per sample, and specimen preservation is moderate to good. The assemblages are similar to those from central and southern California but differ in that they are dominated by long-ranging, high-latitude species.

The Roseburg and Lookingglass Formations were deposited during early Eocene time and are assigned on the basis of planktonic foraminifers to Zone P7-8 of the standard tropical zonation. The planktonic foraminiferal faunas of these units are indistinguishable. The Flournoy Formation is assigned to lower middle Eocene Zone P10. Planktonic foraminifers and calcareous nannofossils indicate that the unconformity between the Lookingglass and Flournoy Formations represents an interval of approximately 1 to 2 m.y.

Despite reports of Paleocene microfossils occurring in the Roseburg and Lookingglass Formations, no microfossils definitely of that age were found during this study. Cretaceous microfossils were recovered from a small outlier currently included in the Roseburg Formation. Although the rocks containing these fossils closely resemble those of the Roseburg Formation, they probably belong to a separate unit.

INTRODUCTION

Objectives

Calcareous plankton, including planktonic foraminifers and calcareous nannofossils, are the most widely used fossils for interregional and intercontinental correlation of Cenozoic sediments and sedimentary rocks. The paucity of calcareous planktonic microfossils in some Oregon Tertiary marine sedimentary rocks has led paleontologists to neglect these groups of fossils and to base biostratigraphic conclusions almost entirely on mollusks and benthic foraminifers. The collection and examination of large numbers of samples show, however, that planktonic microfossils occur in several western Oregon Paleogene units in quantities sufficient for biostratigraphic analysis (McKeel, 1972; Robertson, 1972).

The objectives of this investigation are (1) to determine if the recently defined Roseburg, Lookingglass, and Flournoy Formations (Baldwin, 1974, 1975) can be separated from each other by using planktonic foraminifers, (2) to determine whether or not planktonic foraminifers can be used to divide each formation into zones, and (3) to correlate these units to the stratigraphic sequences of California, Trinidad, and Europe. Because of the large area of investigation, the study is necessarily reconnaissance in scope.

The samples and stratigraphic sections discussed herein represent only a small part of my investigation. Readers interested in examining species check lists, locality information, systematics, and discussions of additional samples and sections are encouraged to consult Miles (1977).

William G. Siesser examined 23 calcareous nannofossil samples from the Umpqua Group. His biostratigraphic determinations were used only for comparison with determinations based on planktonic foraminifers, and no attempt was made to examine the calcareous nannofossils in more detail.

Location

The study area lies at the southern end of a large north-trending belt of Cenozoic sedimentary and volcanic rocks extending from the Klamath Mountains in Oregon northward to British Columbia. It extends from lat 42°30′N to

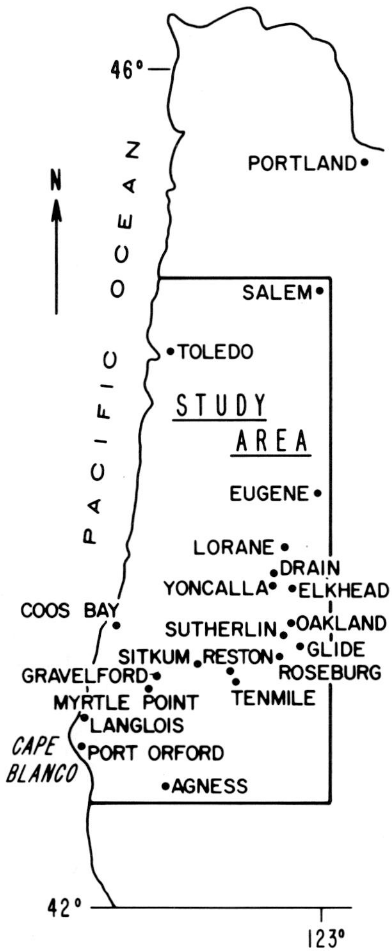

Figure 1. Location map.

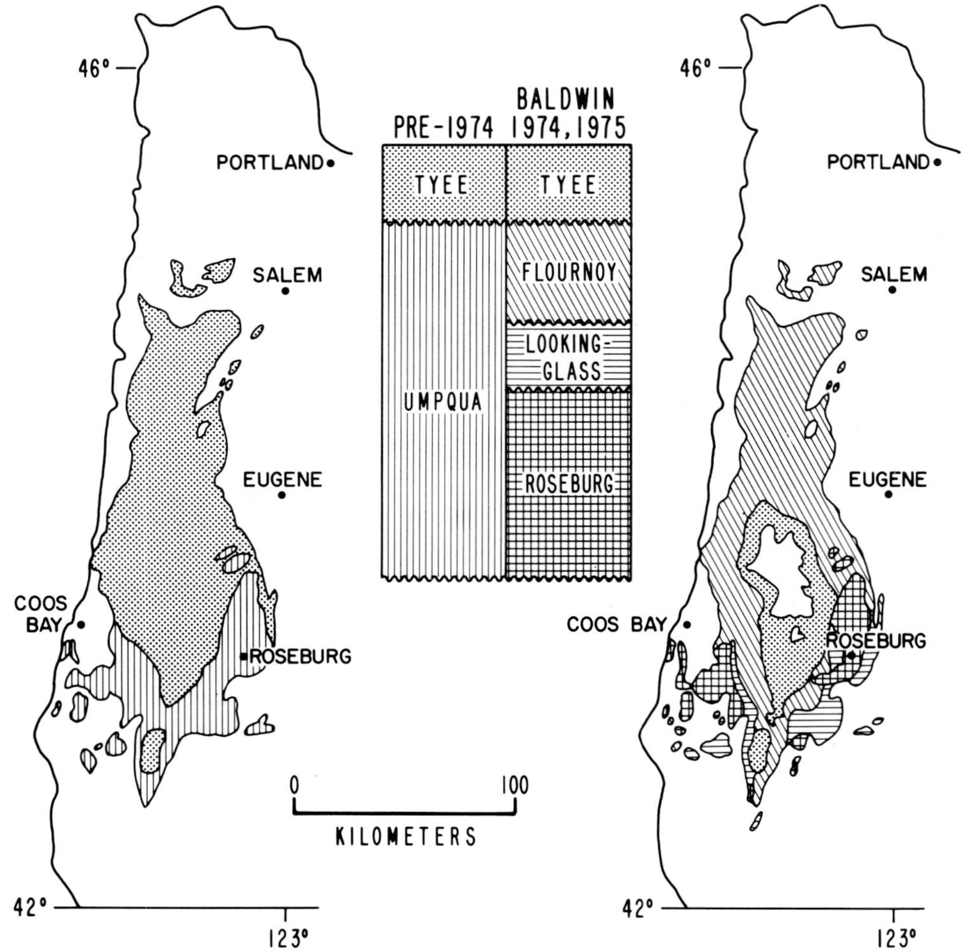

Figure 2. Geologic maps of the Umpqua Group and Tyee Formation showing revisions of Baldwin (1974, 1975).

lat 45°N and is bounded on the east approximately by long 123°W and on the west by the Pacific Ocean (Fig. 1). The Roseburg, Lookingglass, and Flournoy Formations are most widely distributed in the southern part of the Oregon Coast Range, but they also occur in parts of the Klamath Mountains, western Cascade Mountains, and Willamette Valley (Fig. 2; Baldwin, 1973, 1974, 1975, geologic maps).

Samples

Field work was conducted in the summer and fall of 1976 and consisted of collecting 168 spot samples from roadcuts and outcrops along stream beds. Mobil Oil Corporation, the state of Oregon Department of Geology and Mineral Industries, and Richard E. Thoms of Portland State University loaned 137 additional samples to me.

Most exposures of Paleogene fine-grained sedimentary rocks in southwestern Oregon are small and discontinuous because of the dense vegetation and thick mantle of weathered rocks in this region. None of the formations comprising the Umpqua Group is completely exposed, and stratigraphic sections extensive enough to warrant detailed measuring and collecting for biostratigraphic purposes are uncommon. Because of the large area of investigation and the practical limits on the number of samples that could be collected, processed, and examined, more effort was made to examine spot samples from numerous key areas than to examine numerous samples from a few stratigraphic sections. During this investigation I examined samples from nine measured sections collected by Mobil Oil Corporation and Richard E. Thoms (see Miles, 1977); three of these sections are discussed here.

SUMMARY OF UMPQUA GROUP STRATIGRAPHY

Near the turn of the century, Diller (1898) named and described the Umpqua Formation for exposures along the Umpqua River near Roseburg. Similar strata exposed in the vicinity of Coos Bay were later recognized as belonging to the same unit (Allen and Baldwin, 1944). The first attempt to subdivide the Umpqua Formation was made by Turner (1938), who separated the unit into lower and upper members. Later work by Baldwin (1961, 1963, 1965, 1973) led to a tripartite division of the Umpqua Formation into lower, middle, and upper members. In 1974, he elevated these members to formation status and introduced the names Roseburg, Lookingglass, and Flournoy (Fig. 2). Major features of Baldwin's revisions (Baldwin, 1974, 1975) are (1) all the volcanic rocks and the lowermost sedimentary rocks of the former Umpqua Formation have been assigned to the Roseburg Formation, (2) the middle part of the former Umpqua Formation is assigned to the Lookingglass Formation, and (3) the upper part of the former Umpqua Formation and many of the rocks previously mapped as Tyee Formation are now assigned to the Flournoy Formation.

The Roseburg Formation consists of a volcanic sequence overlain by rhythmically bedded sandstone and siltstone. The Lookingglass Formation rests unconformably upon the Roseburg Formation and consists of a lower conglomerate member, a middle rhythmically bedded sandstone and siltstone member, and an upper conglomerate and pebbly sandstone member. The Flournoy Formation, which is divided into a lower sandstone member and an upper siltstone member, rests unconformably upon the Lookingglass Formation and is overlain unconformably by the Tyee Formation.

PREVIOUS WORK

Parts of the Umpqua Group were known to be Eocene nearly a century ago (Diller, 1893). Turner (1938) studied the megainvertebrate fauna of the "Umpqua Formation" and correlated most of that unit with the Capay Formation of California. The Glide section (Lookingglass Formation) was interpreted as possibly intermediate in age between the Capay and Domengine Formations (Turner, 1938, p. 32). The Capay and Domengine provincial molluscan stages (Weaver and others, 1944) are approximately correlative to the lower Eocene Series and the lower part of the middle Eocene Series (Fig. 3).

Thoms (1965, 1975) examined the benthic foraminiferal faunas of the Roseburg and Lookingglass Formations and assigned some of them to the Penutian Stage of Mallory (1959). Other faunas were not particularly age diagnostic and were assigned to the undifferentiated Penutian-Ulatisian Stages. Thoms found that subdivision of the Umpqua Group on the basis of benthic foraminifers was difficult.

Planktonic foraminifers from the Roseburg, Lookingglass, and Flournoy Formations have thus far received only limited attention (Thoms, 1965, 1975; Bird, 1967; McKeel, 1972; principal results published in McKeel and Lipps, 1972, 1975). Thoms and Bird were engaged primarily in studies of benthic foraminifers, and McKeel examined only a few samples from the Umpqua Group during his study of the planktonic foraminifers of several units in the central and southern Oregon Coast Range.

Little progress has been made since the work of Thoms (1965, 1975) and McKeel (1972) toward defining the age relationships of the Roseburg, Lookingglass, and Flournoy Formations. The recent determination of Paleocene and Cretaceous ages from strata included in the Roseburg Formation (see Baldwin, 1974, 1975) differs from the findings of Turner (1938) and Thoms (1965, 1975) and has important implications for interpreting the depositional history of this unit.

PLANKTONIC FORAMINIFERAL ZONATION

The planktonic foraminiferal zonation used in this study (Fig. 3) is the "N" and "P" zone system of Banner and Blow (1965), Blow (1969), and Blow and Berggren (Berggren, 1971, 1972; Berggren and Van Couvering, 1974; Hardenbol and Berggren, 1978). This zonal scheme is a modication of the zonation developed by Bolli (1957a, 1957b, 1957c; 1966).

Although the Blow-Banner-Berggren zonation has been applied successfully to stratigraphic sequences in many parts of the world, it has been used with limited success in high latitudes. Tropical Cenozoic planktonic foraminiferal zonations are difficult to apply to other regions because faunas north of lat 35° to 40° N and south of lat 35° to 40°S usually have few species in common with assemblages from equatorial areas. Despite this problem, the Blow-Banner-Berggren zonation has been adopted for use in this study because it is widely known by students of planktonic foraminifers, and because no specific zonation exists for the Paleogene of the northeast Pacific region.

Schmidt (1975) has developed a zonation for the Paleocene to middle Eocene Series in central and southern California. His scheme closely follows that of Bolli (1966) but, because of the absence of some tropical zonal taxa in the California Paleogene, uses different nominate taxa for some zones. Schmidt's zonal scheme, or some modification thereof, may ultimately prove to be the most useful zonation for the lower Paleogene of Oregon.

The designation "Zone P7-8," used frequently in this study, refers to the *Morozovella formosa formosa* and *Morozovella aragonensis* Zones combined. According to Bolli (1966), the *M. formosa formosa* Zone (= Zone P7) is distinguished from the *M. aragonensis* Zone (= Zone P8) by the first appearance of "*Globigerina*" *taroubaensis* Bolli and "*Globigerina*" *turgida* Finlay at the base of the *M. aragonensis* Zone. Neither of these species has been documented from the lower Eocene Series of the West Coast. Several workers, unable to distinguish the two zones, have combined them (Bolli and Cita, 1960; Schmidt, 1970, 1975; Postuma, 1971; Gibson, 1976).

AGE AND CORRELATION

Introduction

Planktonic foraminiferal and calcareous nannofossil assemblages examined during this study indicate that the Roseburg and Lookingglass Formations were deposited during the middle part of the early Eocene Epoch and that the Flournoy Formation was mostly or entirely deposited during the early part of the middle Eocene Epoch. Extensive sampling has revealed no definite Paleocene strata in

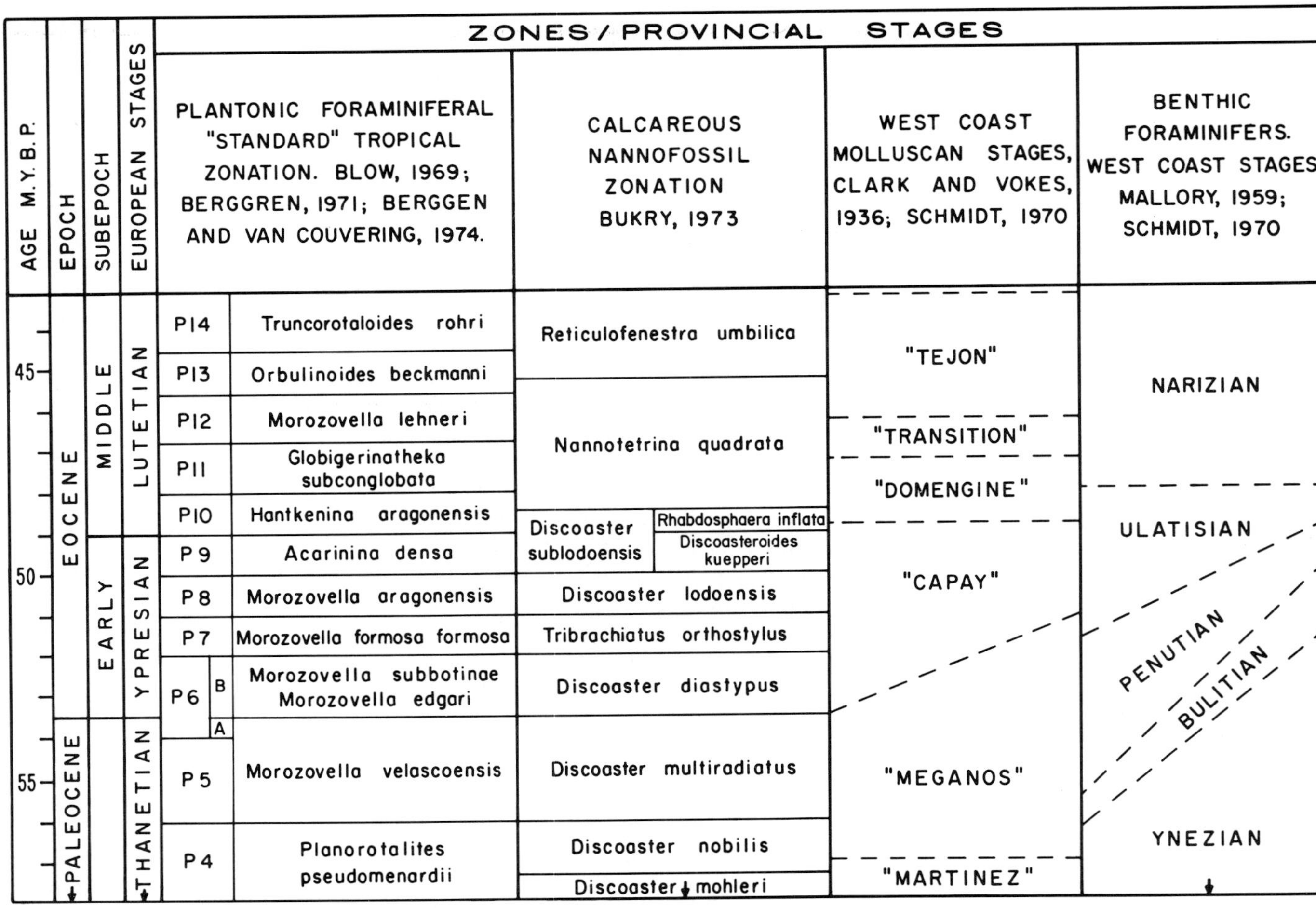

Figure 3. Correlation of the "standard" planktonic foraminiferal zones with the European stages, current calcareous nannofossil zonation, West Coast molluscan stages, and West Coast benthic foraminiferal stages. Names of planktonic foraminifer zones are from Hardenbol and Berggren (1978).

the Umpqua Group. Moreover, some Roseburg Formation strata previously though to be Cretaceous are actually Eocene, and some Cretaceous rocks previously mapped as Roseburg Formation apparently belong to a separate unit.

Figure 4 shows the relative stratigraphic positions of the measured sections and spot samples discussed below. Because of the poor stratigraphic control in many areas, the relative positions of some samples are not well known. Occurrences of planktonic foraminifers in the Roseburg, Lookingglass, and Flournoy Formations are summarized in Figure 5. Age-diagnostic species are illustrated in Plates 1 through 6.

Cretaceous Age Assignments

Baldwin (1965, p. 95) discussed samples collected near Myrtle Point and Gravelford which were assigned to the Upper Cretaceous System on the basis of benthic foraminifers. He stated that the strata "cannot be separated lithologically or structurally from the lower Umpqua" (Baldwin, 1965, p. 97). One of the three samples collected at the Myrtle Point locality (Miles, 1977) yielded well-preserved planktonic foraminifers typical of the lower to middle part of the lower Eocene Series. Age-diagnostic species include *Truncorotaloides nitidus* (Martin), *Truncorotaloides primitivus* (Finlay), and *Morozovella subbotinae* (Morozova). Samples collected from the vicinity of Gravelford are less diagnostic, but the occurrence of *Truncorotaloides wilcoxensis* (Cushman and Ponton) suggests an early Eocene age.

Baldwin (1966, p. 190) mentioned turbidites of probable Cretaceous age which crop out at Five Mile Point, south of Cape Arago in Coos County. Samples collected at this locality yielded no planktonic foraminifers, but Baldwin (1974, p. 9) collected a sample from Five Mile Point which David Bukry of the U.S. Geological Survey, La Jolla, California, determined to be lower Eocene on the basis of calcareous nannofossils.

In addition to strata incorrectly assigned to the Cretaceous System, other rocks definitely of Cretaceous age have been mapped as Roseburg Formation (see Baldwin, 1974, 1975). These rocks almost certainly belong to another unit, perhaps the Upper Cretaceous Hunters Cove Formation of Dott (1971). Although these Cretaceous strata are similar in appearance to those of the Roseburg Formation, they should probably be regarded as a separate unit be-

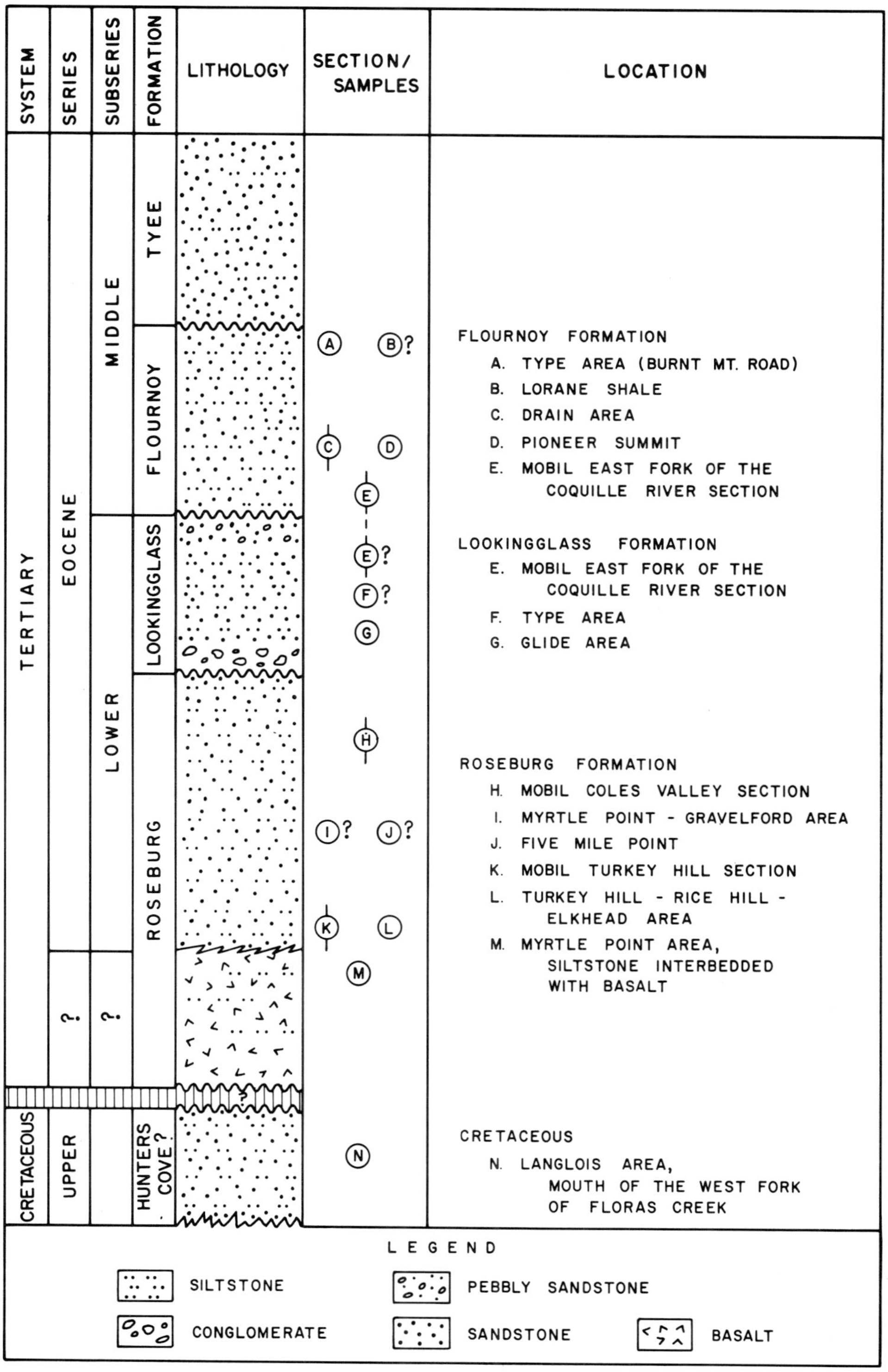

Figure 4. Schematic section of the Tyee Formation, Umpqua Group, and underlying Cretaceous strata showing the approximate stratigraphic positions of the measured sections and spot samples discussed in the text. Geologic relationships of the units are after Baldwin (1974, 1975).

Species	FLOURNOY	LOOKINGGLASS	ROSEBURG
M. aequa			x
M. formosa formosa			x
T. broedermanni			x
T. convexus			x
T. mckannai			x
T. soldadoensis angulosus			x
G. pseudoscitula			x
Gu. columbiana			x
S. varianta		x	x
Tu. pseudomayeri		x	x
M. lensiformis		x	x
M. subbotinae		x	x
T. esnaensis		x	x
T. nicoli		x	x
T. nitidus		x	x
T. pseudotopilensis		x	x
T. wilcoxensis		x	x
S. eocaena	x	x	x
M. aragonensis aragonensis	x	x	x
M. aragonensis caucasica	x	x	x
T. collacteus	x	x	x
T. primitivus	x	x	x
T. soldadoensis soldadoensis	x	x	x
G. planoconica	x	x	x
G. cf. G. planoconica	x	x	x
P. wilcoxensis	x	x	x
Gt. (?) senni	x		
S. frontosa	x		
Tu. wilsoni	x		
T. spinuloinflatus	x		
T. mattseensis wartsteinensis	x		
T. pentacameratus	x		
C. eocanica	x		
P. micra	x		

Figure 5. Distribution of planktonic foraminifers in the Roseburg, Lookingglass, and Flournoy Formations. The eight species and subspecies found only in the Roseburg Formation are extremely rare or, in the case of *M. aequa*, highly localized. *C.* = *Clavigerinella*, *G.* = *Globanomalina*, *Gt.* = *Globigerinatheka*, *Gu.* = *Guembelitria*, *M.* = *Morozovella*, *P.* = *Pseudohastigerina*, *S.* = *Subbotina*, *T.* = *Truncorotaloides*, *Tu.* = *Turborotalia*.

cause the sedimentary rocks of the Roseburg Formation are no older than middle early Eocene. Strata at Blacklock Point, north of Cape Blanco, and along Edson Creek, approximately 10 km northwest of Port Orford, contain Mesozoic fossils (Dott, 1962, 1971; Lent, 1969). Brownfield (1972) collected a sample from strata at the mouth of the West Fork of Floras Creek, approximately 10 km west of Langlois, which he sent to Amoco Production Company in Denver, Colorado, for palynological analysis. An age of "Danian or Paleocene" was determined. Samples collected at this locality during my investigation (Miles, 1977), however, yielded large numbers of Cretaceous calcareous nannofossils. W. G. Siesser examined the floras and reported the presence of *Watznaueria barnesae* (Black), *Parhabdolithus embergeri* (Noel), and *Cretarhabdus crenulatus* Bramlette and Martini. Siesser (1977, written commun.) stated: "Preservation is poor, and the Cretaceous species could, of course, be reworked; however, the abundance of *W. barnesae* makes me doubt this."

In summary, Cretaceous or presumed Cretaceous strata currently included in the Roseburg Formation fall into two categories: (1) those actually of Eocene age (Five Mile Point, Myrtle Point, and Gravelford), and (2) those of Cretaceous age which probably belong to a separate stratigraphic unit (Blacklock Point, Edson Creek, and mouth of the West Fork of Floras Creek).

Paleocene Age Determinations

Faunas of supposed Paleocene age in the Roseburg Formation have been reported in three exposures in the Myrtle Point area (Baldwin, 1965, p. 97; 1969). One of these outcrops, located 3 km northeast of Gravelford, and another at the northern base of Cribbins Hill are now deeply weathered and can no longer be collected. The third outcrop is located approximately 8 km southeast of Myrtle Point along Oregon Highway 42. Samples collected from siltstone interbedded with basalt at this locality yielded sparse assemblages of poorly preserved arenaceous foraminifers which are not age diagnostic. Although the presence of Paleocene strata in the Myrtle Point area cannot be confirmed or denied at this time, their presence is doubted. All age-diagnostic species that I examined from the Myrtle Point–Gravelford area are Eocene (see Miles, 1977, Appendixes).

Roseburg Formation

The Roseburg Formation was named by Baldwin (1974) for the basalt and overlying interbedded sandstone and siltstone exposed in the Roseburg area. The designated type area is along the North Umpqua River from Frear bridge, 5 km west of Glide, downstream to the confluence of the north and south forks of the Umpqua River. Most of the sedimentary rock sequence of the Roseburg Formation is composed of flyschlike interbeds of sandstone and siltstone. Conglomerate and thickly bedded sandstone are present in some parts of the formation. According to Baldwin (1965, 1974, 1975) the Roseburg Formation was uplifted prior to deposition of the Lookingglass Formation, and much of the original upper Roseburg was removed by erosion. The uneroded top of this unit has not been reported to date.

During this investigation 218 microfossil samples from the Roseburg Formation were examined. All age-diagnostic samples suggest assignment of this unit to lower Eocene Zone P7-8 on the basis of planktonic foraminifers and to the interval from the *Tribrachiatus orthostylus* Zone to the *Discoasteroides kuepperi* Subzone (equivalent to Zones P7-8 and P9 combined) on the basis of calcareous nannofossils. Baldwin (1974, 1975) proposed that the Roseburg Formation was highly folded, thrust faulted,

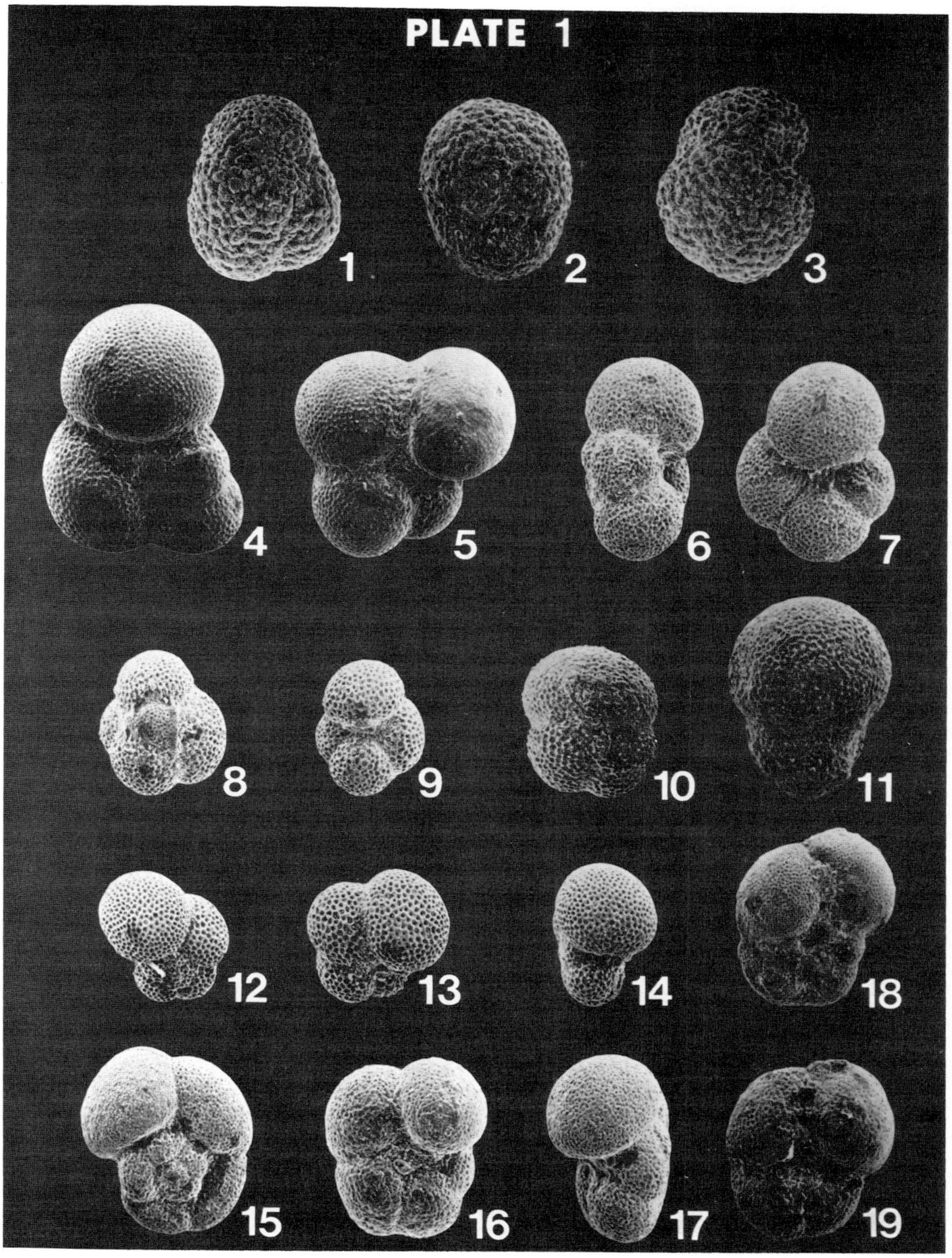

PLATE 1

Figure	
1-3	*Globigerinatheka(?) senni* (Beckmann), Flournoy Formations X93.
4-7, 9	*Subbotina eocaena* (Gumbel), Roseburg Formation, X77.
8	*Subbotina eocaena* (Gumbel), Flournoy Formation, X77.
10-11	*Subbotina frontosa* (Subbotina), Flournoy Formation, X93.
12-14	*Subbotina varianta* (Subbotina), Lookingglass Formation, X93.
15-17	*Turborotalia pseudomayeri* (Bolli), Roseburg Formation, X93.
18-19	*Turborotalia wilsoni* (Cole), Flournoy Formation, X93.

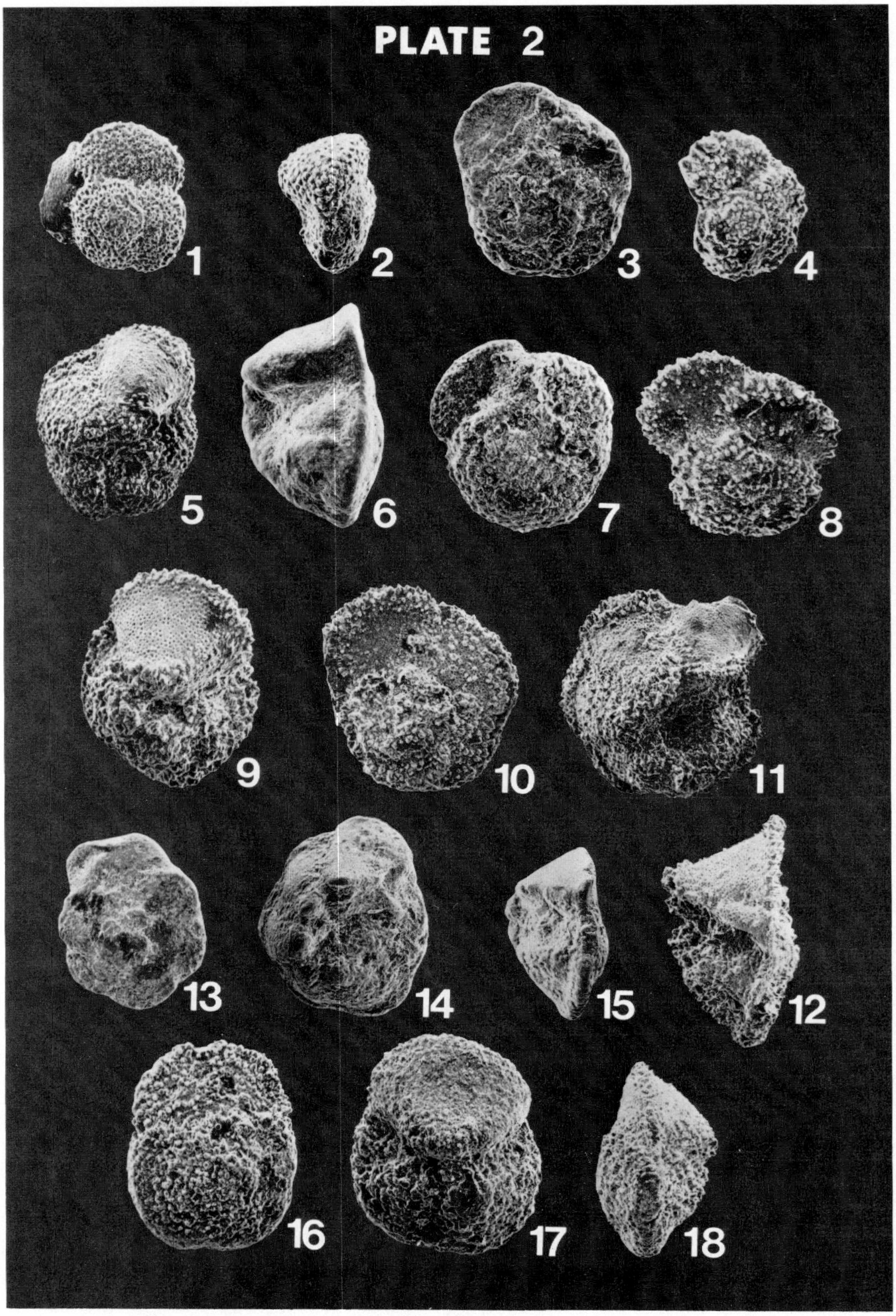

PLATE 2

Figure	
1-2	*Morozovella aequa* (Cushman and Renz), Roseburg Formation, X93.
3	*Morozovella aragonensis aragonensis* (Nuttall), Roseburg Formation, X77.
4, 7-9	*Morozovella aragonensis aragonensis* (Nuttall), Flournoy Formation, X77.
5	*Morozovella aragonensis aragonensis* (Nuttall), Lookingglass Formation, X77.
6	*Morozovella aragonensis aragonensis* (Nuttall), Lookingglass Formation, X60.
10	*Morozovella aragonensis caucasica* (Glaessner), Flournoy Formation, X60.
11-12	*Morozovella aragonensis caucasica* (Glaessner), Roseburg Formation, X60.
13	*Morozovella formosa formosa* (Bolli), Roseburg Formation, X60.
14-15	*Morozovella formosa formosa* (Bolli), Roseburg Formation, X47.
16-18	*Morozovella lensiformis* (Subbotina), Roseburg Formation, X77.

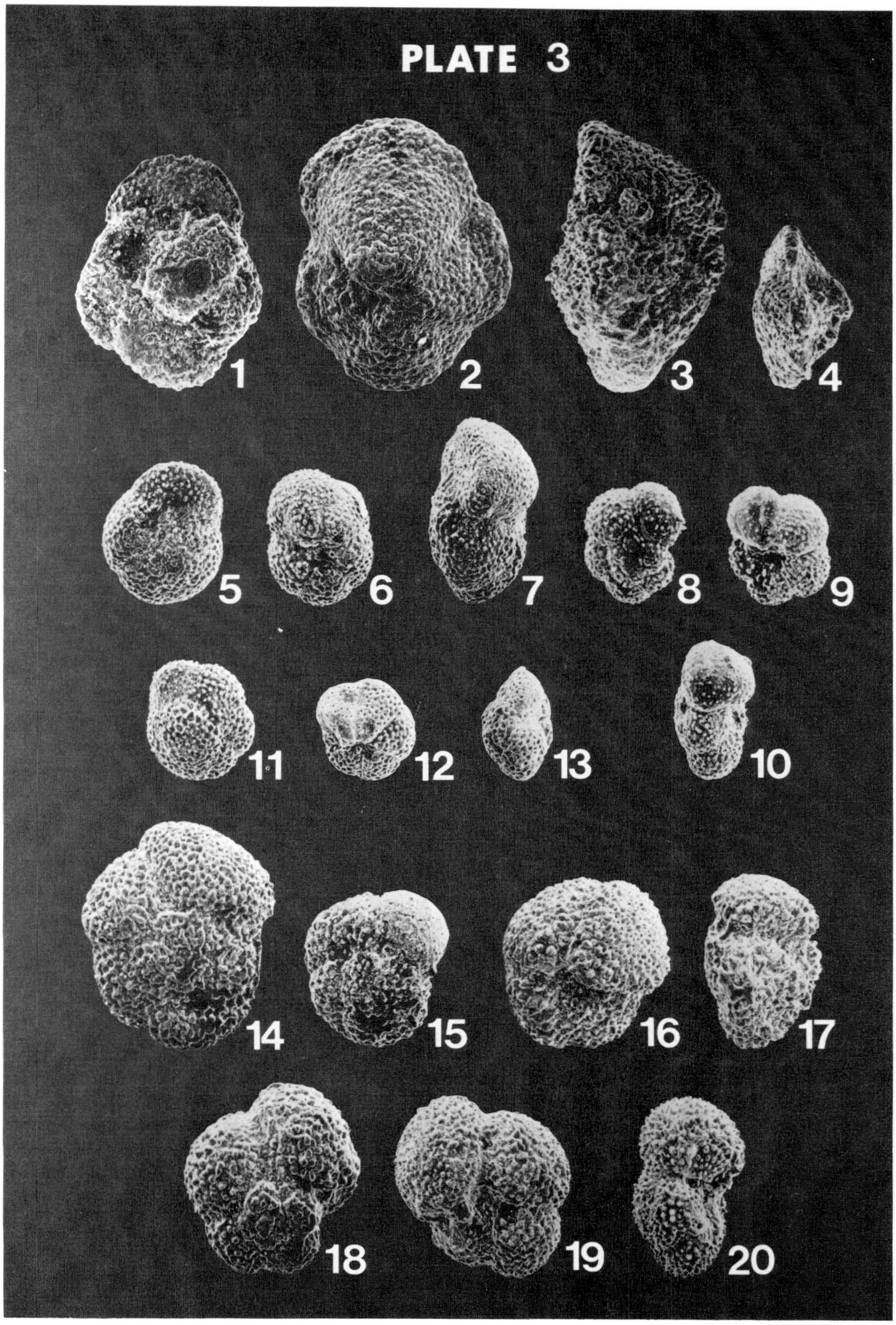

PLATE 3

Figure
1-4 *Morozovella subbotinae* (Morozova), Roseburg Formation, X93.
5-7 *Truncorotaloides broedermanni* (Cushman and Bermudez), Roseburg Formation, X93.
8-10 *Truncorotaloides collacteus* (Finlay), Flournoy Formation, X93.
11-13 *Truncorotaloides convexus* (Subbotina), Roseburg Formation, X93.
14-17 *Truncorotaloides spinuloinflatus* (Bandy), Flournoy Formation, X93.
18-19 *Truncorotaloides esnaensis* (LeRoy), Roseburg Formation, X93.
20 *Truncorotaloides esnaensis* (LeRoy), Lookingglass Formation, X93.

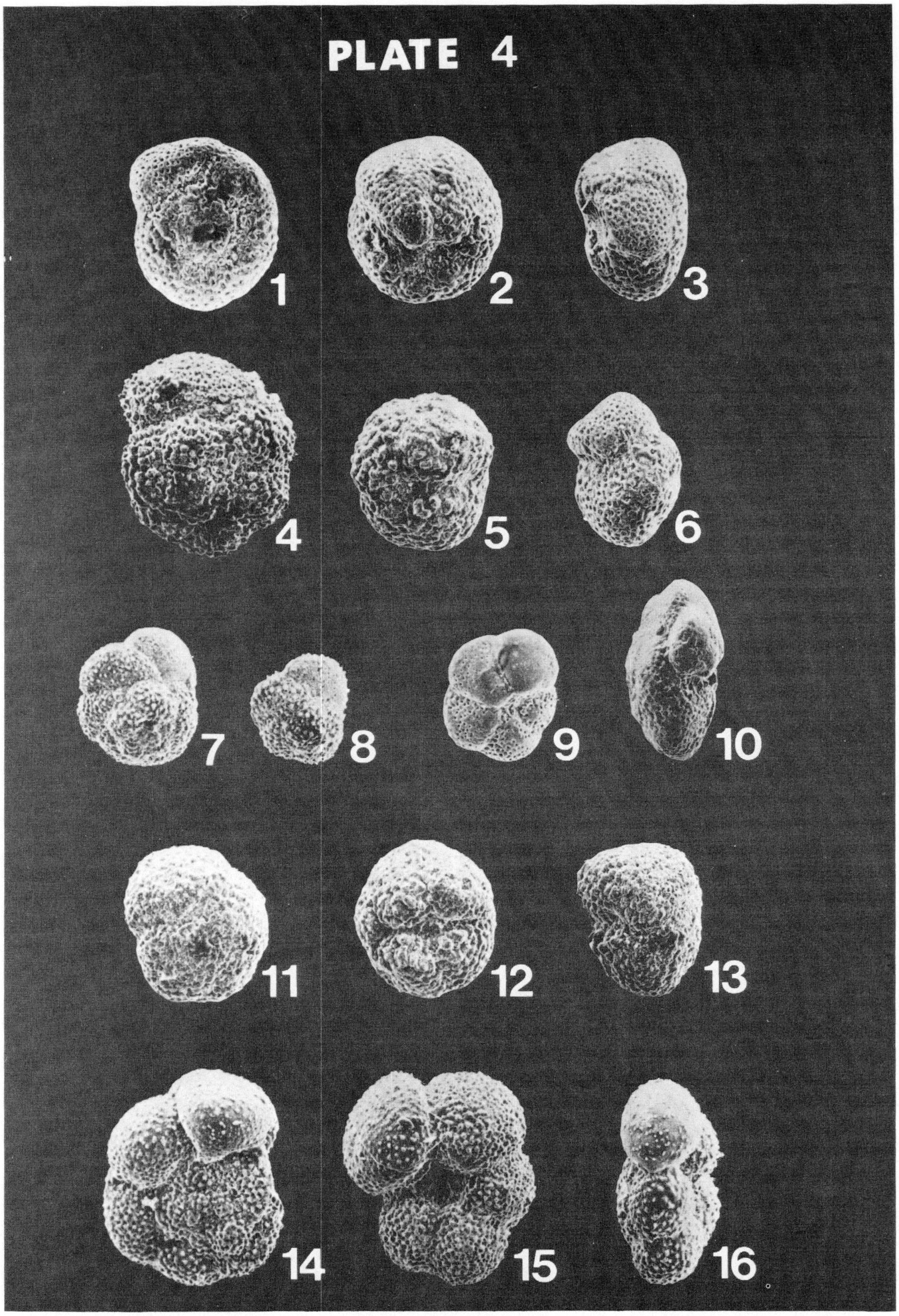

PLATE 4

Figure	
1-3	*Truncorotaloides mattseensis wartsteinensis* (Gohrbandt), Flournoy Formation, X100.
4-6	*Truncorotaloides mckannai* (White), Roseburg Formation, X100.
7-10	*Truncorotaloides nicoli* (Martin), Roseburg Formation, X100.
11-13	*Truncorotaloides nitidus* (Martin), Roseburg Formation, X100.
14-16	*Truncorotaloides pentacameratus* (Subbotina), Flournoy Formation, X100.

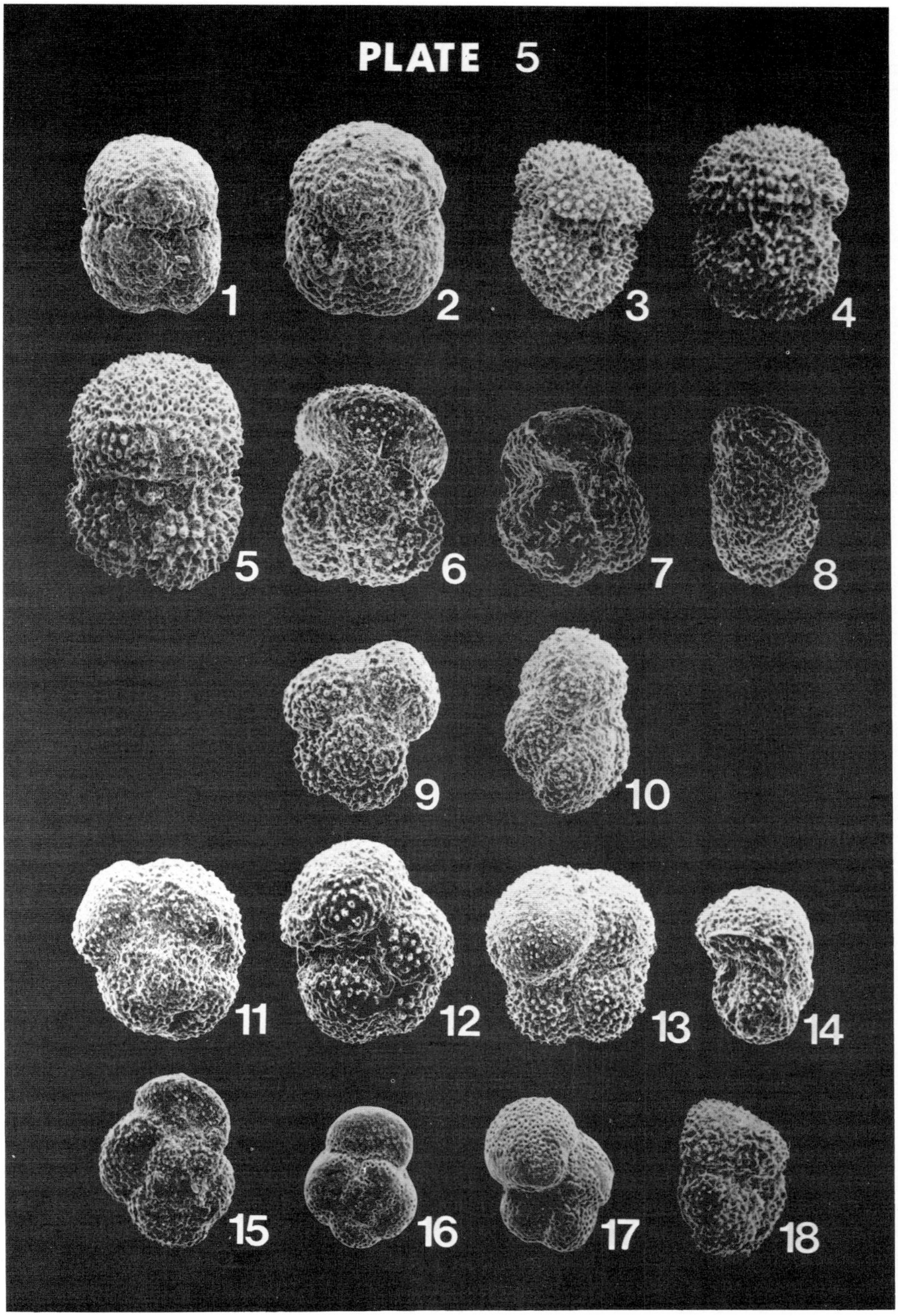

PLATE 5

Figure
1-5 *Truncorotaloides primitivus* (Finlay), Roseburg Formation, X100.
6-8 *Truncorotaloides pseudotopilensis* (Subbotina), Roseburg Formation, X100.
9-10 *Truncorotaloides soldadoensis angulosus* (Bolli), Roseburg Formation, X100.
11-12,14 *Truncorotaloides soldadoensis soldadoensis* (Bronnimann), Lookingglass Formation, X100.
13 *Truncorotaloides* cf. *T. soldadoensis soldadoensis* (Bronnimann), Flournoy Formation, X100.
15-18 *Truncorotaloides wilcoxensis* (Cushman and Ponton), Roseburg Formation, X100.

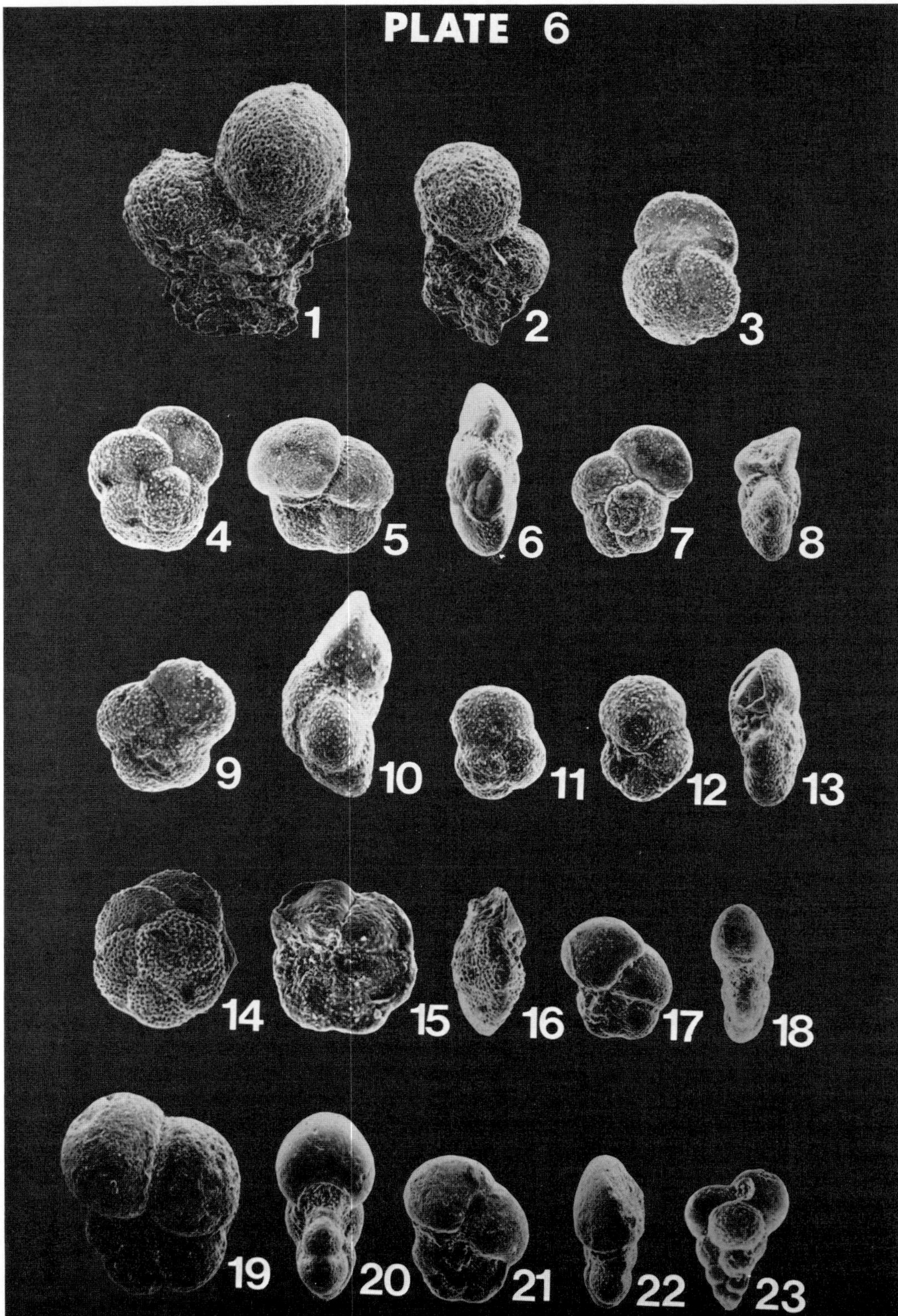

PLATE 6

Figure
1-2 *Clavigerinella eocanica* (Nuttall), Flournoy Formation, X83.
3-8 *Globanomalina planoconica* (Subbotina), Roseburg Formation, X100.
9-10 *Globanomalina planoconica* (Subbotina), Flournoy Formation, X100.
11-13 *Globanomalina* cf. *G. planoconica* (Subbotina), Flournoy Formation, X100.
14-16 *Globanomalina pseudoscitula* (Glaessner), Roseburg Formation, X132.
17-18 *Pseudohastigerina micra* (Cole), Flournoy Formation, X100.
19,21-22 *Pseudohastigerina wilcoxensis* (Cushman and Ponton), Roseburg Formation, X100.
20 *Pseudohastigerina wilcoxensis* (Cushman and Ponton), Flournoy Formation, X100.
23 *Guembelitria columbiana* (Howe), Roseburg Formation, X165.

and eroded before deposition of the Lookingglass Formation. If this interpretation is correct, the Roseburg Formation cannot be stratigraphically higher than Zone P7-8 because the overlying Lookingglass Formation is also assigned to Zone P7-8. The age of the volcanic sequence comprising the lower part of the Roseburg Formation is difficult to establish because the sedimentary rocks interbedded with the basalt lack microfossils which permit a precise age determination. The volcanic sequence may in part be as old as earliest Eocene or even late Paleocene, but this cannot be determined with certainty until diagnostic fossils are found or until accurate radiometric dates become available.

Two measured sections collected by Mobil Oil Corporation—the Turkey Hill section and the Coles Valley section—contain typical Roseburg Formation planktonic foraminiferal assemblages (see Miles, 1977, species check lists in Appendix 2). The Turkey Hill section includes strata lying immediately above the contact with the Roseburg Formation basalt. Assuming that the Roseburg Formation basalt is everywhere older than the sedimentary rocks (except where the two are interbedded), the strata exposed immediately south of the crest of Turkey Hill are among the oldest planktonic foraminifer-bearing rocks within the Roseburg Formation. The Coles Valley section is situated much higher stratigraphically than the Turkey Hill section (Baldwin, 1974, geologic map) but still bears the same planktonic foraminiferal faunas.

Mobil Turkey Hill Section. In 1974, Mobil Oil Corporation collected a section along Interstate 5 on the southern flank of Turkey Hill. The section is situated approximately 9 km north of Oakland in sec. 17, T. 24 S., R. 5 W. Along Interstate 5 at the crest of Turkey Hill, Roseburg Formation sedimentary rocks have been removed by erosion, so that Roseburg Formation basalt is exposed in the core of the Red Hill anticline. Rich assemblages of planktonic foraminifers, including *Morozovella aragonensis aragonensis* (Nuttall), *Morozovella formosa formosa* (Bolli), *Morozovella lensiformis* (Subbotina), *Morozovella subbotinae* (Morozova), *Truncorotaloides wilcoxensis* (Cushman and Ponton), and *Pseudohastigerina wilcoxensis* (Cushman and Ponton), dictate an assignment of Zone P7-8 for these strata. The age of these rocks suggests that deposition of postbasalt Roseburg Formation strata, at least in this area, may have started 1 or 2 m.y. after the beginning of the Eocene Epoch. Moreover, the absence of Zone P6b at this locality suggests that this zone may not be developed anywhere in the Roseburg Formation (althought the basalt may in part be equivalent to Zone P6b).

Standard Oil Company of California collected two samples on the northern flank of Turkey Hill. The fossil assemblages in these samples include rare specimens of *Morozovella aragonensis aragonensis, M. lensiformis, Truncorotaloides wilcoxensis,* and *Pseudohastigerina wilcoxensis* and corroborate at the Zone P7-8 assignment of rock on the south limb of the Red Hill anticline. Additional Standard Oil Company and Mobil samples collected to the north along Interstate 5 between Rice Hill and Yoncalla bear similar assemblages (Miles, 1977, Appendixes).

Numerous spot samples collected along the southern limb of the Red Hill anticline between the Goodrich highway and Elkhead have meager faunas indicative of an early Eocene age. The absence of *M. aragonensis* s.l. is enigmatic but not necessarily indicative of the presence of Zone P6b. This conclusion is corroborated by calcareous nannofossils in one sample that are indicative of the interval from the *Tribrachiatus orthostylus* Zone to the *Discoaster lodoensis* Zone (equivalent to Zone P7-8; W. G. Siesser, 1977, written commun.).

Mobil Coles Valley Section. In 1971, Mobil Corporation collected samples from a measured section along the Umpqua River, approximately 12 km west of Sutherlin. The section extends a distance of nearly 5 km from sec. 25, T. 25 S., R. 7 W. to sec. 5, T. 26 S., R. 6 W. The planktonic foraminiferal faunas are similar throughout the section. The assemblages, typical of the Roseburg Formation, consist of numerous specimens of *Subbotina eocaena* (Gumbel), *Truncorotaloides primitivus* (Finlay), *T. wilcoxensis* (Cushman and Ponton), *Globanomalina planoconia* (Subbotina), *Pseudohastigerina wilcoxensis* (Cushman and Ponton), and rare specimens of *Turborotalia pseudomayeri* (Bolli), *M. subbotinae* (Morozova), *Truncorotaloides broedermanni* (Cushman and Bermudez), *T. esnaensis* (LeRoy), *T. mckannai* (White), *T. nicoli* (Martin), T. *nitidus* (Martin), T. *pseudotopilensis* (Subbotina), *T. soldadoensis soldadoensis* (Bronnimann), and *Globanomalina pseudoscitula* (Glaessner).

Samples collected along strike in strata equivalent to the middle part of the Coles Valley section include *Morozovella aragonensis* s.l., *M. aequa* (Cushman and Renz), *T. broedermanni*, and other species indicative of Zone P7-8.

Lookingglass Formation

The Lookingglass Formation derives its name from Lookingglass Valley, which is situated approximately 15 km west of Roseburg. This unit has been divided into the Bushnell Rock, Tenmile, and Olalla Creek Members and is interpreted by Baldwin (1974) to represent a transgressive-regressive cycle. The Bushnell Rock Member was deposited during the onlap phase of the cycle, the Tenmile Member during an intermediate phase, and the Olalla Creek Member during the offlap phase.

Most of the 52 planktonic foraminiferal samples collected from the Lookingglass Formation bear sparse assemblages that suggest an early Eocene age (see Miles, 1977, species check lists in Appendix 2). Those containing the greatest abundance and diversity of foraminifers indicate that the Lookingglass Formation was deposited during the middle part of the early Eocene Epoch and that this unit is assignable to Zone P7-8. The faunas in these samples are indistinguishable from those of the Roseburg Formation.

Type Area. The type section of the Lookingglass Formation is exposed along Tenmile Creek between Bushnell Rock and Tenmile Butte in western Douglas County (Baldwin, 1974, p. 12). Samples from several exposures of this unit near the type section yielded sparse lower Eocene planktonic foraminiferal assemblages. Foraminifers collected in steeply dipping strata a few kilometres south of Tenmile near Berry Creek (see Baldwin, 1964, Text-Fig. 4) include *Morozovella aragonensis aragonensis, Truncorotaloides soldadoensis soldadoensis,* and other species characteristic of Zone P7-8 assemblages.

Calcareous nannofossils from samples collected near the type area indicate that the strata are assignable to the interval from the *Tribrachiatus orthostylus* Zone to the *Discoasteroides kuepperi* Subzone (equivalent to Zones P7-8 and P9 combined; W. G. Siesser, 1977, written commun.).

Glide Area. A thick sequence of sandstone, siltstone, and mudstone of the Lookingglass Formation is exposed along the North Umpqua River near Glide. Foraminiferal species listed by Stewart (1957, p. 13) and Thoms (1975, p. 546) indicate that planktonic foraminifers are sparse or absent in much of the section. I collected two samples from the lower part of this sequence which contain planktonic foraminifers characteristic of Zone P7-8 assemblages, including *Subbotina eocaena, S. varianta* (Subbotina), *Turborotalia pseudomayeri, Truncorotaloides esnaensis, T. nicoli, T. nitidus, T. wilcoxensis, Globanomalina planoconica,* and *Pseudohastigerina wilcoxensis.* Calcareous nannofossils indicate assignment of these samples to the interval from the *Tribrachiatus orthostylus* Zone to the *Discoasteroides kuepperi* Subzone (equivalent to Zones P7-8 and P9 combined; W. G. Siesser, 1977, written commun.).

Turner (1938, p. 12) proposed that the Glide section may be intermediate in age between the Capay and Domengine Formations of California, based on molluscan faunas, and Thoms (1975, p. 532) assigned this section to the Penutian provincial benthic foraminiferal stage of Mallory (1959). In this part of the Lookingglass Formation, then, planktonic foraminiferal Zone P7-8 overlaps with or is equivalent to the Penutian benthic foraminiferal stage, is intermediate(?) between the Capay and Domengine provincial molluscan stages, and falls within the interval from the *T. orthosylus* Zone to the *D. kuepperi* Subzone.

Mobil East Fork of the Coquille River Section. This section, collected by Mobil Oil Corporation a few kilometres west of Sitkum, extends from sec. 11, T. 28 S., R. 11 W. to sec. 6, T. 28 S., R. 10 W. and includes strata of the Lookingglass and Flournoy Formations. The planktonic foraminiferal faunas of the Lookingglass Formation are sparse, but several samples contain *Truncorotaloides wilcoxensis* which indicates an early Eocene age. The faunal break between this unit and the overlying Flournoy Formation is abrupt. The Flournoy Formation assemblages are characterized by *Truncorotaloides spinuloinflatus* (Brandy), *Pseudohastigerina wilcoxensis,* and numerous specimens of *Truncorotaloides collacteus* (Finlay). The mutual occurrence of these species indicates an early middle Eocene age.

Flournoy Formation

The Flournoy Formation was named by Baldwin (1974) for exposures of sandstone and siltstone in Flournoy Valley southwest of Roseburg. The Formation is divided into a lower White Tail Ridge Member and an upper Camas Valley Member. Each of these members is approximately 450 m thick in the type area.

Comparison of Flournoy Formation planktonic foraminifers with those of the *Globigerapsis senni* Zone (equivalent to Zone P9), *Truncorotaloides densus* Zone (equivalent to Zone P10), and *Truncorotaloides rotundimarginatus* Zone (equivalent to Zones P11 and P12 combined) of California (Schmidt, 1975) reveals that the Flournoy Formation most closely correlates with Zone P10 (Miles, 1977, Fig. 8). Several of the same planktonic foraminiferal species occur in widely separated exposures of the Flournoy Formation; this suggests that the age of this unit is similar throughout (Fig. 6).

I examined 36 samples collected from both members of the Flournoy Formation (see Miles, 1977, check list in Appendix 2). Most of the age-diagnostic samples were concentrated around Pioneer Summit and the Drain area (White Tail Ridge Member) and around the type area and the vicinity of Lorane (Camas Valley Member).

Pioneer Summit. Samples collected near Pioneer Summit, 5 km northeast of Toledo in west-central Lincoln County, yielded rich planktonic foraminiferal assemblages similar to those of Zone P10 in California. The faunas are dominated by the long-ranging species, *Subbotina eocaena,* and include *Globigerinatheka(?) senni* (Beckmann), *Subbotina frontosa* (Subbotina), *Morozovella aragonensis* s.l., *Truncorotaloides collacteus, T. spinuloinflatus, T. pentacameratus* (Subbotina), *T. primitivus, T.* cf. *T. soldadoensis soldadoensis, Globanomalina planoconica,* and *Pseudohastigerina wilcoxensis.* In addition, McKeel and Lipps (1972) reported the occurrence of *Clavigerinella colombiana* (Petters) and *Turborotalia wilsoni* (Cole) (given as *Subbotina bolivariana*) at this locality.

The Zone P10 assignment of the White Tail Ridge Member at the Pioneer Summit locality is corroborated by a calcareous nannofossil flora assigned to *Rhabdosphaera inflata* Subzone (equivalent to lower Zone P10) of the lowermost middle Eocene Series (W. G. Siesser, 1977, written commun.).

Drain Area. A few samples collected from the White Tail Ridge Member in the vicinity of Drain yielded planktonic foraminifers. A sample collected along Oregon Highway 38 about 1 km west of Drain includes *Clavigerinella eocanica* (Nuttall), *Truncorotaloides collacteus,* and *Turborotalia wilsoni.* It is assigned to Zone P10.

Lorane Area. One sample collected from the Lorane Shale (Camas Valley Member of the Flournoy Formation) along Territorial Road, about 4 km north of Lorane, is dominated by *Subbotina eocaena* and includes several specimens of *T. collacteus, T. spinuloinflatus,* and *Globanomalina planoconica.* The Lorane Shale at this locality is assigned to Zone P10. The assemblages are similar to those of the Elkton Formation (Bird, 1967, Pl. 12) and of the type Flournoy Formation on Burnt Mountain Road. Evidence from benthic foraminifers offered by Bird (1967) suggests that the Lorane Shale correlates with the Elkton Formation, but Baldwin (1975) placed the Lorane Shale in

TYPE AREA (BURNT MT. RD.)	LORANE SHALE	DRAIN AREA	PIONEER SUMMIT	EAST FORK COQUILLE RIVER	
			x		Globigerinatheka (?) senni
x	x	x	x	x	Subbotina eocaena
			x		S. frontosa
		x			Turborotalia wilsoni
x			x		Morozovella aragonensis s.l.
x	x	x	x	x	Truncorotaloides collacteus
x	x	x	x	x	T. spinuloinflatus
		x			T. mattseensis wartsteinensis
x			x		T. pentacameratus
x	x		x	x	T. primitivus
x		x	x		T. soldadoensis soldadoensis
		x			Clavigerinella eocanica
x	x	x	x	x	Globanomalina planoconica
x	x	x	x	x	Pseudohastigerina spp.

Figure 6. Comparison of planktonic foraminferal faunas from widely separated exposures of the Flournoy Formation.

the Flournoy Formation on the basis of its stratigraphic position. Calcareous nannofossils from the sample collected on Territorial Road indicate assignment of these rocks to the *Rhabdosphaera inflata* Subzone (equivalent to lower Zone P10; W. G. Siesser, 1977, written commun.).

Type area. Siltstone constituting part of the type Flournoy Formation is exposed along Burnt Mountain Road near Reston. Fresh exposures are uncommon because of the susceptibility of these rocks to rapid and deep weathering. Three samples collected from Burnt Mountain Road exposures yielded planktonic foraminifers indicative of upper Zone 10. Species include *Truncorotaloides collacteus, T. spinuloinflatus, T. pentacameratus, T. primitivus,* and *T.* cf. *T. soldadoensis soldadoensis.* Calcareous nannofossils in these samples are assigned an Eocene age, but further refinement is not possible because of the sparse flora (W. G. Siesser, 1977, written commun.).

Tyee Formation

The Tyee Formation was named by Diller (1898) for exposures on Tyee Mountain, which is situated 12 km west of Sutherlin. The formation assumes a central position in the Coast Range syncline and extends from the vicinity of Agness northward nearly to the Siuslaw River (Baldwin, 1975). It is composed of rhythmically bedded micaceous, arkosic sandstone with thin interbeds of siltstone. The Tyee Formation is similar in lithology to the White Tail Ridge Member of the underlying Flournoy Formation and can be readily distinguished from it only when the unconformity between the two formations is visible. Most of the strata formerly mapped as Tyee Formation (see Wells and Peck, 1961, and references therein) are now considered to belong to the Flournoy Formation (Baldwin, 1975; Fig. 2).

Because the Tyee Formation lacks age-diagnostic fossils, the exact age is unknown. The stratigraphic position of this unit between the lower middle Eocene Flournoy and Elkton Formations, however, dictates an early middle Eocene age (Miles, 1977, p. 18–20).

Correlations with West Coast Paleogene Strata

The use of fossil calcareous plankton to correlate the Roseburg, Lookingglass, and Flournoy Formations with other Paleogene units in the Pacific Northwest is difficult because few studies on fossil plankton are available for this region.

On the basis of benthic foraminifers, Thoms (1975, p. 538) correlated the Roseburg Formation with the Siletz River Formation of northwestern Oregon and with the Metchosin Volcanics, Crescent Formation, and lower "Scow Bay" Formation of western Washington. The benthic foraminifers of the Flournoy Formation have not yet been studied in detail, but Bird (1967, p. 86) and Rau (Baldwin, 1974, p. 19) indicated that parts of this unit are assignable to the Ulatisian Stage of Mallory (1959). Bird (1967, p. 79) correlated the Elkton Formation (including the Lorane Shale and "Sacchi Beach Beds," which are here considered to be within the Flournoy Formation) with the Yamhill Formation of northwestern Oregon and the Aldwell and McIntosh Formations of western Washington.

Planktonic foraminiferal assemblages of the Umpqua Group are similar to age-equivalent faunas from California but are less diverse and have a more temperate aspect. Planktonic foraminiferal and calcareous nannofossil data indicate that the Roseburg and Lookingglass Formations are correlative with the middle part of the Llajas Formation and the middle to upper parts of the Anita and Lodo Formations (Fig. 7; California sections from Schmidt, 1970). The Flournoy Formation is correlative with the Matilija Formation, Domengine Sandstone, the uppermost part of the Llajas and Anita Formations, and the basal part of the "Tejon" Formation (Point of Rocks Member).

Correlations Outside Western North America

Correlations to key stratigraphic sections and stages outside western North America are indicated in Figure 8. The sections selected for this diagram are particularly important to planktonic foraminiferal micropaleontologists because their faunas have been studied in detail and provide the bases for global correlation.

Attempts to correlate the classical European Eocene stages to

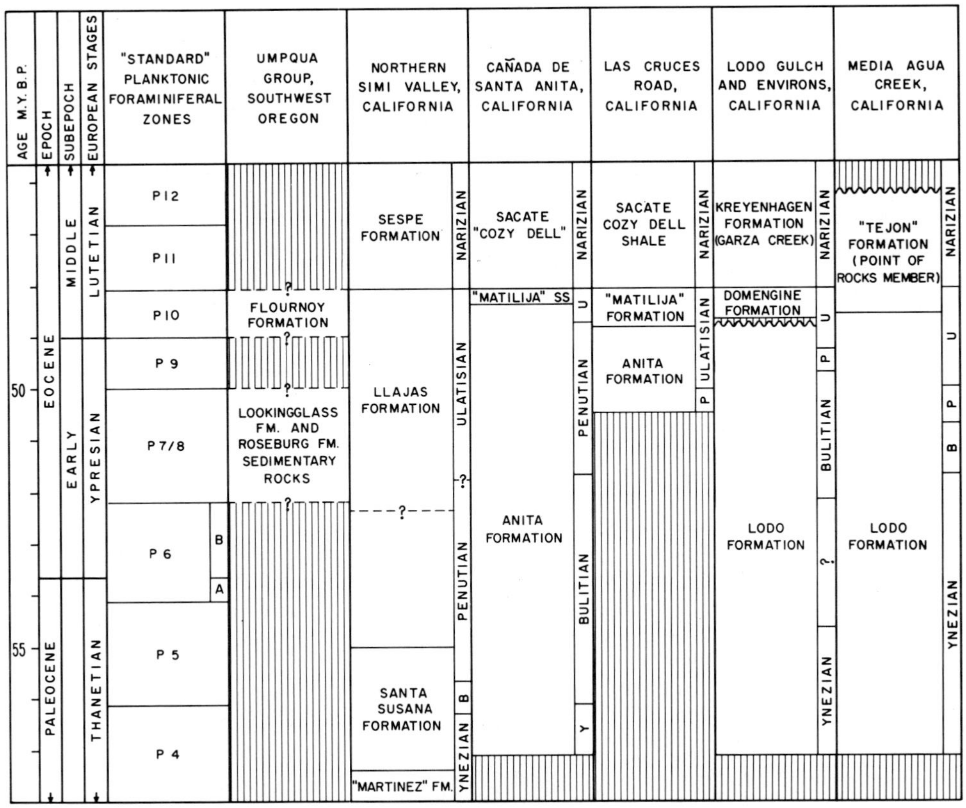

Figure 7. Correlation of the Roseburg, Lookingglass, and Flournoy Formations with the "standard" planktonic foraminiferal zones and selected lower Tertiary sections from California. California sections from Schmidt (1970).

existing calcareous plankton zonations have met with limited success because the European stratotypes are closely allied to transgressive and regressive phases of sedimentary cycles. The lower and middle Eocene European type sections are composed of nonmarine and shallow-marine sedimentary rocks, thus diminishing their usefulness for direct intercontinental correlation.

The correlations indicated on Figure 8 have been made indirectly by tying the European stratotypes and the Umpqua Group to the "standard" planktonic foraminiferal zonations. The necessity of using the "standard" zonation for this purpose stresses the importance of these zones as global datums.

SUMMARY AND CONCLUSIONS

Summary

The principal purposes of this investigation were to examine the biostratigraphic usefulness of planktonic foraminifers from the Roseburg, Lookingglass, and Flournoy Formations and to correlate these assemblages with those of well-known stratigraphic sequences in California, Trinidad, and Europe. Of the 305 samples that were examined during this study, 244 contain planktonic foraminifers. I recovered 33 species and subspecies from the Roseburg, Lookingglass, and Flournoy Formation; 23 of these are reported from the Pacific Northwest for the first time.

Principal Conclusions

The planktonic foraminiferal faunas of the Roseburg and Lookingglass Formations are indistinguishable. These units are assigned to Zone P7-8 of the "standard" tropical zonation, which is equivalent to the *Morozovella aragonensis* Zone in California (Schmidt, 1975). Calcareous nannofossil assemblages from these formations are assigned to the interval from the *Tribrachiatus orthosylus* Zone to the *Discoasteroides kuepperi* Subzone (equivalent to Zones P7-8 and P9 combined).

The faunas of the Flournoy Formation can be distinguished from those of the underlying Roseburg and Lookingglass Formations by the persistent occurrence of *Truncorotoloides spinuloinflatus* (Bandy) and numerous specimens of *Truncorotaloides collacteus* (Finlay). This unit is assigned to Zone P10, which is equivalent to the *Truncorotaloides densus* Zone in California (Schmidt, 1975). Calcareous nannofossils are assigned to the *Rhabdosphaera inflata* Subzone (equivalent to lower Zone P10). The unconformity between the Lookingglass and Flournoy Formations represents an interval of 1 to 2 m.y.

Cretaceous or presumed Cretaceous strata presently

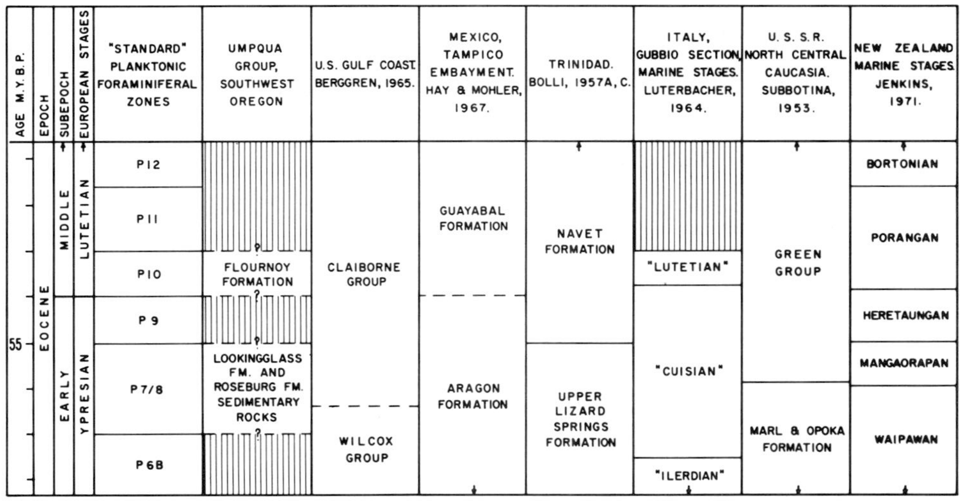

Figure 8. Correlation of the Roseburg, Lookingglass, and Flournoy Formations with key stratigraphic sections and stages outside western North America.

mapped as Roseburg Formation (Baldwin, 1974) fall into two categories: (1) strata of Eocene age which previously have been erroneously assigned to the Cretaceous System (localities near Myrtle Point and Gravelford in Coos County), and (2) strata of Cretaceous age which almost certainly belong to a separate unit (outcrops at Blacklock Point, Dott, 1962; Edson Creek, Lent, 1969; and mouth of the West Fork of Floras Creek, Brownfield, 1972).

Despite numerous published and unpublished reports of strata of Paleocene age occurring within the Roseburg and Lookingglass Formations, I found no faunas definitely assignable to the Paleocene Series.

The scarcity of key stratigraphic markers in the Roseburg Formation hinders interpretations of age and zone assignments in many areas. The most pressing problem is whether or not Zone P6b, the basal zone of the Eocene Series, is developed in the Roseburg Formation. This zone occupies the interval between the first appearance of species of *Pseudohastigerina* and the first appearance of *Morozovella aragonensis* (Nuttall) s.l. Unfortunately, *M. aragonensis* is rare and spotty in the Umpqua Group and, although its presence indicates Zone P7-8 or higher, its absence is not particularly meaningful. In order to determine if Zone P6b occurs in the Roseburg Formation, I examined samples from a measured section on the south limb of the Red Hill anticline north of Oakland, Oregon. The section was collected immediately above Roseburg Formation basalt in the core of the anticline and should represent the oldest planktonic foraminifer-bearing rocks examined during this study. The samples contain several specimens of *M. aragonensis* and *Morozovella formosa formosa* (Bolli); this indicates a Zone P7-8 assignment. This assignment suggests that Zone P6b is not developed in the Roseburg Formation and that all postbasalt Roseburg Formation sedimentary rocks are assignable to Zone P7-8.

None of the single-formation sections I examined (Miles, 1977) could be subdivided by using planktonic foraminifers or calcareous nannofossils together with existing calcareous plankton zonations.

The Roseburg and Lookingglass Formations are correlative with the middle part of the Llajas Formation and the upper parts of the Lodo and Anita Formations of California. They are also equivalent to the middle part of the Upper Lizard Springs Formation of Trinidad and the middle part of the Ypresian Stage of Europe. The Flournoy Formation correlates with the Matilija, Domengine, and lower "Tejon" Formations of California, the lower part of the Navet Formation of Trinidad, and the lower Lutetian Stage of Europe.

ACKNOWLEDGMENTS

I express my sincere thanks to the following people for providing helpful ideas and suggestions: W. N. Orr and E. M. Baldwin, University of Oregon; W. G. Siesser, University of Cape Town; R. E. Thoms, Portland State University; and W. O. Seeley and J. M. Armentrout, Mobil Oil Corporation.

Costs of scanning electron microscopy were met through a grant from the University of Oregon Department of Geology Graduate Student Research Fund; Exxon Company, U.S.A., kindly provided funds for plate reproduction and for drafting the figures used in this study.

REFERENCE CITED

Allen, J. E., and Baldwin, E. M., 1944, Geology and coal resources of the Coos Bay quadrangle: Oregon Department of Geology and Mineral Industries Bulletin 27, 153 p.

Baldwin, E. M., 1961, Geologic map of the lower Umpqua River area, Oregon: U.S. Geological Survey Oil and Gas Investigations Map OM-204, scale 1:62,500.

——1963, Lower and middle Eocene formations of southwestern

Oregon: Geological Society of America Special Paper 76, p. 188-189.
Baldwin, E. M., 1964, Thrust faulting in the Roseburg area, Oregon: The Ore Bin, v. 26, p. 178-184.
——1965, Geology of the south end of the Oregon Coast Range Tertiary basin: Northwest Science, v. 39, no 3. p. 93-103.
——1966, Some revisions of the geology of the Coos Bay area, Oregon: The Ore Bin, v. 28, p. 189-203.
——1969, Geologic map of the Myrtle Point area, Coos County, Oregon: U.S. Geological Survey, Mineral Investigations Map MF-302, scale 1:62,500.
——1973, Geologic map of Coos County, *in* Baldwin, E. M., and Beaulieu, J. D., Geology and mineral resources of Coos County, Oregon: Oregon Department of Geology and Mineral Industries Bulletin 80, scale 1:62,500.
——1974, Eocene stratigraphy of southwestern Oregon: Oregon Department of Geology and Mineral Industries Bulletin 83, 40 p.
——1975, Revision of the Eocene stratigraphy of southwestern Oregon, *in* Paleogene symposium and selected technical papers: American Association of Petroleum Geologists-Society of Economic Paleontologists-Society of Economic Paleontologists and Mineralogists-Society of Economic Geologists, Pacific Sections, Annual Meeting, Long Beach, California, p. 49-64.
Banner, F. T., and Blow, W. H., 1965, Progress in the planktonic foraminiferal biostratigraphy of the Neogene: Nature, v. 208, p. 1164-1166.
Berggren, W. A., 1965, Some problems of Paleocene-lower Eocene planktonic foraminiferal correlations: Micropaleontology, v. 11, p. 278-300.
——1971, Multiple phylogenetic zonations of the Cenozoic based on planktonic foraminifera: International Conference on Planktonic Microfossils, 2nd, Rome, 1970, Proceedings, v. 1, p. 41-56.
——1972, A Cenozoic time scale—Some implications for regional geology and paleobiogeography: Lethaia, v. 5, p. 195-215.
Berggren, W. A., and Van Couvering, J.A., 1974, The late Neogene: Biostratigraphy, geochronology and paleoclimatology of the last 15 million years in marine and continental sequences: Amsterdam, Elsevier, 219 p.
Bird, K. J., 1967, Biostratigraphy of the Tyee Formation (Eocene), southwestern Oregon [Ph.D. thesis]: Madison, University of Wisconsin, 209 p.
Blow, W. H., 1969, Late middle Eocene to Recent planktonic foraminiferal biostratigraphy: International Conference on Planktonic Microfossils, 1st, Geneva, 1967, v. 1, p. 199-421.
Bolli, H. M., 1957a, The genera *Globigerina* and *Globorotalia* in the Paleocene-lower Eocene Lizard Springs Formation of Trinidad, B.W.I.: U.S. National Museum Bulletin 215, p. 61-81.
——1957b, Planktonic foraminifera from the Oligocene-Miocene Cipero and Lengua Formations of Trinidad, B.W.I.: U.S. National Museum Bulletin 215, p. 97-123.
——1957c, Planktonic foraminifera from the Eocene Navet and San Fernando Formations of Trinidad, B.W.I.: U.S. National Museum Bulletin 215, p. 155-172.
——1966, Zonation of Cretaceous to Pliocene marine sediments based on planktonic foraminifera: Asociación Venezolana de Geología, Minería y Petróleo Boletín Informativo, v. 9, no. 1, p. 3-32.
Bolli, H. M., and Cita, M. B., 1960, Globigerine e Globorotalie del Paleocene di Paderno d'Adda (Italia) [Globigerinids and globorotaliids of the Paleocene of Paderno d'Adda (Italy)]: Rivista Italiana de Paleontologia e Stratigrafia, v. 66, no. 3, p. 361-402.
Brownfield, M. E., 1972, Geology of the Floras Creek drainage, Langlois quadrangle, Oregon [M.S. thesis]: Eugene, University of Oregon, 104 p.
Bukry, D., 1973, Coccolith stratigraphy, Leg 10, Deep Sea Drilling Project, *in* Worzel, J. L., Bryant, W., and others, Initial reports of the Deep Sea Drilling Project, Volume 10: Washington, D.C., U.S. Government Printing Office, p. 385-406.
Clark, B. L., and Vokes, H. E., 1936, Summary of marine Eocene sequence of North America: Geological Society of America Bulletin, v. 47, p. 851-878.
Diller, J. S., 1893, Cretaceous and early Tertiary of northern California and Oregon: Geological Society of America Bulletin, v. 4, p. 205-224.
——1898, Description of the Roseburg quadrangle, Oregon: U.S. Geological Survey Geologic Atlas, Roseburg Folio, no 49.
Dott, R. H., Jr., 1962, Geology of the Cape Blanco area, southwest Oregon: The Ore Bin, v. 24, p. 121-133.
——1971, Geology of the southwestern Oregon coast west of the 124th meridian: Oregon Department of Geology and Mineral Industries Bulletin 69, 63 p.
Gibson, J. M., 1976, Distribution of planktonic foraminifera and calcareous nannoplankton, Late Cretaceous and early Paleogene, Santa Ynez Mountains, California: Journal of Foraminiferal Research, v. 6, no. 2, p. 87-106.
Hardenbol, J., and Berggren, W. A., 1978, A new Paleogene numerical time scale: American Association of Petroleum Geologists Special Studies in Geology, no. 6, p. 213-234.
Hay, W. W., and Mohler, H. P., 1967, Calcareous nannoplankton from early Tertiary rocks at Pont Labau, France, and Paleocene-early Eocene correlations: Journal of Paleontology, v. 41, no. 6, p. 1505-1541.
Jenkins, D. G., 1971, New Zealand Cenozoic planktonic foraminifera: New Zealand Geological Survey Paleontological Bulletin 42, 278 p.
Lent, R. L., 1969, Geology of the southern half of the Langlois quadrangle, Oregon [Ph.D. thesis]: Eugene, University of Oregon, 189 p.
Luterbacher, H., 1964, Studies in some *Globorotalia* from the Paleocene and lower Eocene of the central Apennines: Eclogae Geologicae Helvetiae, v. 57, p. 631-730.
Mallory, V. S., 1959, Lower Tertiary biostratigraphy of the California Coast Ranges: Tulsa, Oklahoma, American Association of Petroleum Geologists, 416 p.
McKeel, D. R., 1972, Systematics, biostratigraphy, and paleooceanography of Tertiary planktonic foraminifera assemblages, Oregon Coast Range [M.S. thesis]: Davis, University of California, 118 p.
McKeel, D. R., and Lipps, J. H., 1972, Calcareous plankton from the Tertiary of Oregon: Paleoceanography, Paleoclimatology, Paleoecology, v. 12, p. 75-93.
——1975, Eocene and Oligocene planktonic foraminifera from the central and southern Oregon Coast Range: Journal of Foraminiferal Research, v. 5, no. 4, p. 249-269.
Miles, G. A., 1977, Planktonic foraminifera of the lower Tertiary Roseburg, Lookingglass, and Flournoy Formations, southwest Oregon [Ph.D. thesis]: Eugene, University of Oregon, 359 p.
Postuma, J. A., 1971, Manual of planktonic foraminifera: Amsterdam, Elsevier, 420 p.
Robertson, R. D., 1972, An early Tertiary pelagic microfauna from the Coast Range, western Washington County, Oregon [M.S. thesis]: Eugene, University of Oregon, 100 p.

Schmidt, R. R., 1970, Planktonic foraminifera from the lower Tertiary of California [Ph.D. thesis]: Los Angeles, University of California, 339 p.

——1975, Upper Paleocene-middle Eocene planktonic biostratigraphy from the Great Valley of California and adjacent areas, and correlation to the West Coast microfaunal stages, *in* Paleogene symposium and selected technical papers: American Association of Petroleum Geologists-Society of Economic Paleontologists-Society of Economic Geologists, Pacific Sections, Annual Meeting, Long Beach, California p. 439-463.

Stewart, R. E., 1957, Stratigraphic implications of some Cenozoic foraminifera from western Oregon: The Ore Bin, v. 19, p. 11-15.

Subbotina, N. N., 1953, Globigerinidae, Hantkeninidae and Globorotaliidae: Fossil foraminifera of the U.S.S.R. (in Russian): VNIGRI, Trudy, new ser., no. 76, p. 1-296.

Thoms, R. E., 1965, Biostratigraphy of the Umpqua Formation, southwest Oregon [Ph.D. thesis]: Berkeley, University of California, 215 p.

——1975, Biostratigraphy of the Umpqua Group, southwestern Oregon, *in* Paleogene symposium and selected technical papers: American Association of Petroleum Geologists-Society of Economic Paleontologists and Mineralogists-Society of Economic Geologists, Pacific Sections, Annual Meeting, Long Beach, California, p. 563-587.

Turner, F. E., 1938, Stratigraphy and mollusca of the Eocene of western Oregon: Geological Society of America Special Paper 10, 130 p.

Weaver, C. E., and others, 1944, Correlation of the marine Cenozoic formations of western North America: Geological Society of America Bulletin, v. 55, p. 569-598.

Wells, F. G., and Peck, D. L., 1961, Geologic Map of Oregon west of the 121st meridian: U.S. Geological Survey, Miscellaneous Geologic Investigations Map I-325, scale 1:500,000

MANUSCRIPT RECEIVED BY THE SOCIETY MARCH 14, 1980
MANUSCRIPT ACCEPTED MAY 7, 1980

Printed in U.S.A.

Geological Society of America
Special Paper 184
1981

Calcareous plankton biostratigraphy of the type Bastendorff Formation, southwest Oregon

A. D. WARREN
Anderson, Warren & Associates, Inc., 11526 Sorrento Valley Road, Suite G, San Diego, California 92121

J. H. NEWELL
Superior Oil Company, Geoscience Laboratory, 12401 Westheimer Road, Houston, Texas 77077

ABSTRACT

Seventy-six samples from the type Bastendorff Formation exposed near Coos Bay, Oregon, were examined to determine if significant calcareous plankton were present. Planktonic foraminifera and/or calcareous nannofossils are recorded from 24 of those samples.

Planktonic foraminiferal zones believed present are the *Globigerinatheka semiinvoluta* and *Globorotalia cerroazulensis* s.l. Zones of Bolli and Zones P15, P16, and P17 of Blow.

Nannoplankton zones recognized are the *Chiasmolithus oamaruensis* and *Isthmolithus recurvus* Subzones of Bukry; these are considered equivalent to Zone NP18 and Zones NP19 and NP20 of Martini, respectively.

The middle and upper parts of the Bastendorff—assigned by Tipton to the provincial uppermost part of the Narizian Stage and the Refugian Stage, respectively, on the evidence of benthic foraminiferal distribution—are of late Eocene age based on calcareous plankton.

INTRODUCTION

The type Bastendorff Formation is located at Bastendorff Beach, near Coos Bay, Oregon, where, as late as 1950, approximately 671 m (2,200 ft) of the formation were well exposed, and an additional 183 to 213 m (600 to 700 ft) of lowermost Bastendorff were presumed to be concealed beneath alluvium at the mouth of Miner Creek (Fig. 1). Today, most of this section has been covered by dense brush, so that further measurements and detailed collection of samples are impossible. Fortunately, this type section was measured and collected in 1950 by John Browning and William Grier; their samples and data were reposited at the Museum of Paleontology, University of California, Berkeley, to be studied at a later date.

The purpose of this paper is to document the occurrence of significant calcareous microplankton in the type Bastendorff Formation and to interpret the age and zonal assignments within that unit according to criteria currently in wide usage by specialists working with calcareous nannofossils and planktonic foraminifera. The Browning-Grier samples used in this study are the same as those reported on by Tipton (1975) and McKeel and Lipps (1975); thus, calcareous plankton zonations and age assignments may be directly correlated with the provincial zonal and stage assignments of Tipton (1975).

PREVIOUS STUDIES

The predominantly shale section lying between the Coaledo and Tunnel Point Formations at Bastendorff Beach was named "Bassendorf shale" by Schenck (1927). He considered the unit to be "Lower Oligocene" in age.

Cushman and Schenck (1928) described and illustrated many of the benthic foraminiferal species of the type "Bassendorf" and Keasey shales. A few planktonic specimens, assigned to *Globigerina conglomerata,* were noted but not illustrated. They considered the Bastendorff and Keasey Formations to be correlative and concluded that they are "probably lowest Oligocene."

In erecting the Refugian Stage, Schenck and Kleinpell (1936) included the "Bassendorf beach" section as an important part of their concept in establishing that stage which they considered to be late Eocene or early Oligocene in age.

Allen and Baldwin (1944), in their report on the geology and coal resources of the Coos Bay quadrangle, established the correct spelling as "Bastendorff" for the formation in accordance with the spelling of the Bastendorff family name.

Detling (1946), in a study of the foraminiferal assem-

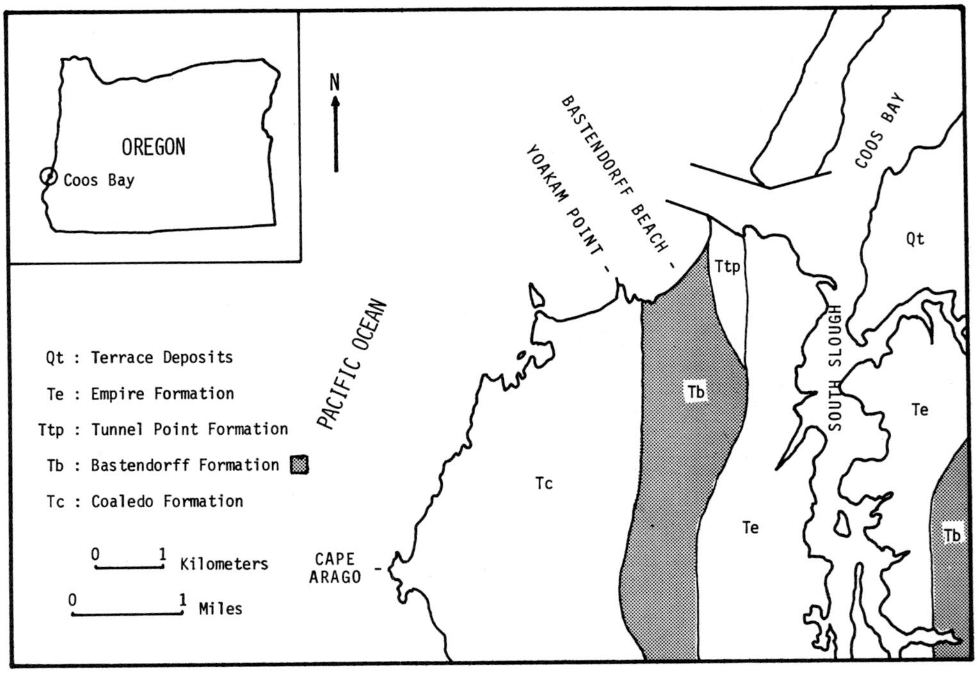

Figure 1. Location map, type Bastendorff Formation. Geologic contacts after Baldwin (1973).

blages of the Coaledo and "Bastendorf" Formations, listed occurrences of *Globigerina triloculinoides* in five sample stations distributed from near the middle to the top of the "Bastendorf" Formation. She remarked on the considerable variation in these specimens, so it is probable that more than one taxon was involved. On the basis of observed benthic foraminiferal associations, Detling assigned an Eocene or Oligocene age to the "Bastendorf" Formation.

In one of a series of short papers on studies of foraminifera from western Oregon, Stewart (1956) assigned the upper 213 m (700 ft) of the type Bastendorff to the Refugian Stage and considered that section to be early Oligocene in age. The middle and unexposed lower parts of the Bastendorff were assigned to the upper Eocene by Stewart. He listed *Globigerina* cf. *G. bulloides* as having been observed in the middle Bastendorff.

Rooth (1974), utilizing both benthic and planktonic foraminifera in biostratigraphic analysis, conducted a comprehensive study of the biostratigraphy and paleoecology of the Coaledo and Bastendorff Formations. He listed *Globigerapsis index, Globigerina prasaepis?, Globigerina senilis,* and *Subbotina linaperta* from the middle Bastendorff and *Chiloguembelina cubensis, Globigerina prasaepis?, Globigerina tripartita tripartita, Globorotalia gemma, Globorotalia* cf. *G. gemma, Globorotalia increbescens, Globorotalia opima nana,* and *Globorotalia* sp. from the upper Bastendorff at the stratotype locality. He concluded (p. 93) that the Coaledo and Bastendorff Formations are late Eocene in age and that they can be correlated confidently with the Runangan Stage of New Zealand and tentatively with the "P.17" partial-range zone of Blow (1969) and others. Rooth's description and illustration of *Globorotalia* sp. from the uppermost Bastendorff Formation (22 m below the Tunnel Point Formation) are sufficient to permit identification of the two specimens found as *Globorotalia insolita,* a late Eocene index species originally described from New Zealand by Jenkins (1966). In addition, the specimens identified *and* illustrated by Rooth as *Globorotalia gemma* and *Globorotalia* aff. *G. gemma* also appear to be specimens of *Globorotalia insolita.*

Tipton (1975) detailed the vertical distribution of foraminifera within the type Bastendorff. She established the presence of approximately 180 m (600 ft) of lower Refugian and 60 m (200 ft) of upper Refugian sediments at the top of that formation; the lower and middle Bastendorff were assigned to the upper Narizian Substage. Tipton's placement of the Eocene/Oligocene boundary was shown to coincide questionably with the boundary between the lower and upper Refugian Substages. *Globigerina* cf. *G. wilsoni, Globigerina* spp., and *Globigerinatheka index index* are listed as elements of the foraminiferal assemblages within the type Bastendorff Formation.

McKeel and Lipps (1975) recorded *Chiloguembelina cubensis, Globigerina praebulloides, Globigerina* sp., *Globigerinatheka index, Globorotaloides* aff. *G. pseudokugleri,* and *Subbotina wilsoni* in three samples from the type Bastendorff Formation. Their Table 2 shows the Eocene/Oligocene boundary placed questionably at a point between the middle and upper parts of the Bastendorff Formation.

Warren and Newell (1980), in a broader study of the calcareous plankton biostratigraphy of the Refugian and adjoining stages, demonstrated that the Refugian Stage of the Pacific Coast Tertiary may be equated with the *Discoaster barbadiensis* Zone, *Isthmolithus recurvus* Subzone of Bukry (1975), and with planktonic foraminiferal Zones P16 and P17 of Blow (1969); thus the Refugian Stage may also be correlated with Zones NP19 and NP20 of Martini (1971) and with the *Globigerinatheka semiinvoluta* (part) and *Globorotalia cerroazulensis* s.l. Zones of Bolli (1972). They concluded that the Refugian Stage is late Eocene in age and correlative to all or parts of the Jacksonian Stage of the Atlantic and Gulf Coastal province (U.S.A.), the Priabonian Stage of Europe, and the Kaiatan and Runangan Stages of New Zealand.

ADDITIONAL DATA, PLANKTONIC FORAMINIFERA

Seventy-six samples from the Browning-Grier collection of the type Bastendorff Formation were examined in this study to determine the distribution of planktonic foraminifera in that section. Well-preserved planktonics were found to be present in rather low frequencies in eight of the samples examined. A few additional samples contained rare specimens that were too poorly preserved to permit identification beyond the fact that they were the remains of planktonic foraminifera. Occurrences noted here are listed in Figure 2, which also lists occurrences of planktonic foraminiferal species reported by Rooth (1974), Tipton (1975), and McKeel and Lipps (1975).

Owing to low frequency and diversity, planktonic foraminiferal zonal boundary assignments can only be tentatively drawn within the middle and upper Bastendorff Formation. However, the presence of *Globigerinatheka index index, G. index tropicalis, Globorotalia cerroazulensis pomeroli, G. increbescens,* and *G. insolita* and their vertical distribution within the section permit gross recognition of the *Globigerinatheka semiinvoluta* and *Globorotalia cerroazulensis* s.l. Zones of Bolli (1957, 1966, 1972) and Zones P15, P16 and P17 of Blow (1969).

ADDITIONAL DATA, CALCAREOUS NANNOPLANKTON

Twenty of the Browning-Grier samples from the type Bastendorff were found to contain sufficient numbers and diversity of nannofossils to permit analysis. Thirty species and their vertical distribution are listed in Figure 3.

Nine samples from the middle Bastendorff (late Narizian) are assigned to the *Chiasmolithus oamaruensis* Subzone, *Discoaster barbadiensis* Zone of Bukry (1975), or to Zone NP18 of Martini (1971), on the basis of occurrences of *Chiasmolithus oamaruensis* prior to the earliest occurrence of *Isthmolithus recurvus.*

Nine samples from the upper Bastendorff (early and late Refugian) are confidently assigned to the *Isthmolithus recurvus* Subzone, *Discoaster barbadiensis* Zone of Bukry (1975), or to Zones NP19 and NP20 of Martini (1971). This assignment is based on the occurrence of *Isthmolithus recurvus* with rare and intermittent occurrences of *Discoaster barbadiensis* and *Discoaster saipanensis.*

The two samples from the uppermost Bastendorff (late Refugian) cannot be assigned with certainty to any single zone or subzone. However, these samples cannot be younger than the *Coccolithus subdistichus* Subzone of the *Helicopontosphaera reticulata* Zone of Bukry (1973) [=NP21 Zone of Martini (1971)]. We, taking note of the extreme rarity of *Discoaster barbadiensis* and *Discoaster saipanensis* in high-latitude areas, consider the uppermost Bastendorff Formation to be most probably a continuation of the *Isthmolithus recurvus* Subzone or Zones NP19 and NP20 even though *Discoaster barbadiensis* and *Discoaster saipanensis* were not recorded from that section in this study.

SUMMARY AND CONCLUDING REMARKS

Documented planktonic foraminiferal and calcareous nannoplankton data from the type Bastendorff Formation support the following conclusions (see Figs. 2, 3, 4):

1. The precise age of the unexposed lower Bastendorff Formation is indeterminate, as no sample material is available for study. Tipton (1975) was able to assign this section to the provincial upper Narizian Substage as it is underlain by the upper Coaledo Formation and overlain by the middle Bastendorff Formation, both of which are—on the basis of benthic foraminifera— in the *Amphimorphina jenkinsi* Zone (late Narizian). The only conclusion that can be drawn from data available at this time is that the lower Bastendorff may be as old as late middle Eocene and as young as early late Eocene.
2. The middle Bastendorff, assigned to the uppermost Narizian Stage and *Amphimorphina jenkinsi* Zone by Tipton (1975), is early late Eocene in age. The *Globigerinatheka semiinvoluta* Zone of Bolli (1972) and planktonic Zones P15 and lower P16 of Blow (1969) are tentatively recognized in this part of the formation. Assignment of the upper half of the middle Bastendorff to the *Chiasmolithus oamaruensis* Subzone of Bukry (1975) or Zone NP18 of Martini (1971) is without reservation.
3. The upper Bastendorff, assigned to the lower and upper Refugian and to the *Valvulineria tumeyensis* and *Uvigerina vicksburgensis* Zones by Tipton (1975, 1980), is considered here to be middle late Eocene to late late Eocene in age. This section is tentatively assigned to the *Globorotalia cerroazulensis s.l.* Zone of Bolli (1972) or to Zones P16 and P17 of Blow (1969). The *Isthmolithus recurvus* Subzone of Bukry (1975), equivalent to Zones NP19 and NP20 of Martini (1971), is definitely present within the upper Bastendorff, although the uppermost 39 m of the formation could not be assigned with certainty to that subzone.

Special note should be taken that here the uppermost part of the Narizian Stage is of early late Eocene age. Other locations where the uppermost part of the Narizian Stage has been shown to be of early late Eocene age are the upper Nestucca Formation of northwest Oregon, the Sacate Formation of Arroyo el Bulito, and the San Lorenzo Formation of Santa Cruz County, California (Warren and Newell, 1980). However, most of the lower and upper parts of the Narizian Stage have been shown to be of middle Eocene age (Bandy and Kolpack, 1963; Bukry and others,

CHECKLIST OF PLANKTONIC FORAMINIFERA REPORTED FROM THE TYPE BASTENDORFF FORMATION BY ROOTH (1974), TIPTON (1975), MCKEEL AND LIPPS (1975) AND WARREN AND NEWELL (THIS PAPER)				SELECTED PLANKTONIC FORAMINIFERA																		MODIFIED AFTER ROOTH (1974) AND TIPTON (1975, IN PRESS)		
SUBSERIES	Planktonic Foraminiferal Zonation of Bolli (1957, 1966, 1972) Zones	Planktonic Foraminiferal Zonation of Blow (1969) Zones	Locality Numbers of the Museum of Paleontology, U. C. Berkeley and of Rooth (1974)	Chiloguembelina cubensis	Globigerina praebulloides	G. senilis	G. tripartita tripartita	G. (Subbotina) angiporoides angiporoides	G. linaperta	G. wilsoni	Globigerinatheka index s.l.	G. index index	G. index tropicalis	Globorotalia cerroazulensis pomeroli	G. gemma	G. cf. G. gemma	G. increbescens	G. insolita	G. opima nana	Globorotaloides aff. G. pseudokugleri	G. suteri	Meters below the base of the Tunnel Point Formation	Formation	Provincial Substage and Benthonic Foraminiferal Zonal Assignments
UPPER EOCENE	Globorotalia cerroazulensis s.l.	P.17 Gl. gortanii gortanii-Globorotalia centralis	A 8079		F																	0.05	upper Bastendorff	Late Refugian, Uvigerina vicksburgensis Zone
			Rooth 98	R											C			R	R			21.5		
		P.16 Cribrohantkenina inflata	Rooth 96													C						125.0		Early Refugian, Valvulineria tumeyensis Zone
			A 8058						R													175.6		
			A 8047											R								214.6		
	Globigerinatheka semiinvoluta		A 8046										R									216.6		
			Rooth 90													R						280.5		Late Narizian, Amphimorphina jenkinsi Zone
			Rooth 88	R		R	R				R						F					334.0	middle Bastendorff	
		P.15 Globigerapsis mexicana	A 8029	C				R		F	F	C	R							R	R	357.0		
			A 8028	C				R		F	F	F	R							R	R	357.9		
			A 8024									R										360.0		
			A 8023									R										361.9		
			Rooth 87						F													389.0		
			A 8014								C	F	F									446.3		
			Rooth 82								C											578.0	lower Bastendorff	

Figure 2. Occurrence of selected planktonic foraminifera (for references, see App.1) in the type Bastendorff Formation as reported by Rooth (1974, locality descriptions, p. 239-241 and Pl. 12), Tipton (1975, localities descriptions, p. 580-582 and Fig. 6), McKeel and Lipps (1975), and Warren and Newell (1980 and this paper). R, rare; F, frequent; C, common.

SUBSERIES	Calcareous Nannoplankton Zonation of Bukry (1973, 1975): Zones	Calcareous Nannoplankton Zonation of Bukry (1973, 1975): Subzones	Calcareous Nannoplankton Zonation of Martini (1971): Zones	Locality Numbers of Museum of Paleontology, U.C., Berkeley	Braarudosphaera bigelowi	irregularis	Chiasmolithus altus	oamaruensis	cf. C. oamaruensis	Coccolithus eopelagicus	pelagicus	Cyclococcolithina formosa	Dictyococcites bisectus	scrippsae	Discoaster barbadiensis	distinctus	nodifer	saipanensis	tani	Discolithina distincta	panaria	Helicopontosphaera euphratis	intermedia	reticulata	Isthmolithus recurvus	Neococcolithes dubius	Reticulofenestra hillae	umbilica	Rhabdosphaera crebra	tenuis	Sphenolithus cf. S. abies	moriformis	Transversopontis pulcher	Zygrhablithus bijugatus	Meters Below Base Tunnel Point Formation	Substages of Bastendorff Formation after Tipton (1975)
UPPER EOCENE TO LOWER OLIGOCENE	Helicopontosphaera reticulata to Discoaster barbadiensis	Coccolithus subdistichus to Isthmolithus recurvus	NP 21 to NP 19/NP 20	A 8079	R		R						R	R						R					R		R	R							.05	Upper Refugian
				A 8076	R		R				R	R	R	F						R					F			R							38.4	
UPPER EOCENE	Discoaster barbadiensis	Isthmolithus recurvus	NP 19/NP 20	A 8075			R			R	R	R	R	F				R		R					F	R	R	F						R	39.9	
				A 8074	R		R				R		R	F											F			F				R			42.0	
				A 8073	R		R			R			R	F											R			R			R	R			43.9	
				A 8071							R		R	F				R							R		R	R				R			81.7	Lower Refugian
				A 8070						R	R	R	R	F											R		R	R							87.8	
				A 8069			R				R	R	R	F											R			R							90.2	
				A 8065									R	F											R			R							96.6	
				A 8058						R			R	F			R		R				R		R			R				R			175.6	
				A 8057			R			R		R	R	C	R				R						R		R	R							176.2	
		Chiasmolithus oamaruensis	NP 18	A 8039						R	R			R													R	R							306.0	Upper Narizian
				A 8038	R	F				R	R		R	F		R											R	F							307.6	
				A 8036						R	R		R	F	R			R				R													311.5	
				A 8035							R	R	R	F	R					R								R							312.5	
				A 8029				R		R	R	R	R	C						R									R	R				R	357.0	
				A 8028				R		R	R	R		C						F										R		R		R	357.9	
				A 8027				R		R	R	R		C						R										R	R	R			358.8	
				A 8025				R		R	R			C						R	R		R			R							R		359.7	
				A 8014			F		R	R	R	R	R	C								R		R		R		R				R	R		446.3	

Figure 3. Occurrence of calcareous nannoplankton (for references, see App. 2) in the middle and upper parts of the type Bastendorff Formation (locality descriptions, Tipton, 1975, p. 580-582 and Fig. 6). R, rare; F, frequent; C, common.

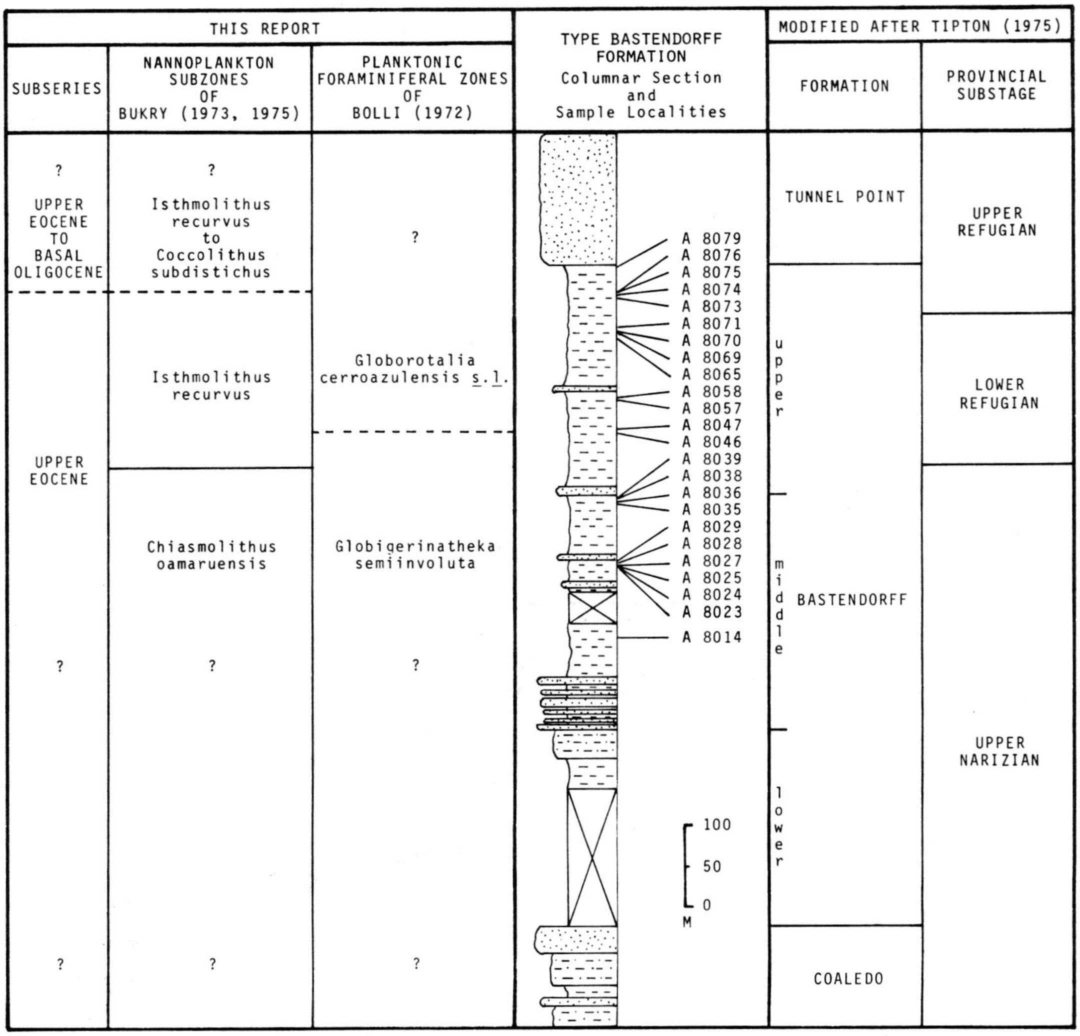

Figure 4. Relationships of nannoplankton and planktonic foraminiferal zones within the type Bastendorff Formation and stratigraphic allocation of Browning-Grier samples containing calcareous microplankton (locality descriptions, Tipton, 1975, p. 580-582 and Fig. 6).

1977; Schmidt, 1975; Steineck and Gibson, 1971; Warren and Newell, 1980; and others). Thus, the Narizian Stage in its entirety includes sediments ranging from early middle Eocene to early late Eocene in age.

Evidence developed from this study of the Bastendorff Formation and from previous studies of several other formations in Oregon and California (Brabb and others, 1971; Lipps, 1967; Warren and Newell, 1976, 1980) indicates that the Refugian Stage in its entirety contains sediments ranging from middle late Eocene to late late Eocene in age.

Plankton workers should be aware of the presense of *Globorotalia insolita* in Refugian sediments of the Pacific Northwest. This tiny, distinctive, and stratigraphically important taxon was first recognized in the subsurface Refugian of northwest Washington by R. S. Boettcher (1971, personal commun.). *Globorotalia insolita,* recorded as *Globorotalia* sp., *G. gemma,* and *G.* cf. *G. gemma* from the uppermost Bastendorff by Rooth (1974), has recently been observed in the Keasey Formation of northwest Oregon by K. McDougall (1978, personal commun.). Formations which may yet be found to contain occurrences of that species are the Alsea and Nestucca Formations of Oregon, the Lincoln Creek, Twin River, and Blakely Formations of Washington, and the Poul Creek Formation of Alaska. *Globorotalia insolita* was originally described from the Kaiatan Stage of New Zealand by Jenkins (1966) and has been recorded in a deep-sea core from the eastern equatorial Pacific by Jenkins and Orr (1972), where it was associated with *Globorotalia gemma* and other planktonic species of late Eocene age. No occurrences have been recorded to date in surface or subsurface sections in California.

ACKNOWLEDGMENTS

We are grateful to the late Gordon Hornaday, Museum of Paleontology, University of California, Berkeley, and to Ann Tipton Donnelly for making available the Browning-

Grier samples and locality descriptions. John Armentrout, Kristin McDougall, Ronald Morin, and S. F. Percival kindly contributed their time in reviewing this paper prior to publication. Thanks are also due to Anderson, Warren & Associates, Inc., for subsidizing all expenses involved in preparing this report.

APPENDIX 1. REFERENCES FOR SPECIES OF FORAMINIFERA IN FIGURE 2

Chiloguembelina cubensis (Palmer) = *Gümbelina cubensis* Palmer, 1934, Cubana Natural History Society Memoir, v. 8, p. 74, Text-Figs, 1-6.

Globigerina praebulloides Blow, 1959, American Paleontology Bulletins, v. 39, no. 178, p. 180, Pl. 8, figs. 47a-c, Pl. 9, fig. 48.

Globigerina senilis Bandy = *Globigerina ouachitaensis senilis* Bandy, 1949, American Paleontology Bulletins, v. 32, no. 131, p. 121, Pl. 22, figs. 5a-c.

Globigerina tripartita tripartita Koch = *Globigerina bulloides tripartita* Koch, 1926, Eclogae Geologicae Helvetiae,v. 19, no. 3, p. 746, Text-Figs. 21a-b.

Globigerina (Subbotina) angiporoides angiporoides Hornibrook = *Globigerina angiporoides* Hornibrook, 1965, New Zealand Journal of Geology and Geophysics, v. 8, no. 5, p. 834, Figs. 1, 2.

Globigerina (Subbotina) linaperta Finlay = *Globigerina linaperta* Finlay, 1939, Royal Society of New Zealand Transactions, v. 69, no. 1, p. 125, Pl. 23, figs. 54-57.

Globigerina (Subbotina) wilsoni Cole = *Globigerina wilsoni* Cole, 1927, American Paleontology Bulletins, v. 14, no. 51, p. 34, Pl. 4, figs. 8-9.

Globigerinatheka index s.l. (Finlay) = *Globigerinoides index* Finlay, 1939, Royal Society of New Zealand Transactions, v. 69, pt. 1, p. 125, Pl. 14, figs. 85-88.

Globigerinatheka index index (Finlay) of Bolli, 1972, Journal of Foraminiferal Research, v. 2, no. 3, p. 124, Pl. 1, figs. 1-4, 6, 7; Text-Figs. 51-57, 63, 64.

Globigerinatheka index tropicalis (Blow and Banner) of Bolli, 1972, Journal of Foraminiferal Research, v. 2, no. 3, p. 127, Pl. 3, figs. 1-24; Pl. 4, figs. 7-12; Text-Figs. 58-60.

Globorotalia cerroazulensis pomeroli Toumarkine and Bolli, 1970, Revue de Micropaleontologie, v. 13, no. 3, p. 140-142, Pl. 1, figs. 10-18; Pl. 2, Figs. 1, 2, 11-19.

Globorotalia gemma Jenkins, 1966, New Zealand Journal of Geology and Geophysics, v. 8, no. 6, p. 1115, Fig. 11, no. 97-103, Pl. 7, figs. 29-31.

Globorotalia cf. *G. gemma* Jenkins of Rooth, 1974 = *Globorotalia insolita* Jenkins, 1966, New Zealand Journal of Geology and Geophysics, v. 8, no. 6, p. 1120, Fig. 13, no. 113-118.

Globorotalia increbescens (Bandy) = *Globigerina increbescens* Bandy, 1949, American Paleontology Bulletins, v. 32, no. 131, p. 120, Pl. 23, figs. 3a-c.

Globorotalia insolita Jenkins, 1966, New Zealand Journal of Geology and Geophysics, v. 8, no. 6, p. 1120, Fig. 13, no. 113-118.

Globorotalia opima nana Bolli, 1957, U.S. National Museum Bulletin 215, p. 118, Pl. 28, figs. 3a-c.

Globorotaloides aff. *G. pseudokugleri* (Blow) = aff. *Globorotalia pseudokugleri* Blow, 1969, International Conference of Planktonic Microfossils, 1st, Geneva, p. 391, Pl. 10, figs. 4-6.

Globorotaloides suteri Bolli, 1957, U.S. National Museum Bulletin 215, p. 117, Pl. 27, figs. 9a-13b.

APPENDIX 2. REFERENCES FOR SPECIES OF NANNOPLANKTON IN FIGURE 3

Braarudosphaera bigelowi (Gran and Braarud) = *Pontosphaera bigelowi* Gran and Braarud, 1935, Biological Board of Canada Journal, v. 1, p. 388, Fig. 67.

Chiasmolithus altus Bukry and Percival, 1971, Tulane Studies in Geology and Paleontology, v. 8, no. 3, p. 126, Pl. 2, figs. 1, 2.

Chiasmolithus oamaruensis (Deflandre) = *Tremalithus oamaruensis* Deflandre in Deflandre and Fert, 1954, Annals of paleontology, v. 40, p. 154, Pl. 11, fig. 22; Text-Figs. 72-74.

Coccolithus eopelagicus (Bramlette and Riedel) = *Tremalithus eopelagicus* Bramlette and Riedel, 1954, Journal of Paleontology, v. 28, p. 392, Pl. 38, figs. 2a,b.

Coccolithus pelagicus (Wallich) = *Coccosphaera pelagica* Wallich, 1877, Annual Magazine of Natural History, ser. 4, v. 19, p. 348, Figs. 1, 2, 5, 11, 12.

Cyclococcolithina formosa (Kamptner) = *Cyclococcolithus formosus* Kamptner, 1963, Annalen Naturhistorisches Museum Wien, v. 66, p. 163, Pl. 2, fig. 8; Text-Fig. 20.

Dictyococcites bisectus (Hay, Mohler and Wade) = *Syracosphaera bisecta* Hay, Mohler and Wade, 1966, Eclogae Geologicae Helvetiae, v. 59, p. 393, Pl. 10, figs. 1-6.

Dictyococcites scrippsae Bukry and Percival, 1971, Tulane Studies in Geology and Paleontology, v. 8, no. 3, p. 128, Pl. 2, figs. 7, 8.

Discoaster barbadiensis Tan, 1927, Koninkljke Nederlandse Akademie van Wetenschappen Proceedings, v. 30, p. 145 (not figured).

Discoaster distinctus Martini, 1958, Senckenbergiana Lethaea, v. 39, p. 363, Pl. 4, figs. 17a, 17b.

Discoaster nodifer Bramlette and Riedel = *Discoaster tani nodifer* Bramlette and Riedel, 1954, Journal of Paleontology, v. 28, no. 4, p. 397, Pl. 39, fig. 2.

Discoaster saipanensis Bramlette and Riedel, 1954, Journal of Paleontology, v. 28, no. 4, p. 298, Pl. 39, fig. 4.

Discoaster tani Bramlette and Riedel, 1954, Journal of Paleontology, v. 28, no. 4, p. 397, Pl. 39, fig. 1.

Discolithina distincta (Bramlette and Sullivan) = *Discolithus distinctus* Bramlette and Sullivan, 1961, Micropaleontology, v. 7, no. 2, p. 141, Pl. 2, figs. 8a, 8b, 9a-c.

Discolithina panaria (Deflandre) = *Discolithus panarium* Deflandre in Deflandre and Fert, 1954, Annals of Paleontology, v. 40, p. 141, Text-Figs. 39, 40.

Helicopontosphaera euphratis (Haq) = *Helicosphaera euphratis* Haq, 1966, Stockholm Contributions in Geology, v. XV, no. 3, p. 33, Pl. II, figs. 1, 3 (= *Helicosphaera parallela* Bramlette and Wilcoxon).

Helicopontosphaera intermedia (Martini) = *Helicosphaera intermedia* Martini, 1965, Colston Research Society Symposium, 17th, Proceedings, v. 17, p. 404, Pl. 35, figs. 1, 2.

Helicopontosphaera reticulata (Bramlette and Wilcoxon) = *Helicosphaera reticulata* Bramlette and Wilcoxon, 1967, Tulane Studies in Geology and Paleontology, v. 5, no. 3, p. 106, Pl. 6, fig. 15.

Isthmolithus recurvus Deflandre and Fert, 1954, Annals of Paleontology, v. 40, p. 169, Pl. 12, figs. 9-13, Text-Figs. 119-122.

Neococcolithes dubius (Deflandre) = *Zygolithus dubius* Deflandre in Deflandre and Fert, 1954, Annals of Paleontology, v. 40, p. 149, Text-Figs. 43-44, 68.

Reticulofenestra hillae Bukry and Percival, 1971, Tulane Studies in Geology and Paleontology, v. 8, no. 3, p. 136, Pl. 6, figs. 1-3.

Reticulofenestra umbilica (Levin) = *Coccolithus umbilicus* Levin, 1965, Journal of Paleontology, v. 39, p. 265, Pl. 41, fig. 2.

Rhabdosphaera crebra (Deflandre) = *Rhabdolithus creber* Deflandre *in* Deflandre and Fert, 1954, Annals of Paleontology, v. 40, p. 157, Pl. 12, figs. 31-33, Text-Figs. 81, 82.

Rhabdosphaera tenuis Bramlette and Sullivan, 1961, Micropaleontology, v. 7, no. 2, p. 147, Pl. 5, figs. 14a-b.

Sphenolithus cf. *S. abies* Deflandre in Deflandre and Fert, 1954, Annals of Paleontology, v. 40, p. 164, Pl. 10, figs. 1-4.

Sphenolithus moriformis (Bronnimann and Stradner) = *Nannoturbella moriformis* Bronnimann and Stradner, 1960, Erdoel-Zeitschrift, v. 76, p. 368, Figs. 11-16.

Transversopontis pulcher (Bramlette and Sullivan) = *Discolithus pulcher* Bramlette and Sullivan, 1961, Micropaleontology, v. 7, no. 2, p. 143, Pl. 3, figs. 8a-c.

Zygohablithus bijugatus (Deflandre) = *Zygolithus bijugatue* Deflandre, and Fert, Annals of Paleontology, v. 40, p. 148, Pl. 11, fig. 21 (not Text-Fig. 59).

REFERENCES CITED

Allen, J. E., and Baldwin, E. M., 1944, Geology and coal resources of the Coos Bay quadrangle: Oregon Department of Geology and Mineral Industries Bulletin 27, 153 p.

Baldwin, E. M., 1973, Geology and mineral resources of Coos County, Oregon: Oregon Department of Geology and Mineral Industries Bulletin 80, p. 1-82.

Bandy, O. L., and Kolpack, R. L., 1963, Foraminiferal and sedimentological trends in the Tertiary section of Tecolote Tunnel, California: Micropaleontology, v. 9, no. 2, p. 117-170.

Blow, W. H., 1969, Late middle Eocene to Recent planktonic foraminiferal biostratigraphy: International Conference on Planktonic Microfossils, 1st, Geneva, 1967, v. 1, p. 199-421.

Bolli, H. M., 1957, Planktonic foraminifera from the Eocene Navet and San Fernando Formations of Trinidad, B.W.I.: U.S. National Museum Bulletin 215, p. 155-172.

——1966, Zonation of Cretaceous to Pliocene marine sediments based on planktonic foraminifera: Associación Venezolana de Geología Minería y Petróleo Boletín Informativo, v. 9, no. 1, p. 3-32.

——1972, The genus *Globigerinatheka* Bronnimann; Journal of Foraminiferal Research, v. 2, no. 3, p. 109-136.

Brabb, E. E., Bukry, D., and Pierce, R. L., 1971, Eocene (Refugian) nannoplankton in the Church Creek Formation near Monterey, central California: U.S. Geological Survey Professional Paper 750, p. C44-C57.

Bukry, D., 1973, Low-latitude coccolith biostratigraphic zonation, *in* Initial reports of the Deep Sea Drilling Project, Volume 15: Washington, D.C., U.S. Government Printing Office, p. 685-703.

——1975, Coccolith and silicoflagellate stratigraphy, northwestern Pacific Ocean. Leg 32, *in* Initial reports of the Deep Sea Drilling Project Volume 32: Washington, D.C., U.S. Government Printing office, p. 677-701.

Bukry, D., Brabb, E. E., and Vedder, J. G., 1977, Correlation of Tertiary nannoplankton assemblages from the Coast and Peninsular Ranges of California: Congreso Latino-americano do Geologia, 2nd, Caracas, 1973, Memoria, tme 3, p. 1461-1483.

Cushman, J. A., and Schenck, H. G., 1928, Two foraminiferal faunules from the Oregon Tertiary: University of California Publications, Department of Geological Sciences Bulletin, v. 17, no. 9, p. 304-324.

Detling, M. R., 1946, Foraminifera of the Coos Bay lower Tertiary, Coos County, Oregon: Journal of Paleontology, v. 20, no. 4, p. 348-361.

Jenkins, D. G., 1966, Planktonic foraminiferal zones and new taxa from the Danian to lower Miocene of New Zealand: New Zealand Journal of Geology and Geophysics, v. 8, no. 6, p. 1080-1126.

Jenkins, D. G., and Orr, W., 1972, Planktonic foraminiferal biostratigraphy of the eastern equatorial Pacific, Leg 9. *in* Initial reports of the Deep Sea Drilling Project, Volume 9: Washington, D.C., U.S. Government Printing Office, p. 1059-1193.

Lipps, J. H., 1967, Planktonic foraminifera, intercontinental correlation and age of California mid-Cenozoic microfaunal stages: Journal of Paleontology, v. 41, no. 4, p. 994-999.

Martini, E., 1971, Standard Tertiary and Quaternary calcareous nannoplankton zonation: International Conference on Planktonic Microfossils, 2d, Rome, 1970, v. II, p. 739-786.

McKeel, D. R., and Lipps, J. H., 1975, Eocene and Oligocene planktonic foraminifera from the central and southern Oregon Coast Range: Journal of Foraminiferal Research, v. 5, p. 249-269.

Rooth, G. H., 1974, Biostratigraphy and paleoecology of the Coaledo and Bastendorff Formations, southwestern Oregon [Ph.D. dissert.]: Corvallis, Oregon State University, 270 p.

Schenck, H. G., 1927, Marine Oligocene of Oregon: California University Publications, Department of Geological Sciences Bulletin, v. 16, p. 449-460.

Schenck, H. G., and Kleinpell, R. M., 1936, Refugian Stage of the Pacific coast Tertiary: American Association of Petroleum Geologists Bulletin, v. 20, p. 215-225.

Schmidt, R. R., 1975, Upper Paleocene-middle Eocene planktonic biostratigraphy from the Great Valley of California and adjacent areas, and correlation to the West Coast microfaunal stages, *in* Paleogene symposium and selected Technical Papers: American Association of Petroleum Geologists-Society of Economic Paleontologists and Mineralogists-Society of Exploration Geophysicists, Pacific Sections, Annual Meeting, Long Beach, California, p. 439-455.

Steineck, P. L., and Gibson, J. M., 1971, Age and correlation of the Eocene Ulatisian and Narizian Stages, California: Geological Society of American Bulletin, v. 82, p. 477-480.

Stewart, R. E., 1956, Stratigraphic implications of some Cenozoic foraminifera from western Oregon: The Ore Bin, v. 18, no. 7, p. 57-63.

Tipton, A., 1975, Foraminiferal biostratigraphy of the late Eocene and early Oligocene type Bastendorff Formation, near Coos Bay, Oregon, *in* Paleogene symposium and selected technical papers: American Association of Petroleum Geologists-Society of Exploration Geophysicists, Pacific Sections, Annual Meeting, Long Beach, California, p. 563-585.

——1980, Foraminiferal zonation of the Refugian Stage, uppermost Eocene of California, *in* O. L. Bandy Memorial Volume: Cushman Foundation for Foraminiferal Research, Special Publication 18 (in press).

Warren, A. D., and Newell, J. H., 1976, Nannoplankton biostratigraphy of the upper Sacate and Gaviota Formations, Arroyo el Bulito, Santa Barbara County, California [abs.]: American Association of Petroleum Geologists-Society of Exploration Geophysicists-Society of Economic Paleontologists and Mineralogists Program, Pacific Sections, Annual Meeting, San Francisco, p. 19-20.

——1980, Plankton biostratigraphy of the Refugian and adjoining stages of the Pacific Coast Tertiary, *in* O. L. Bandy memorial volume: Cushman Foundation for Foraminiferal Research Special Publication 18, p. 233-251.

MANUSCRIPT RECEIVED BY THE SOCIETY MARCH 14, 1980
MANUSCRIPT ACCEPTED MAY 7, 1980

Printed in U.S.A.

Geological Society of America
Special Paper 184
1981

Marine diatom biostratigraphy of the Montesano Formation near Aberdeen, Washington

JOHN A. BARRON
U.S. Geological Survey, MS 15, 345 Middlefield Road, Menlo Park, California 94025

ABSTRACT

The upper part of the Montesano Formation near Aberdeen, Washington, contains early late Miocene marine diatoms of Schrader's North Pacific Diatom Zones XIII and XIV and Subzone b of combined Zones XV–XVI as modified by Barron. Silicoflagellates are referable to Bukry's *Distephanus pseudofibula* Zone and support the diatom correlations.

The type Empire Formation at Coos Bay, Oregon, and Diller's Empire Formation near Cape Blanco, Oregon, contain a somewhat younger (late late Miocene) diatom assemblage.

Diatoms from the uppermost part of the Astoria(?) Formation of Rau, which underlies the Montesano Formation along the Middle Fork of the Wishkah River, are indicative of the middle Miocene North Pacific Diatom Zone XXI. The middle Miocene/upper Miocene boundary lies either in the diatom-poor lower Montesano or at the unconformity with the underlying Astoria(?) Formation.

INTRODUCTION

In recent years marine diatoms have become valuable tools for Neogene correlation in the North Pacific area. Nearly continuous Neogene sections of diatomaceous sediments recovered by the Deep Sea Drilling Project (DSDP) have allowed the establishment of Neogene diatom biostratigraphic zonations that have widespread application.

Schrader (1973) defined 25 North Pacific diatom zones for lower Miocene through Holocene sediments recovered at DSDP site 173 off northern California. Schrader (1973) recognized his zones in DSDP holes as far north as the Gulf of Alaska, and Barron (1976a, 1976b) used them with slight modification in middle Miocene through lower Pliocene correlations of onshore stratigraphic sections in both northern and southern California.

Koizumi (1973, 1975) also established a marine diatom biostratigraphic zonation for the Neogene of the Bering Sea and western North Pacific. Koizumi's zones can be recognized in Neogene sediments throughout the North Pacific area, including Japan, and are especially useful for the Pliocene and Quaternary (Burckle and Opdyke, 1977). Schrader's (1973) zones for the middle Miocene and lower upper Miocene, however, are considerably more numerous and allow refinement of the biostratigraphy where they can be recognized.

Until now, neither of these diatom zonations has been applied to the scattered diatomaceous sediments in the Neogene of Oregon and Washington. Diatoms are reported in the Montesano and Quinault Formations of western Washington and in the Empire Formation and an isolated outcrop near the town of Bandon in southwestern Oregon (Orr, 1972; Orr and Conley, 1976). Studies by Orr (1972) and Orr and others (1971) on the diatoms from the Bandon locality predate the establishment of the diatom zonations and list only a few biostratigraphically diagnostic diatom species.

The Montesano Formation, which is exposed in the Grays Harbor area of western Washington (Fig. 1), contains rare to abundant diatoms according to Fowler (1965). The co-occurrence of mollusks and benthic foraminifers in the same stratigraphic section allows for correlation of diatom biostratigraphy with the provincial biostratigraphic frameworks used by molluscan and benthic foraminiferal workers. The benthic foraminifers of the Montesano Formation have been assigned to the Mohnian and Delmontian benthic foraminiferal stages of Kleinpell (1938; Fowler, 1965; Rau, 1967; Bergen and Bird, 1972). In international usage, the Mohnian is considered to be middle to upper Miocene, whereas the "Delmontian" interval is assigned to the lower Pliocene and possibly the uppermost Miocene (Berggren and Van Couvering, 1974; Barron, 1976c; Bukry and others, 1977). Molluscan biostratigraphy, which is updated and summarized by Addicott (1976), places the Montesano in the provincial upper Miocene of the West Coast. Fowler (1965) stated that the low diversity and the small size of most of the Montesano planktonic foraminifers make correlation difficult; they can at best set a lower limit of upper middle Miocene for the Montesano Formation.

MATERIALS AND METHODS

Twenty-two samples were collected from Fowler's (1965) type section of the Montesano Formation along the Middle Fork of the Wishkah River north of Aberdeen, Washington (Fig. 2). According to Fowler (1965), the Montesano Formation is best exposed and least tectonically disturbed in this section. Addicott's (1976) stratotype for his molluscan Graysian Stage was also established in this section.

The Montesano Formation at this locality is 759 m (2,490 ft) thick (Fowler, 1965; Fig. 3). It rests unconformably on the Astoria(?) Formation of Rau (1967) and is truncated at its top by an erosion surface overlain by the sands and gravels of the nonmarine Satsop Formation of Pleistocene age (Fowler, 1965).

Fowler (1965) divided the Montesano Formation in this area into two members and seven units. Along the Middle Fork of the Wishkah River, the lower member is 445 m (1,460 ft) thick and can be divided into five units (Fig. 3). In general, massive, sandy siltstone (units 2, 4) alternates with fine- to medium-grained sandstone (units 1, 3). Unit 5 is composed of very fine to fine-grained sandstone with minor thin interbeds of mudstone. The upper member contains massive mudstone (unit 6) overlain by poorly bedded sandy siltstone (unit 7; Fowler, 1965).

Fowler (1965) reported that diatoms are very rare in the lower five units, most abundant in unit 6, and rare in unit 7. Consequently, sampling was concentrated in the upper two units (Fig. 3).

In addition to the Montesano samples, one sample (USGS micropaleontology reference collection loc. R1499) was taken from the underlying Astoria(?) Formation of Rau (1967) approximately 10 m below the base of the Montesano Formation (Figs. 2, 3).

Approximately 5 g of sample was processed for siliceous microfossils using the procedure outlined in Barron (1976c). At least two strewn slides per sample were examined in entirety at 500× in the light microscope. In addition, several traverses at 1,250× were made per sample. Taxa were recorded as abundant, if two or more specimens occurred in one field of view at 500×; common, if one specimen was present in two fields of view; few, if one specimen was encountered during each vertical traverse of a coverslip; and rare, for scarcer occurrences.

MONTESANO BIOSTRATIGRAPHY

Table 1 gives the occurrence of the prominent and stratigraphically significant marine diatoms and silicoflagellates in the Montesano Formation samples. All marine diatom taxa observed in the Montesano Formation are listed in Appendix 1. Rare to few fresh-water diatom taxa were also encountered but they were not identified. In Plate 1 selected Montesano diatom taxa are illustrated.

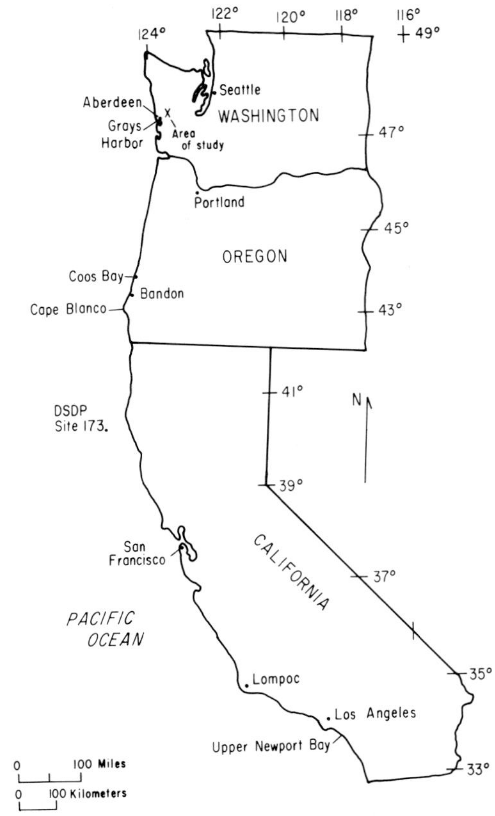

Figure 1. Sketch map of Washington, Oregon, and California, showing the Montesano Formation area of study near Grays Harbor, Washington, and other localities discussed in the text.

Abundance and preservation of diatoms and silicoflagellates are best in samples from USGS localities R1505, R1506, and R1507. Samples from USGS localities R1500, R1501, R1503, and R1521 are barren of marine diatoms and silicoflagellates.

Diatom Biostratigraphy

Figure 4 shows the ranges of the stratigraphically important marine diatoms and silicoflagellates of the Montesano Formation relative to the North Pacific diatom zonation of Schrader (1973) as modified by Barron (1976a). This chart was compiled from the data of Schrader (1973) at DSDP site 173 off northern California and the studies of Barron (1975, 1976a, 1976b) on onshore sections in California.

Coscinodiscus lineatus var. *leptopus* Grunow in the sample from USGS locality R1502 (Table 1) indicates that that sample is no older than Barron's (1976a) Subzone b of North Pacific Diatom Zones XV-XVI (Fig. 4). *Denticula dimorpha* Schrader in USGS locality R1505 and the absence of younger taxa imply that this sample is no younger than Subzone b of Zone XV-XVI (Figs. 4, 5). Barron (1976a) placed this subzone in the upper Miocene.

The interval from USGS localities R1506 through R1508 is correlated with the lower part of North Pacific Diatom

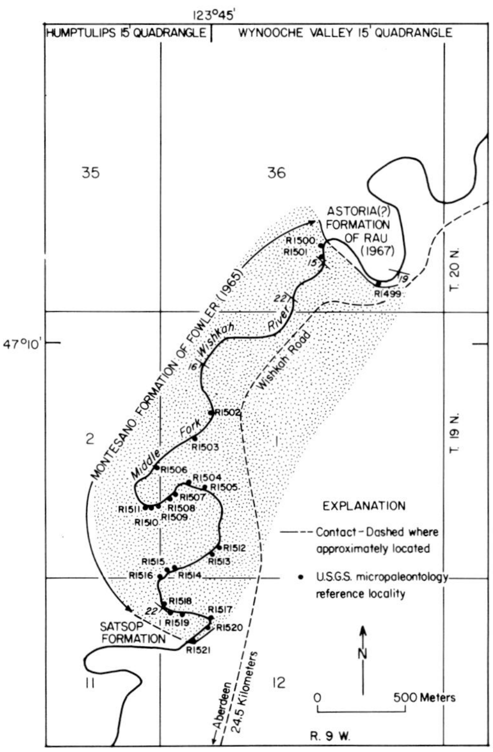

Figure 2. Map of the Middle Fork of the Wishkah River in the Humptulips and Wynoochee Valley 15-minute quadrangles, Washington, showing microfossil localities. Samples from USGS localities R1500 to R1521 are from the Montesano Formation. The contact with the Astoria(?) Formation of Rau (1967) is shown in the north, and the contact with the Satsop Formation is shown in the south. (Modified after Fowler, 1965).

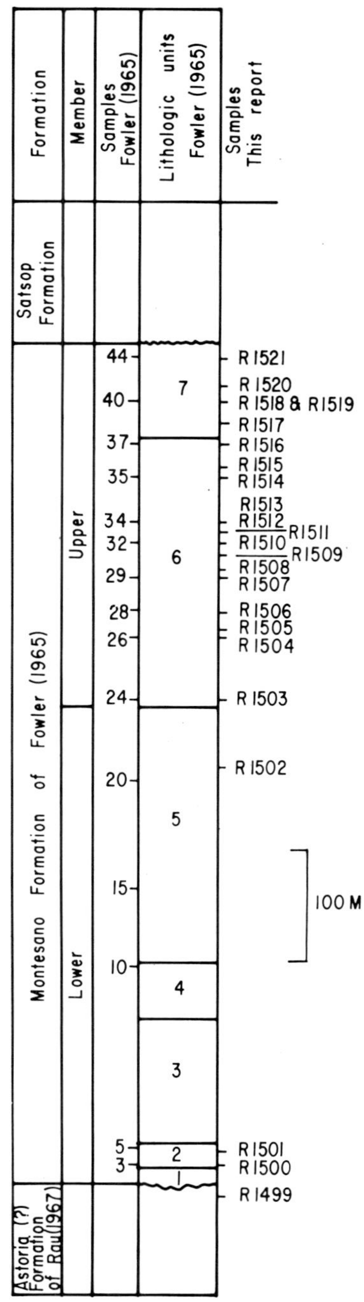

Figure 3. Stratigraphic section of the Montesano Formation (Fowler, 1965) along the Middle Fork of the Wishkah River showing the stratigraphic position of the samples studied.

Zones XIII-XIV (presumably Zone XIV) on the basis of the presence of *Thalassiosira lineata* Jouse in the sample from USGS locality R1506, of *Rhizosolenia miocenica* Schrader and *Coscinodiscus endoi* Kanaya in sample from USGS locality R1507, and of *Cosmiodiscus elegans* Greville in the sample from USGS locality R1508 (Figs. 4, 5). Diatom assemblages are poor above USGS locality R1508; however, rare *Stephanopyxis schenckii* Kanaya through USGS locality R1518 and the absence of younger species such as *Nitzschia fossilis* (Frenguelli) Kanaya and *Thalassiosira* sp. cf. *T. decipiens* (Grunow) Joergensen suggest that this interval is no younger than North Pacific Diatom Zone XIII (Fig. 4). Correlation of this interval (samples from USGS locs. R1509 to R1518) can at best be regarded as tentative.

The Montesano Formation samples from above and below USGS localities R1518 and R1502, respectively, lack diagnostic diatom species and cannot be zoned (Table 1; Fig. 5).

The North Pacific Diatom Zone XV/XIV boundary is stratigraphically just below the top of Koizumi's (1973, 1975) *Denticula hustedtii-Denticula lauta* Zone (Barron, 1976a). On the basis of radiometric dates in Japan, Koizumi (1977) assigned an absolute age of 9.5 m.y. to this boundary.

Silicoflagellate Biostratigraphy

Many of the Montesano samples also contain silicoflagellates (Table 1). The presence of *Distephanus pseudofibula* (Schulz) Bukry in samples from USGS localities

TABLE 1. OCCURRENCE CHART OF SELECTED DIATOMS AND SILICOFLAGELLATES FROM THE MONTESANO FORMATION

	Stratigraphic interval (m) above contact with Astoria(?) Formation																				
	17	28	375	437	494	500	515	547	552	568	578	583	587	595	635	646	667	683	703	703	720	742
	Sample numbers*																					
	R1500	R1501	R1502	R1503	R1504	R1505	R1506	R1507	R1508	R1509	R1510	R1511	R1512	R1513	R1514	R1515	R1516	R1517	R1518	R1519	R1520	R1521
Diatoms																						
<u>Actinocyclus</u> <u>ingens</u>	-	-	-	-	R	R	R	R	R	R	R	-	-	R	R	R	-	-	R	-	-	-
<u>Actinoptychus</u> <u>undulatus</u>	-	-	-	-	-	R	R	R	R	-	R	-	R	R	-	-	-	-	-	-	-	-
<u>Coscinodiscus</u> <u>endoi</u>	-	-	-	-	-	R	R	R	-	-	-	-	-	-	-	-	-	-	-	-	-	-
<u>C</u>. <u>lineatus</u> var. <u>leptopus</u>	-	-	R	-	-	R	R	R	-	-	-	-	-	-	R	-	-	-	-	-	-	-
<u>C</u>. <u>marginatus</u>	-	-	-	-	R	R	R	R	-	R	R	-	R	R	-	R	-	-	-	-	R	-
<u>C</u>. <u>temperei</u>	-	-	-	-	-	-	-	-	-	-	-	-	-	R	-	-	-	-	-	-	-	-
<u>Cosmiodiscus</u> <u>elegans</u>	-	-	-	-	-	R	-	R	R	-	-	-	-	-	-	-	-	-	-	-	-	-
<u>Denticula</u> <u>dimorpha</u>	-	-	-	-	-	R	-	-	-	-	-	-	-	-	-	-	-	-	-	-	-	-
<u>D</u>. <u>hustedtii</u>	-	-	-	-	R	F	R	F	-	R	R	R	R	R	-	R	R	-	R	R	-	-
<u>Melosira</u> <u>sulcata</u>	-	-	R	-	R	R	R	F	-	R	R	R	R	R	R	R	R	R	R	R	R	-
<u>Rhaphoneis</u> <u>amphiceros</u> var. <u>gemmifera</u>	-	-	-	-	-	R	R	R	-	-	R	-	-	R	-	R	R	-	-	-	-	-
<u>Rhizosolenia</u> <u>barboi</u>	-	-	-	-	-	R	R	R	-	-	-	R	-	R	-	R	-	-	-	-	-	-
<u>R</u>. <u>miocenica</u>	-	-	-	-	-	-	-	R	-	-	-	-	-	-	-	-	-	-	-	-	-	-
<u>Rouxia</u> <u>fusiformis</u>	-	-	-	-	-	R	-	-	-	-	-	-	-	-	-	-	-	-	-	-	-	-
<u>Stephanopyxis</u> <u>schenckii</u>	-	-	-	-	-	-	R	-	-	R	-	-	-	R	R	R	-	R	R	-	-	-
<u>S</u>. <u>turris</u>	-	-	-	-	-	-	R	R	R	-	R	-	-	-	-	R	-	R	R	-	-	-
<u>Thalassionema</u> <u>nitzschioides</u>	-	-	R	-	R	A	F	C	R	R	R	R	R	F	R	F	R	R	R	-	R	-
<u>Thalassiosira</u> <u>lineata</u>	-	-	-	-	-	-	R	-	-	-	-	-	-	-	-	-	-	-	R	-	-	-
<u>T</u>. sp. 1	-	-	-	-	-	F	-	-	-	-	-	-	-	R	-	-	-	-	-	-	-	-
Silicoflagellates																						
<u>Distephanus</u> <u>pseudocrux</u>	-	-	-	-	-	R	-	-	-	-	-	-	-	-	-	-	-	-	-	-	-	-
<u>D</u>. <u>pseudofibula</u>	-	-	-	-	-	F	R	R	R	R	R	-	-	R	-	-	R	-	-	-	-	-
<u>Mesocena</u> <u>hexagona</u>	-	-	-	-	-	R	-	-	-	-	-	-	-	-	-	-	-	-	-	-	-	-

<u>Note</u>: R, rare; F, few; C, common; A, abundant.

*Sample numbers refer to USGS micropaleontology reference collection localities.

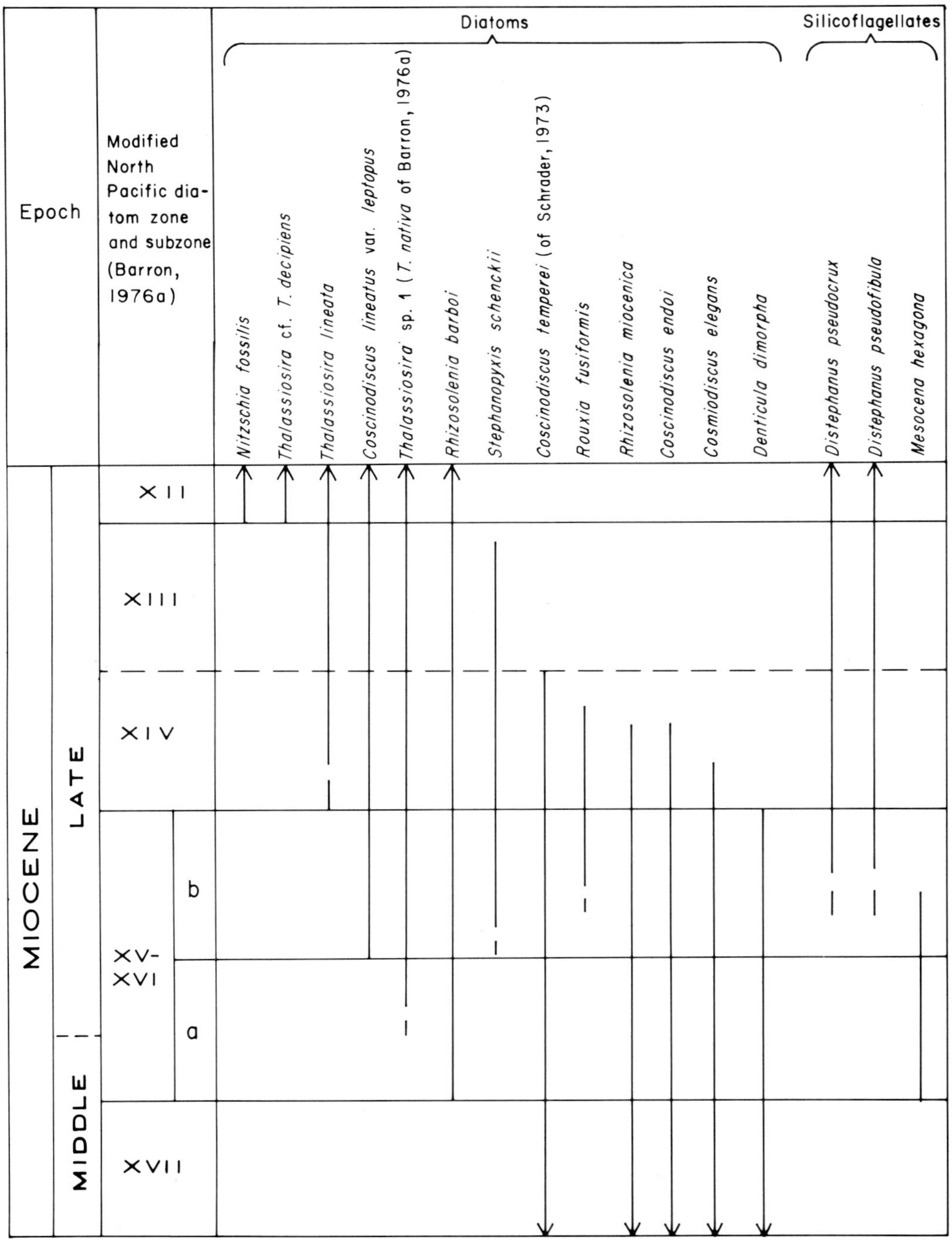

Figure 4. Ranges of selected diatoms and silicoflagellates relative to the North Pacific diatom zones as modified by Barron (1976a). Compiled from Schrader (1973) and Barron (1975, 1976a, 1976c).

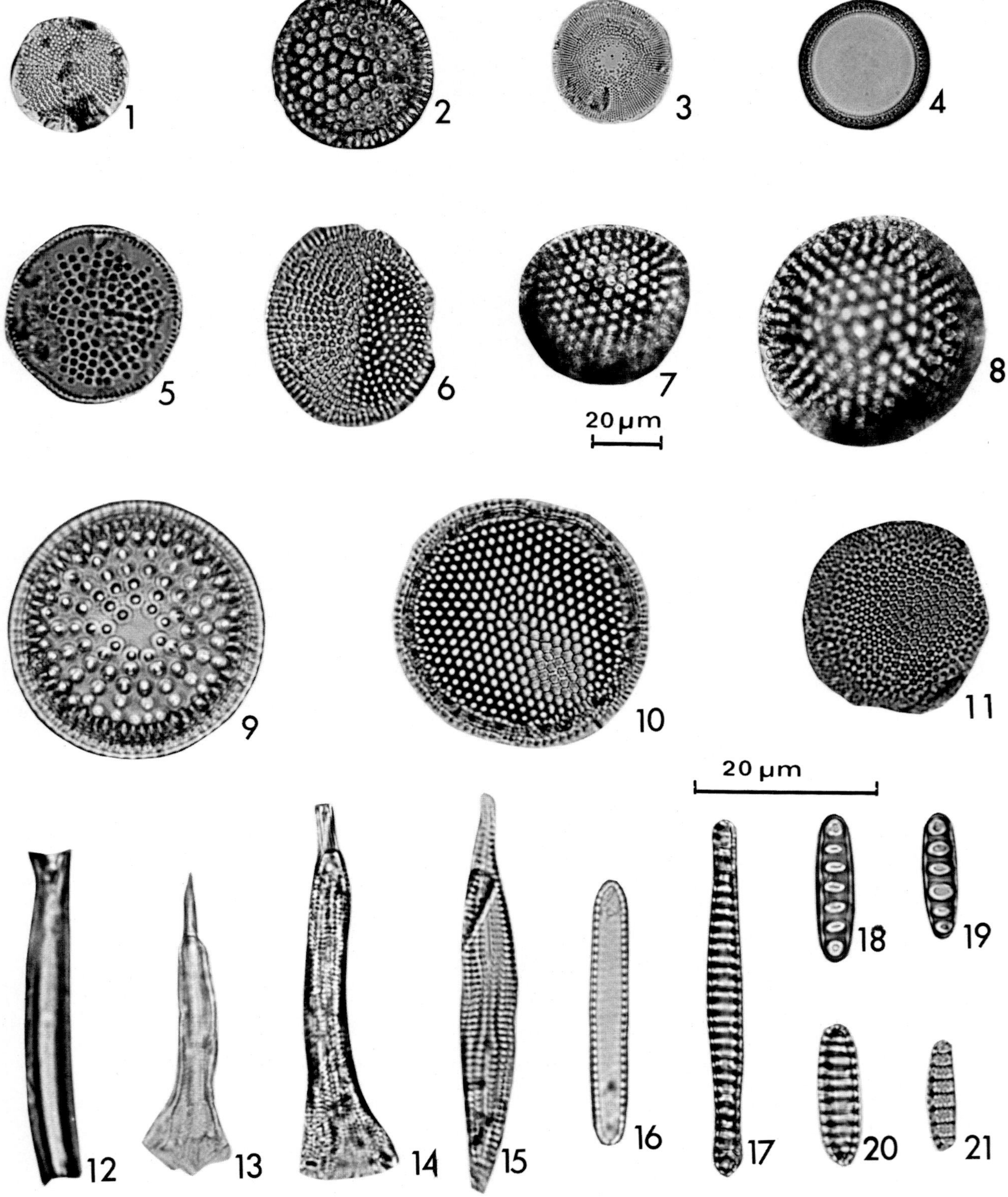
1
2
3
4
5
6
7
20 μm
8
9
10
11
20 μm
12
13
14
15
16
17
18
19
20
21

PLATE 1 (facing page)

Montesano Formation diatoms. Scale bars represent 20 μm (figs. 1-4, 7, 8, small scale bar; others, large scale bar)

Figure

1, 5. ***Coscinodiscus endoi*** **Kanaya. Montesano Formation, sample R 1505.**

2. ***Coscinodiscus marginatus*** **Ehrenberg. Montesano Formation, sample R1505.**

3. ***Cosmiodiscus elegans*** **Greville. Montesano Formation, sample R1505.**

4. ***Melosira sulcata*** **(Ehrenberg) Kutzing. Montesano Formation, sample R1505.**

6. ***Coscinodiscus temperei*** **Brun. Montesano Formation, sample R1513.**

7, 8. ***Stephanopyxis schenckii*** **Kanaya. Montesano Formation, sample R1513.**

9. ***Actinocyclus ingens*** **Rattray. Montesano Formation, sample R1505.**

10. ***Coscinodiscus lineatus*** **var.** ***leptopus*** **Grunow. Montesano Formation, sample R1505.**

11. ***Thalassiosira*** **sp. 1 (*T. nativa* of Barron, 1976a). Montesano Formation, sample R1505.**

12. ***Rhizosolenia barboi*** **Schrader. Montesano Formation, sample R1506.**

13, 14. ***Rhizosolenia miocenia*** **Schrader. Montesano Formation, sample R1507.**

15. ***Rouxia fusiformis*** **Tsumura. Montesano Formation, sample R1505.**

16. ***Thalassionema nitzschioides*** **Grunow. Montesano Formation, sample R1505.**

17, 20. ***Denticula hustedtii*** **Simonsen and Kanaya. Montesano Formation, sample R1505.**

18, 19. ***Denticula dimorpha*** **Schrader. Montesano Formation, sample R1505.**

21. ***Denticula*** **sp. Montesano Formation, sample R1505.**

R1505 to R1516 indicates Bukry's (1973) upper Miocene *Distephanus pseudofibula* Zone. *Distephanus pseudocrux* (Schulz) Bukry in sample from USGS locality R1505 is supportive, and *Mesocena hexagona* Haeckel in that sample suggests equivalence with the lower part of that zone or the underlying *Distephanus longispinus* Zone of Bukry (1973) (Barron, 1976c). Barron (1976a) correlated the *D. pseudofibula* Zone with the interval from Subzone b of North Pacific Diatom Zones (XV-XVI) to Zone XI. Thus, silicoflagellate biostratigraphy agrees with the diatom correlations.

Molluscan and Benthic Foraminferal Biostratigraphy

Addicott (1976) designated the upper two-thirds of the Montesano Formation along the Middle Fork of the Wishkah River the stratotype for his upper Miocene Graysian (molluscan) Stage. He assigned the lower third to his upper Miocene Wishkahan Stage (Fig. 5).

Fowler (1965) correlated the benthic foraminifers from the uppermost 27 m (300 ft) of the Montesano Formation along the Middle Fork of the Wishkah River to the *Rotorbinella*(?) *garveyensis* Zone of Kleinpell's (1938) Delmontian Stage (Fig. 5). He placed the next stratigraphically lower 27 m (300 ft) in the *Bolivina obliqua* Zone of the Delmontian Stage and the remainder of the Montesano Formation into the Mohnian Stage.

Rau (1967) agreed with Fowler and assigned the Montesano Formation at this locality to the Mohnian and Delmontian benthic foraminiferal stages (Fig. 5).

Bergen and Bird (1972) examined benthic foraminifers from the subsurface Montesano Formation near Ocean City, Washington, approximately 29 km (18 mi) west of the area of this study. They assigned these assemblages to the provincial upper Miocene of the West Coast and stated that the lack of most of the key species prevented assignment to the California stages and zones of Kleinpell (1938).

At Upper Newport Bay and Lompoc in southern California, North Pacific Diatom Zones XIII and XIV contain upper Mohnian benthic foraminifers (Barron, 1976a). The type *Bolivina obliqua* Zone of the Delmontian Stage in southern California contains substantially younger diatoms correlative with North Pacific Diatom Zones X and IX and is probably uppermost Miocene to lower Pliocene (Barron, 1976c). The correspondence of Fowler's (1965) *Bolivina obliqua* Zone with upper Miocene North Pacific Diatom Zones XIII-XIV diatoms in the upper Montesano (Fig. 5), therefore, suggests a discrepancy. The diatom correlation of this interval is tentative; however, Bukry and others (1977) pointed out that many of the benthic foraminiferal stages and zones of Kleinpell (1938) appear to be time-transgressive. Perhaps diatom studies in other sections of the Montesano Formation will resolve this problem.

Montesano Correlations

Addicott (1976) correlated molluscan assemblages in the Montesano Formation with assemblages in the upper part of Diller's (1903) Empire Formation at Cape Blanco, Oregon, and the type Empire Formation at Coos Bay, Oregon.

A sample from the upper part of Diller's (1903) Empire Formation was provided for this study by Warren Addicott. This sample was collected at USGS locality M3755 from the SW¼SW¼ of sec. 1, T. 32 S., R. 16 W., Cape Blanco 15-minute quadrangle, Curry County, Oregon. The diatom assemblage of this sample correlates with North Pacific Diatom Zone XI and is younger than the assemblage from the upper member of the Montesano Formation along the Middle Fork of the Wishkah River (Fig. 5). Stratigraphic-

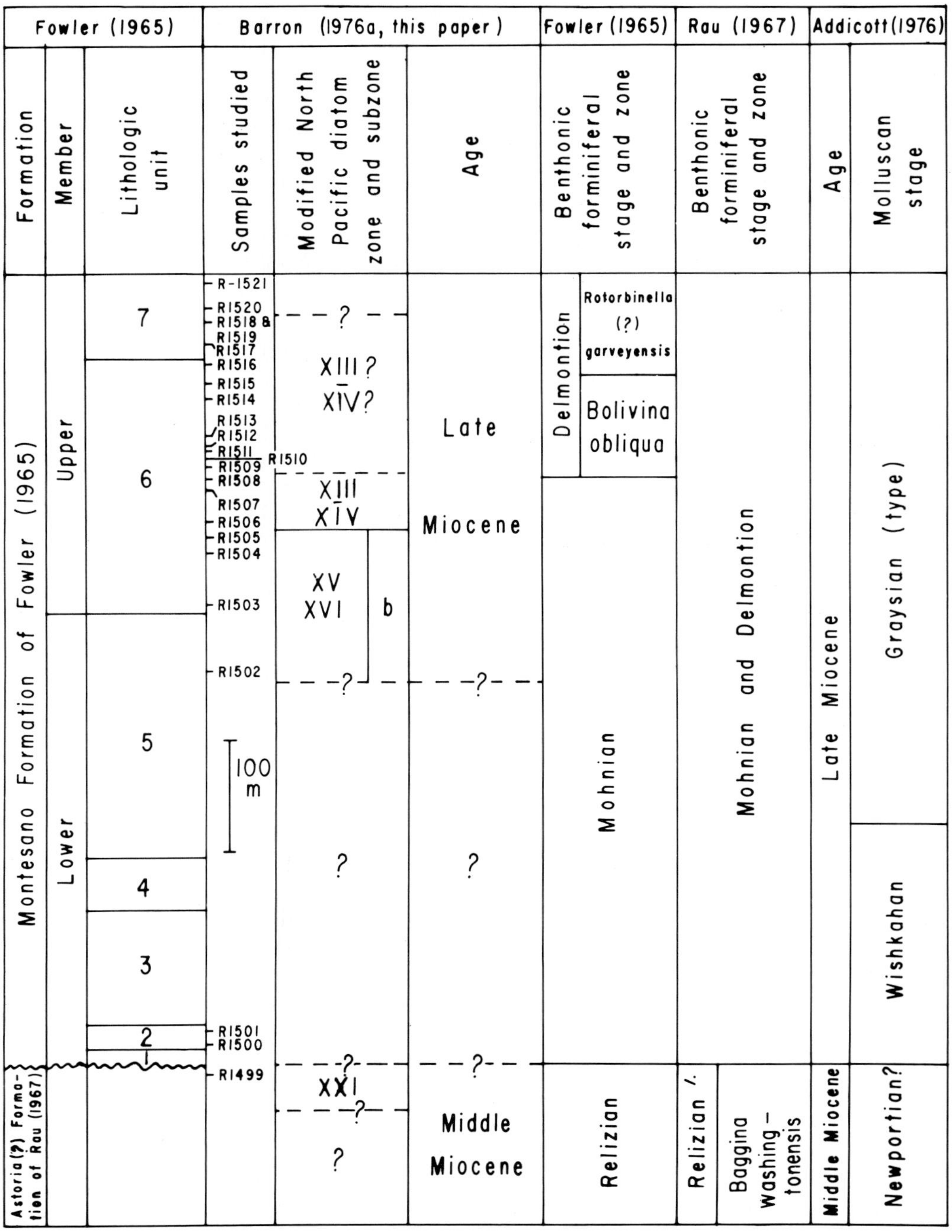

Figure 5. Stratigraphic section along the Middle Fork of the Wishkah River zoned by the modified North Pacific diatom zones of Barron (1976a), benthic foraminifers (Fowler, 1965; Rau, 1967), and mollusks (Addicott, 1976). Relizian of Rau (1967) possibly includes upper part of Saucesian Stage and lower part of Luisian Stage.

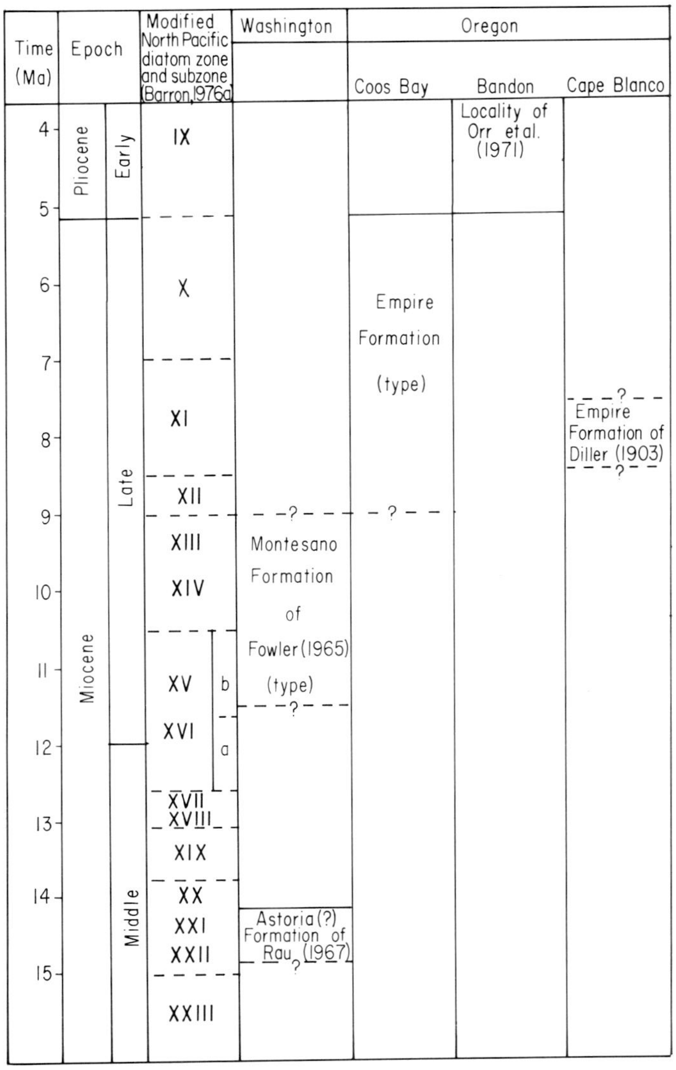

Figure 6. Correlation of the formations discussed with the modified North Pacific diatom zones of Barron (1976a) and with the time scale of Ryan and others (1974). Correlation of diatom zones to time scale after Barron (1976a) with modifications suggested by Burckle and Opdyke (1977) and Burckle (1978).

ally significant diatoms in this sample include *Actinoptychus minutus, A. Vulgaris* var. *monicae* Grunow, *Coscinodiscus lineatus* var. *leptopus, Denticula hustedtii, Stephanopyxis schenckii, Thalassiosira lineata,* and *T. antiqua.* The diatom correlation is supported by the presence of the silicoflagellates *Distephanus pseudocrux* and *D. pseudofibula.*

Samples from the type section of the Empire Formation at Coos Bay, Oregon, were provided for study by John Armentrout. Although two samples from the lower 130 m of section were barren of diatoms, the upper 309 m contained diatoms that are assigned to the upper upper Miocene North Pacific Diatom Zones XII to Zone X (Fig. 6). Stratigraphically significant diatoms observed in this sequence include *Actinocyclus ingens, Coscinodiscus symbolophorus* Grunow, *Denticula hustedtii* Simonsen and Kanaya, *D.* cf. *D. kamtschatica* Sabelina, *Hemiaulus polymorphus* Grunow, *Lithodesmium asketogonium* Barron, *N.* cf. *N. rolandii* Schrader, *Rouxia californica* Peragallo, *Thalassiosira antiqua* (Grunow) Cleve, *T. burckliana* Schrader, *T.* cf. *T. decipiens* (of Barron, 1976a), and *T.* sp. 1 (of this paper).

Addicott (1976) assigned the type Empire Formation and Diller's (1903) Empire Formation to his Wishkahan Stage. The type section of his Graysian Stage, which overlies the Wishkahan Stage along the Middle Fork of the Wishkah River, contains early late Miocene diatoms (Subzone b of North Pacific Diatom Zones XV-XVI to North Pacific Diatom Zone XIII; Fig. 6). Late late Miocene diatoms in Diller's (1903) Empire Formation (North Pacific Diatom Zone XI) and the type section of the Empire Formation (North Pacific Diatom Zone XII to X) suggest that the mollusks should correlate with the Graysian Stage or younger (Figs. 5, 6). This problem bears closer examination, and I plan a more detailed study of the diatoms of the type Empire Formation in the near future.

An isolated exposure of diatomaceous siltstone on the Oregon coast near Bandon was interpreted to be upper Miocene by Orr and others (1971). A sample from this locality was provided by Warren Addicott. This sample is from USGS locality M3631, Bandon State Park, north side of lagoon at the mouth of China Creek 270 m (900 ft) north of southeast corner sec. 11, T. 29 S., R. 15 W., Bandon 15-minute quadrangle. The diatoms in this sample are correlative with early Pliocene North Pacific Diatom Zone IX and are substantially younger than the Montesano assemblage. The diatoms include *Thalassiosira oestrupii* (Ostenfeld) Proshkina-Lavrenko, *T. antiqua* (Grunow) Cleve-Euler, *T. lineata* Jouse, *T. kryophilia* (of Barron, 1975), and *Lithodesmium cornigerum* Grunow. The presence of the silicoflagellate *Distephanus boliviensis jimlingii* Bukry supports this correlation (Barron, 1976a).

ASTORIA(?) FORMATION

On the Middle Fork of the Wishkah River, the Montesano rests unconformably on sandstone and interbedded mudstone of the Astoria(?) Formation of Rau (1967). Fowler (1965) reported that benthic foraminifers from the Astoria(?) just below the base of the Montesano at this locality correlated with faunas from Kleinpell's (1938) Relizian Stage in California. Rau (1967) assigned benthic foraminifers from this locality to his *Baggina washingtonensis* Zone, which he says represents the Relizian Stage and possibly the upper part of the underlying Saucesian Stage and a lower part of the overlying Luisian Stage (Fig. 5).

Fowler (1965) also studied the planktonic foraminifers from this locality and reported *Globorotalia scitula praescitula* Blow and *Globoquadrina tripartita tripartita* (Koch). He assigned this assemblage to the lowermost Aquitanian Stage of the lower Miocene.

On the basis of the benthic foraminiferal correlations,

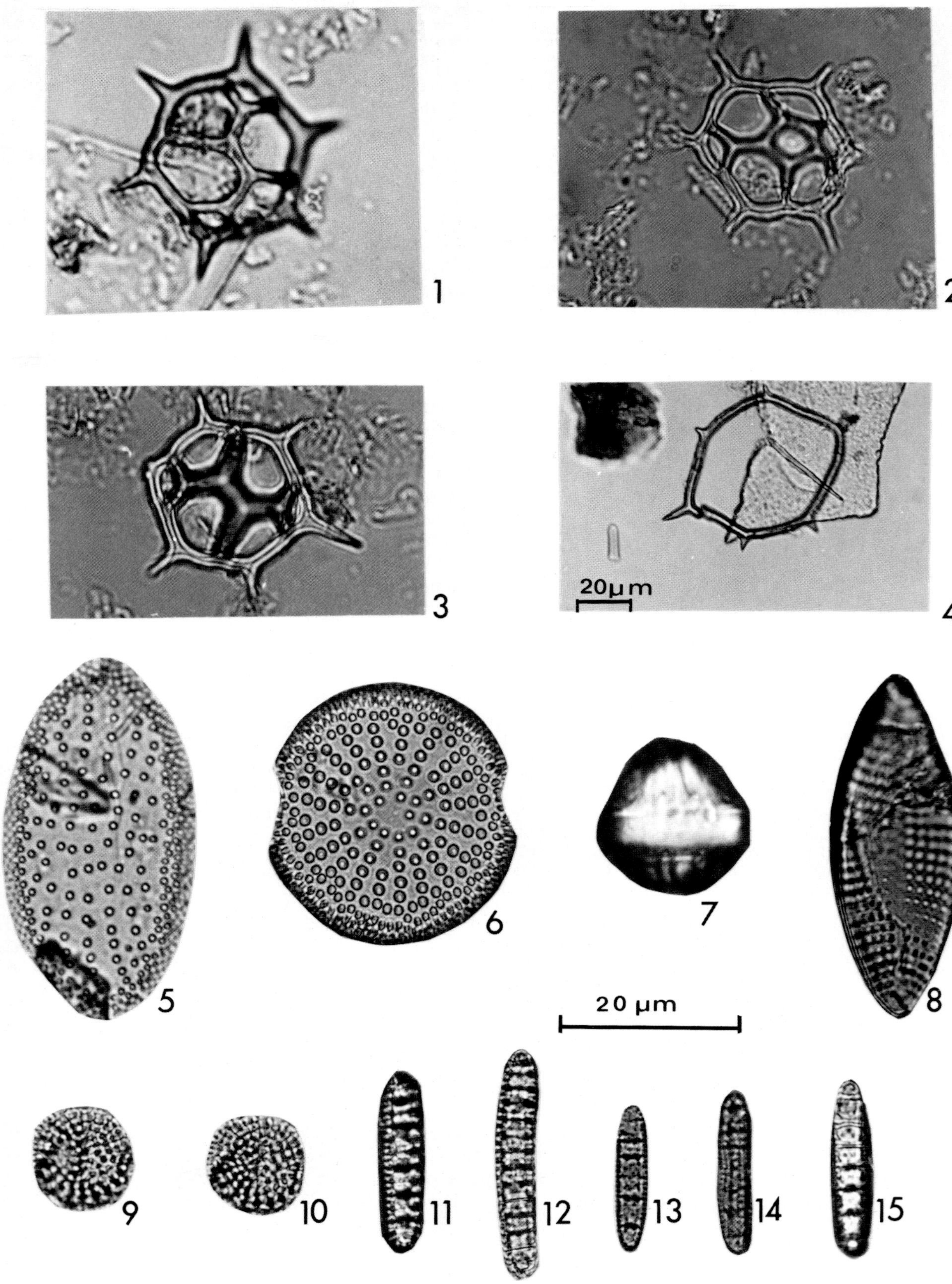
1
2
3
20µm
4
5
6
7
8
20 µm
9
10
11
12
13
14
15

PLATE 2 (facing page)

Montesano Formation silicoflagellates (figs. 1-4) and Astoria(?) Formation diatoms (figs. 5-15). Scale bar represents 20 μm (fig. 4, small scale bar; others, large scale bar).

Figure

1, 3. ***Distephanus pseudofibula*** **(Schulz) Bukry. Montesano Formation, sample R1505.**

2. ***Distephanus pseudocrux*** **(Schulz) Bukry. Montesano Formation, sample R1505.**

4. ***Mesocena hexagona*** **Haeckel. Montesano Formation, sample R1505.**

5. ***Coscinodiscus lewisianus*** **Greville. Astoria(?) Formation, sample R1499.**

6. ***Actinocyclus ingens*** **Rattray. Astoria(?) Formation, sample R1499.**

7. ***Cladogramma dubium*** **Lohman. Astoria(?) Formation, sample R1499.**

8. ***Rhaphoneis elliptica*** **Schrader. Astoria(?) Formation, sample R1499.**

9, 10. ***Coscinodiscus praeyabei*** **Schrader (focus change). Astoria(?) Formation, sample R1499.**

11, 12. ***Denticula hustedtii*** **Simonsen and Kanaya. Astoria(?) Formation, sample R1499.**

13, 14. ***Denticula nicobarica*** **Grunow. Astoria(?) Formation, sample R1499.**

15. ***Denticula punctata*** **Schrader. Astoria(?) Formation, sample R1499.**

Addicott (1976) assigned the Astoria(?) Formation to his middle Miocene Newportian (molluscan) Stage (Fig. 5).

The sample from USGS locality R1499 was collected from the Astoria(?) Formation (Fig. 2) about 10 m below the base of the Montesano Formation on the Middle Fork of the Wishkah River (Fig. 3). Diatoms are relatively rare; however, examination of four microscope slides revealed the following stratigraphically important diatom species: *Actinocyclus ingens* Rattray, *Brunia mirabilis* (Brun) Tempere, *Cladogramma dubium* Lohman, *Coscinodiscus lewisianus* Greville, *C. praeyabei* Schrader, *Denticula hustedtii* Simonsen and Kanaya, *D. lauta* Bailey, *D. nicobarica* Grunow, *D. punctata* Schrader, *Eupodiscus californicus* Grunow, *Rhaphoneis elliptica* Schrader, and *R. parilis* Hanna (Pl. 2).

In the eastern North Pacific, this assemblage is indicative of middle Miocene North Pacific Diatom Zone XXI (Schrader, 1973; Barron, 1976a). At Upper Newport Bay, Zone XXI contains benthic foraminifers assigned to the upper Luisian Stage by Warren (1972) and planktonic foraminifers assigned to Zones N10 to N12 of Banner and Blow (1965) by Lipps (1964, 1967) (Barron, 1967b). My studies in the offshore area of southern California and at DSDP site 173 are supportive of these correlations.

R. Z. Poore (1977, personal commun.) reported that the sample from USGS locality R1499 contains poorly preserved and nondiagnostic planktonic foraminifers. Fowler's (1965) report of the early Miocene *Globoquadrina tripartita tripartita* in the uppermost Astoria(?) Formation at this locality may represent reworking.

CONCLUSIONS

The upper part of the Montesano Formation along the Middle Fork of the Wishkah River in western Washington contains early late Miocene marine diatoms correlative with Subzone b of North Pacific Diatom Zones XVI–XV to Zone XIII. Silicoflagellates in this interval are assigned to Bukry's (1973) *Distephanus pseudofibula* Zone and support the diatom correlations.

Although mollusks suggest that the upper Montesano Formation correlates with Diller's (1903) Empire Formation at Cape Blanco, Oregon, and with the type Empire Formation at Coos Bay, Oregon, the late late Miocene diatoms of these formations are somewhat younger than the diatoms from the Montesano.

The Astoria(?) Formation of Rau (1967), which underlies the Montesano Formation along the Middle Fork of the Wishkah River, contains middle Miocene diatoms correlative with North Pacific Diatom Zone XXI near its top.

The middle Miocene/upper Miocene boundary, therefore, must be in the poorly diatomaceous lower Montesano Formation or must correspond with the unconformity at the top of the Astoria(?) Formation.

Addendum

Subsequent to the completion of this manuscript in 1977, I proposed (Barron, 1980) a modified version of the diatom zonation of Koizumi (1973) for the northeastern Pacific area. Figure 7 shows a correlation of Pacific Northwest diatom localities. Assignment of the type section of the Empire Formation is modified slightly following Barron and Armentrout (1980). The paleomagnetic epochs are after Mankinen and Dalrymple (1979).

ACKNOWLEDGMENTS

I thank Warren O. Addicott, Richard Z. Poore, and Kristin McDougall of the U.S. Geological Survey, Menlo Park, for their helpful discussions. I also thank Lloyd H. Burckle of Lamont-Doherty Geological Observatory, Palisades, New York, and Jonathan Rider of Mobil Oil Exploration and Producing Services, Inc., Dallas, Texas, for reviewing the manuscript.

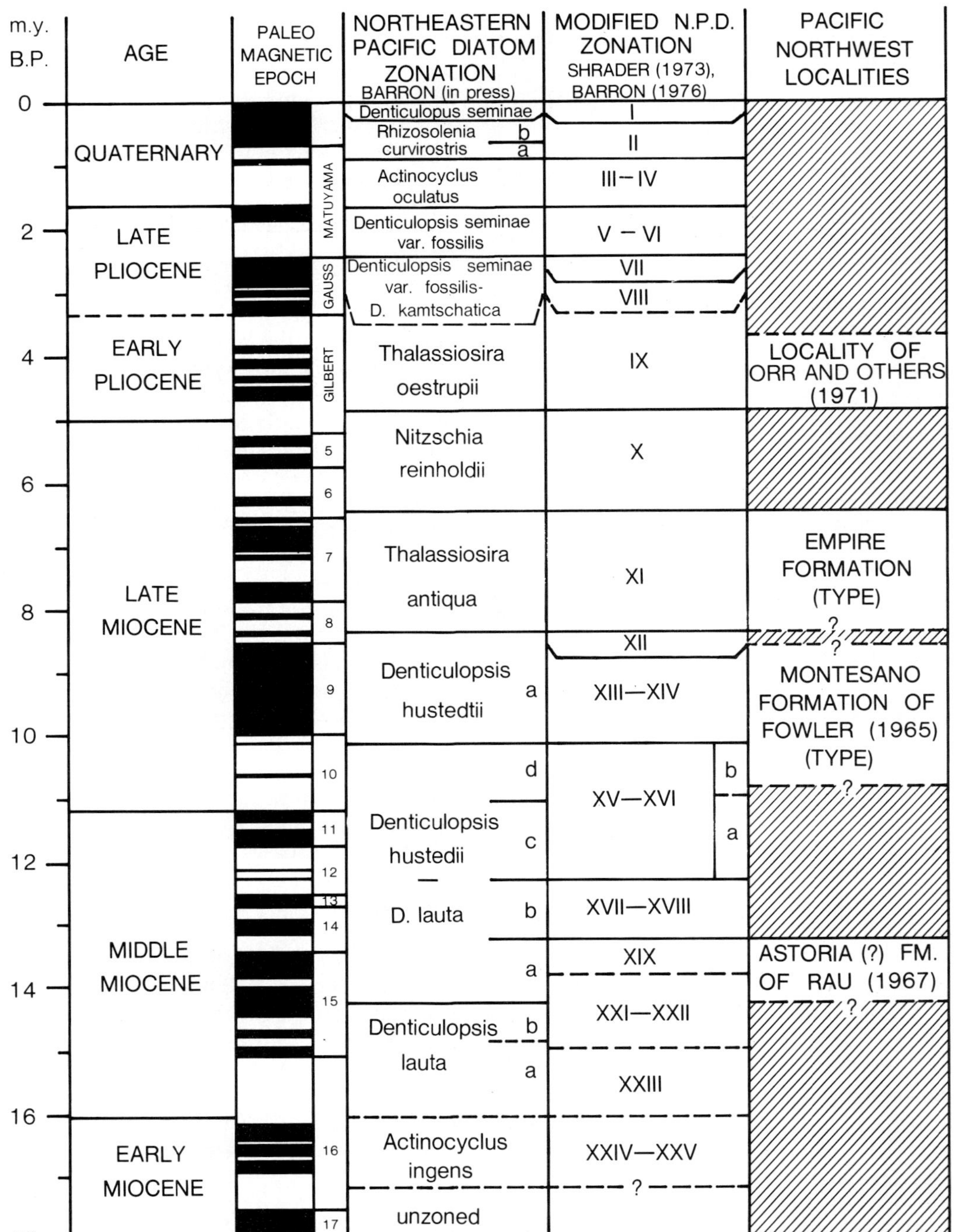

Figure 7. Correlation of Pacific Northwest diatom localities.

APPENDIX 1. MARINE DIATOMS RECORDED FROM THE MONTESANO FORMATION ALONG THE MIDDLE FORK OF THE WISHKAH RIVER

Actinocyclus divisus (Grunow) Hustedt

A. sp. cf. A. divisus (Grunow) Hustedt

A. ehrenbergii Ralfs

A. ellipticus var. moronensis (Deby) Kolbe

A. ingens Rattray

Actinoptychus minutus Greville

A. splendens var. halionyx Grunow

A. undulatus (Bailey) Ralfs

Asteromphalus darwinii Ehrenberg

Aulacodiscus concentricus (Mann) Boyer

Chaetoceros dîcladia Castracane

Cladogramma dubium Lohman

Cocconeis costata Gregory

C. sp. cf. C. distans Gregory

C. scutellum Ehrenberg

Coscinodiscus argus Ehrenberg

C. curvatulus Grunow

C. endoi Kanaya

C. gigas var. diorama (Schmidt) Grunow

C. lineatus var. leptopus Grunow[1]

C. marginatus Ehrenberg

C. monicae (Grunow) Rattray

C. sp. cf. C. pacificus (Grunow) Rattray

C. plicatus Grunow s. ampl.

C. obscurus Schmidt

C. radiatus Ehrenberg

C. stellaris Roper

C. temperei Brun

C. sp. cf. C. vetustissimus Pantocsek

Cosmiodiscus elegans Greville

Cyclotella sp. cf. C. stylorum Brightwell

Denticula dimorpha Schrader

D. hustedtii Simonsen and Kanaya

Diploneis smithii (Brebisson) Cleve

Endictya sp.

Eupodiscus californicus Grunow

Goniothecium rogersii Ehrenberg

Grammatophora angulosa Ehrenberg

G. merletta Hanna and Grant

Hemiaulus polymorphus Grunow

Hemidiscus cuneiformis Wallich

Hyalodiscus sp.

Lithodesmium minisculum Grunow

Melosira sulcata (Ehrenberg) Kützing

Navicula lyra Ehrenberg

N. optima Hanna

Nitzschia sp. cf. N. rolandii Schrader

Rhabdonema japonicum var. sparsicostatum Tempere and Brun

Rhaphoneis amphiceros var. gemmifera Peragallo

R. sachalinensis Sheshukova-Poretzkaya

Rhizosolenia barboi Schrader

R. miocenica Schrader

R. styliformis Brightwell

R. fusiformis Tsumura

R. peragalli Brun and Heribaud

Stephanopyxis schenckii Kanaya

S. turris (Greville and Arnott) Ralfs

S. sp. [= S. lineata of Barron, 1975, non S. lineata (Ehrenberg) Forti]

Synedra jouseana Sheshukova-Poretzkaya

Thalassionema nitzschioides Grunow

Thalassiosira lineata Jouse

T. sp. 1 (= T. nativa of Barron, 1976a, non T. nativa Sheshukova-Poretzkaya)

Thalassiothrix longissima Cleve and Grunow

Triceratium condecorum Brightwell

---

[1]Light microscope studies indicate that C. lineatus var. leptopus probably should be reassigned to the genus Thalassiosira.

REFERENCES CITED

Addicott, W. O., 1976, Neogene molluscan stages of Oregon and Washington, *in* The Neogene symposium: Society of Economic Paleontologists and Mineralogists, Pacific Section, San Francisco, California, April 1976, p. 95-115, Pls. 1-5.

Banner, F. T., and Blow, W. H., 1965, Progress in planktonic foraminiferal biostratigraphy of the Neogene: Nature, v. 208, p. 1164-1166.

Barron, J. A., 1975, Late Miocene-early Pliocene marine diatoms from southern California: Palaeontographica, v. 151, pt. B, p. 97-170, Pls. 1-15.

——1976a, Revised Miocene and Pliocene diatom biostratigraphy of Upper Newport Bay, Newport Beach, California: Marine Micropaleontology, v. 1, p. 27-63.

——1976b, Middle Miocene-lower Pliocene marine diatom and silicoflagellate correlations in the California area, *in* The Neogene symposium: Society of Economic Paleontologists and Mineralogists, Pacific Section, San Francisco, California, April 1976, p. 117-124.

——1976c, Marine diatom and silicoflagellate biostratigraphy of the type Delmontian Stage and the type *Bolivina obliqua* Zone: U.S. Geological Survey Journal of Research, v. 4, p. 339-351.

——1980 Late Cenozoic diatom biostratigraphy and paleoceanography of the middle-latitude eastern North Pacific, DSDP Leg 63, *in* Initial reports of the Deep Sea Drilling Project, Volume 63: Washington, D.C., U.S. Government Printing Office (in press).

Barron, J. A., and Armentrout, J. M., 1980, Late Miocene diatom florules of the Empire Formation, Coos Bay, Oregon: Geological Society of America Abstracts with Programs, v. 12, no. 3, p. 95-96.

Bergen, F. W., and Bird, K. J., 1972, The biostratigraphy of the Ocean City area, Washington, *in* Symposium on Miocene biostratigraphy of California: Society of Economic Paleontologists and Mineralogists, Pacific Section, Bakersfield, California, March 1972, p. 173-191.

Berggren, W. A., and Van Couvering, J. A., 1974, The late Neogene: Biostratigraphy, geochronology and paleoclimatology of the last 15 million years in marine and continental sequences: Palaeogeography, Palaeoclimatology, Palaeoecology, v. 16, p. 1-216.

Bukry, David, 1973, Coccolith and silicoflagellate stratigraphy, Deep Sea Drilling Project Leg 18, eastern North Pacific, *in* Initial reports of the Deep Sea Drilling Project, Volume 18: Washington, D.C., U.S. Government Printing Office, p. 817-831.

Bukry, David, Brabb, E. E., and Vedder, J. G., 1977, Correlation of Tertiary calcareous nannoplankton assemblages from the Coast and Peninsular Ranges of California: Latin American Geological Congress, 2nd, Caracas, 1973, Venezuela, Ministerio Minas Hidrocarburos, Boletin de Geologia, Publicacion Especial 7, p. 1461-1483.

Burckle, L. H., 1978, Early Miocene to Pliocene diatom datum levels for the equatorial Pacific: Geological Research and Development Centre, Republic of Indonesia, Special Publication no. 1, p. 25-44.

Burckle, L. H., and Opdyke, N. D., 1977, Late Neogene diatom correlations in the Circum-Pacific: International Congress on Pacific Neogene Stratigraphy, 1st, Tokyo, Japan, May 1976, Proceedings.

Diller, J. S., 1903, Description of the Port Orford quadrangle, Oregon: U.S. Geological Survey Geologic Atlas, Folio 89, 6 p.

Fowler, G. A., 1965, The stratigraphy, foraminifera, and paleoecology of the Montesano Formation, Grays Harbor County, Washington [Ph.D. thesis]: Los Angeles, University of Southern California, 354 p.

Kleinpell, R. M., 1938, Miocene stratigraphy of California: Tulsa, Oklahoma, American Association of Petroleum Geologists, 450 p.

Koizumi, Itaru, 1973, The late Cenozoic diatoms of sites 183-193, Leg 19, *in* Initial reports of the Deep Sea Drilling Project, Volume 19: Washington, D.C., U.S. Government Printing Office, p. 805-855.

——1975, Neogene diatoms from the western margin of the Pacific Ocean, Leg 31, *in* Initial reports of the Deep Sea Drilling Project, Volume 31: Washington, D.C., U.S. Government Printing Office, p. 779-807.

——1977, Diatom biostratigraphy in the North Pacific region: International Congress on Pacific Neogene Stratigraphy, 1st, Tokyo, Japan, May 1976, Proceedings, Kaiyo Shuppan Co., Ltd. (Tokyo) p. 235-253.

Lipps, J. H., 1964, Miocene planktonic foraminifera from Newport Bay, California: Tulane Studies in Geology, v. 2, p. 109-133.

——1967, Planktonic foraminifera, intercontinental correlation and age of California mid-Cenozoic microfaunal stages: Journal of Paleontology, v. 41, p. 994-999.

Mankinen, E. A., and Dalrymple, G. B., 1979, Revised geomagnetic polarity time scale for the interval 0-5 m.y. B.P.: Journal of Geophysical Research, v. 84 (B2), p. 615-626.

Orr, W. N., 1972, Pacific Northwest siliceous phytoplankton: Palaeogeography, Palaeoclimatology, Palaeoecology, v. 12, p. 95-114.

Orr, W. N., and Conley, Steven, 1976, Siliceous dinoflagellates in the northeast Pacific rim: Micropaleontology, v. 22, no. 1, p. 92-99, Pls. 1-2.

Orr, W. N., Ehlen, J., and Zaitzeff, J. B., 1971, A late Tertiary diatom flora from Oregon: California Academy of Sciences Proceedings, ser. 4, v. 37, no. 16, p. 489-500.

Rau, W. W., 1967, Geology of the Wynoochee Valley quadrangle, Grays Harbor County, Washington: Washington Division of Mines and Geology Bulletin 62, 51 p.

Ryan, W.B.R., and others, 1974, A paleomagnetic assignment of Neogene stage boundaries and the development of isochronous datum planes between the Mediterranean, the Pacific and Indian Oceans in order to investigate the response of the world oceans to the Mediterranean Salinity Crisis: Rivista Italiana di Paleontologia, v. 80, v. 4, p. 631-688.

Schrader, H. J., 1973, Cenozoic diatoms from the northeast Pacific, Leg 18, *in* Initial reports of the Deep Sea Drilling Project, Volume 18: Washington, D.C., U.S. Government Printing Office, p. 673-798.

Warren, A. D., 1972, Luisian and Mohnian biostratigraphy of the Monterey Shale at Newport Lagoon, Orange County, California, *in* Symposium on Miocene biostratigraphy of California: Society of Economic Paleontologists and Mineralogists, Pacific Section, Bakersfield, California, p. 27-36.

MANUSCRIPT RECEIVED BY THE SOCIETY MARCH 14, 1980
MANUSCRIPT ACCEPTED MAY 7, 1980

Printed in U.S.A.

Geological Society of America
Special Paper 184
1981

Planktonic foraminiferal biostratigraphy, paleoceanographic implications, and deep-sea correlation of the Pliocene-Pleistocene Centerville Beach section, northern California

GERTA KELLER
JAMES C. INGLE, JR.
Department of Geology, School of Earth Sciences, Stanford University, Stanford, California 94305

ABSTRACT

Standard planktonic foraminiferal zones developed for tropical faunas can be estimated in mid- to high-latitude faunal assemblages of the northeast Pacific on the basis of recognition of major paleoclimatic-paleoceanographic events, quantitative distribution of *Neogloboquadrina pachyderma* morphotypes, and coiling ratio of this species in addition to the presence of key taxa. The late Neogene foraminiferal assemblages of DSDP site 173, off northern California, have been quantitatively analyzed and zoned by correlation with the central North Pacific DSDP site 310; multiple correlations of other microfossil groups support this correlation. Site 173 can thus serve as a reference section for zonation of sediments exposed at Centerville Beach, northern California, and in general for mid- to high-latitude faunas of the northeast Pacific. Correlation of planktonic foraminiferal assemblages of the Centerville Beach section with assemblages of site 173 indicate that the Eel River and Rio Dell Formations represent lower Pliocene to lower Pleistocene deposits. The base of the Eel River Formation lies within the upper part of foraminiferal zone N19, the Pliocene-Pleistocene boundary lies within the upper part of the middle member of the Rio Dell Formation, and extrapolation suggests that the top of the Rio Dell Formation is no younger than 0.7 m.y. This correlation is supported by analysis of other planktonic microfossil groups and by recent magnetostratigraphic work by Dodd and others and Kodama.

INTRODUCTION

Correlation of North Pacific Neogene deep-sea deposits with much thicker and geologically more complex epicontinental Neogene deposits exposed along the margin of the Pacific Northwest is commonly hindered by (1) the relative paucity of well-preserved planktonic microfossils in coastal sequences due to paleoenvironmental and diagenetic effects and (2) the general low-diversity aspect of subarctic and temperate planktonic assemblages common in the northeastern Pacific region during much of Neogene time. Significantly, the general absence of warm-water planktonic taxa in both deep-sea and marginal marine sequences of this region has precluded precise correlations with the standard low-latitude planktonic foraminiferal zones of Blow (1969) and in turn with the much utilized correlations of these zones with the radiometric-paleomagnetic time scale (Berggren and Van Couvering, 1974). Nevertheless, recent studies have demonstrated that (1) utilization of planktonic foraminiferal expressions of major Neogene paleoclimatic events in the North Pacific together with (2) occurrences of biostratigraphically useful higher latitude taxa and (3) brief paleoceanographically controlled occurrences of warmer water taxa allow confident correlation across temperate and subarctic faunal provinces of the North Pacific (Ingle, 1973a, 1973b; 1977a, 1977b). This report expands and refines the use of these methods to correlate the important Pliocene-Pleistocene Centerville Beach section exposed near Cape Mendocino,

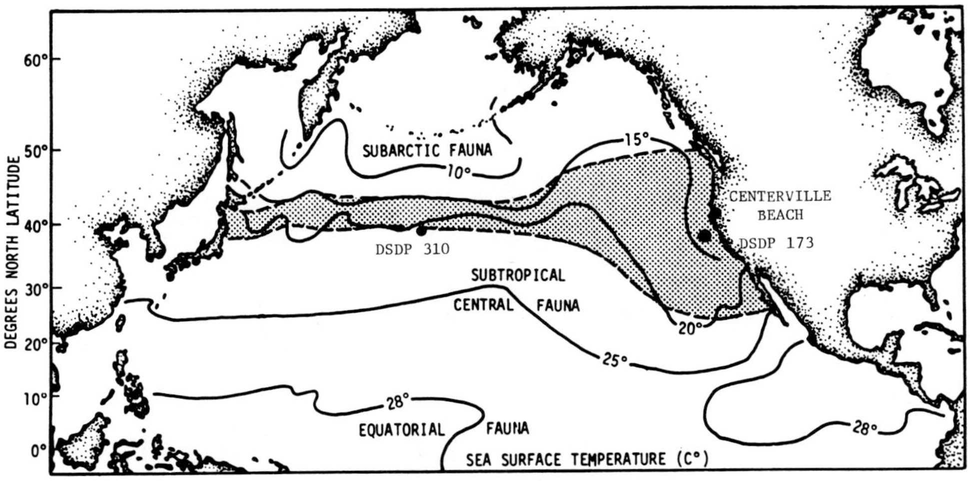

Figure 1. Location map of Deep Sea Drilling Project sites 173 and 310 and Centerville Beach, California, surface isotherms, and distribution of living planktonic foraminiferal assemblages in the North Pacific Ocean. Stippled pattern indicates distribution of transitional fauna (adapted from Ingle, 1967, and Vincent, 1975).

California, with adjacent Deep Sea Drilling Project (DSDP) sites (Figs. 1, 2). In turn, estimated positions of standard Neogene zonal boundaries in the Centerville Beach sequence are aided by recent detailed analyses and correlation of planktonic foraminifera at DSDP site 173 beneath the California Current, with a similar but more diverse late Neogene faunal sequence at DSDP site 310 on the Hess Rise in the central North Pacific (Keller, 1978a, 1978b, 1978c). Although emphasis is placed on quantitative analysis of paleoceanographically important morphotypes of *Neogloboquadrina pachyderma* and occurrences of key transitional water mass species for correlations between DSDP sites 173 and 310 and the Centerville Beach section, proposed correlations are supported by analyses of other planktonic microfossil groups in the DSDP sites and by recent paleomagnetic analyses of the Centerville Beach section (Dodd and others, 1977; Kodama, 1977).

DSDP SITES 173 AND 310

The relatively continuous upper Neogene planktonic foraminiferal sequence at DSDP site 173 in the marginal northeastern Pacific Ocean (Fig. 1) provides an excellent reference section for mid- to high-latitude faunas of the northeastern Pacific (Ingle, 1973b). However, Blow's standard planktonic foraminiferal zones (1969) cannot be identified precisely, except where key evolutionary transitions (that is, the *Orbulina* datum) are present (Ingle, 1973a). Nevertheless, faunal assemblages at site 173 can be correlated with the similar, though warmer, faunal assemblages of site 310 (Keller, 1978c), which can in turn be correlated with the standard Neogene planktonic foraminiferal zones (Vincent, 1975; Keller, 1978b).

Keller (1978c) has correlated site 173 to site 310 by using planktonic foraminiferal expressions of paleoceanographic events in combination with ranges of key index species (Fig. 3). Paleoceanographic-paleoclimatic events have been identified by quantitative distribution of *Neogloboquadrina pachyderma,* coiling ratio curves of this species, the distribution of various *N. pachyderma morphotypes,* and the sporadic but important occurrence of subtropical species. These analyses demonstrate that four distinct cold events can be readily identified at sites 173 and 310 in lower, middle, and upper Pliocene and lower Pleistocene sediments, with intervals of warm surface temperature and faunas connecting these events. Keller's correlation between sites 173 and 310 is supported by independent correlations of other microfossil groups, including diatoms (Koizumi, 1975), radiolarians (Kling, 1973; Foreman, 1975), and calcareous nannofossils (Bukry, 1973; Wise, 1973).

The correlation of planktonic foraminiferal assemblages between DSDP sites 173 and 310, supported by multiple correlations of other microfossil groups, permits extrapolation of the standard planktonic foramineral zonation to site 173. In addition, this zonation can be tied to major paleoclimatic events readily identified in planktonic foraminiferal assemblages in mid- to high-latitude regions of the central and northeastern Pacific Ocean. DSDP site 173 thus serves as a well-documented reference section for the late Neogene of the northeastern Pacific Ocean and, in particular, correlation of subarctic planktonic foraminiferal faunas within the upper Neogene epicontinental sections exposed along the coasts of Alaska, British Columbia, Washington, Oregon, and northern California.

CENTERVILLE BEACH SECTION

The upper Neogene Centerville Beach section exposed in the sea cliffs immediately north of Cape Mendocino, northern California (Figs. 1, 2), forms a portion of the Humboldt

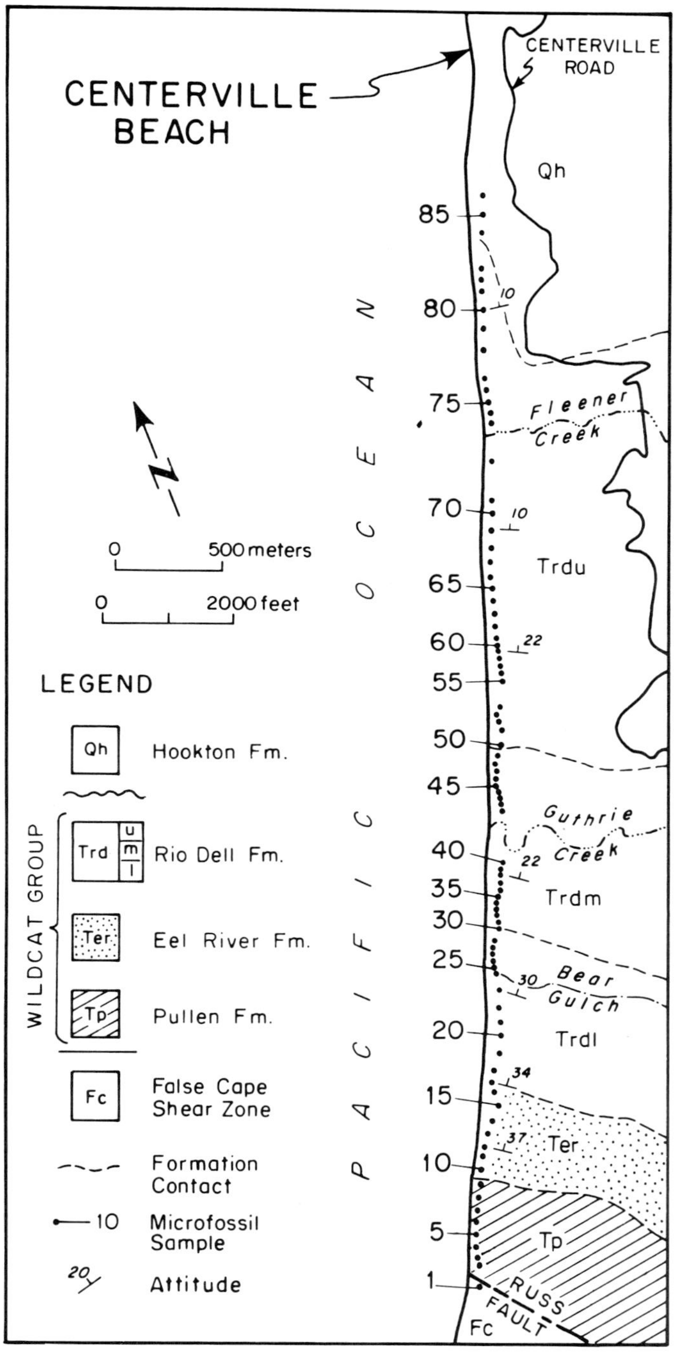

Figure 2. Sample locations and formations in the Pliocene-Pleistocene section (Wildcat Group) exposed south of Centerville Beach, Humboldt County, California (after Ingle, 1976).

Basin deposit representing the southern margin of a northwest-trending basin. The largest portion of this basin is still submerged beneath the continental margin of northern California and southern Oregon (Ogle, 1953; Ingle, 1976). Detailed stratigraphy and depositional history of this basin has been recently discussed by Ingle (1976), and Piper and others (1976). The 1,900 m of sediments exposed in this section (Fig. 2) represent a relatively continuous, though structurally deformed, lower Pliocene to lower Pleistocene sequence which can be considered conterminous with sediments of similar age penetrated at DSDP site 173 on the Delgada Fan (Ingle, 1973a, 1976). The greater sediment thickness at Centerville Beach attests to the extreme rate of sedimentation (70 cm/1,000 yr; Dodd and others, 1977) at the immediate continental margin versus much lower rates of sediment accumulation prevailing at the outer continental slope area of site 173 (Ingle, 1973a, 1973b). Planktonic foraminifera at Centerville Beach are relatively common, but some samples contain few to rare specimens and preservation is commonly poor; in addition, oldest Pleistocene faunas are affected by severe dissolution making quantitative analysis impossible. Nevertheless, faunal assemblages and biofacies trends are sufficiently similar to those of DSDP site 173 to allow for direct correlation between the two sections. This correlation is supported by recent magnetostratigraphic work at Centerville Beach by Dodd and others (1977) and Kodama (1977).

Samples from the Centerville Beach section consist of 200 to 350 g of dry sediments. (For details of collection and preparation of samples, see Ingle (1976.) Quantitative frequency determinations of planktonic foraminifera were made by counting 200 to 400 specimens within the size fraction $>62\ \mu m$.

Age and Biostratigraphy

The Neogene history, age, and biostratigraphy of the Wildcat group (Fig. 2) at Centerville Beach have been discussed by Ingle and Takayanagi (1968), Ingle (1969, 1976, 1977a, 1977b), Orr and Zaitzeff (1971), and Piper and others (1976). These studies indicate that the Pullen, Eel River, and Rio Dell Formations (Fig. 2) were deposited during early Pliocene through early Pleistocene time. Recently described magnetostratigraphy of the Rio Dell Formation (Dodd and others, 1977; Kodama, 1977) and correlation of the planktonic foraminiferal data with the upper Neogene sequence at DSDP site 173 provide new evidence for the correlation and placement of standard Neogene planktonic foraminiferal zones of Blow (1969) and Berggren and Van Couvering (1974) within the Centerville Beach section (Fig. 4). Figure 4 illustrates the cumulative distribution of *Neogloboquadrina pachyderma* morphotypes, coiling ratio of this species, zonation, and paleomagnetic dates (P) in the Centerville Beach sections.

The Eel River and Rio Dell Formations provide sufficient planktonic foraminifera for quantitative analysis. Foraminifera are rare or absent in the Pullen Formation. Correlation of the faunal sequence at the Centerville Beach section with the planktonic reference section established at DSDP site 173 suggests that the base of the Eel River Formation lies within the upper part of the planktonic foraminiferal zone N19. Thus, the N19/N21 boundary lies within the lower part of the Eel River Formation (Fig. 4), and the base of the Rio Dell Formation lies within the lower half of zone N21. It is important to emphasize

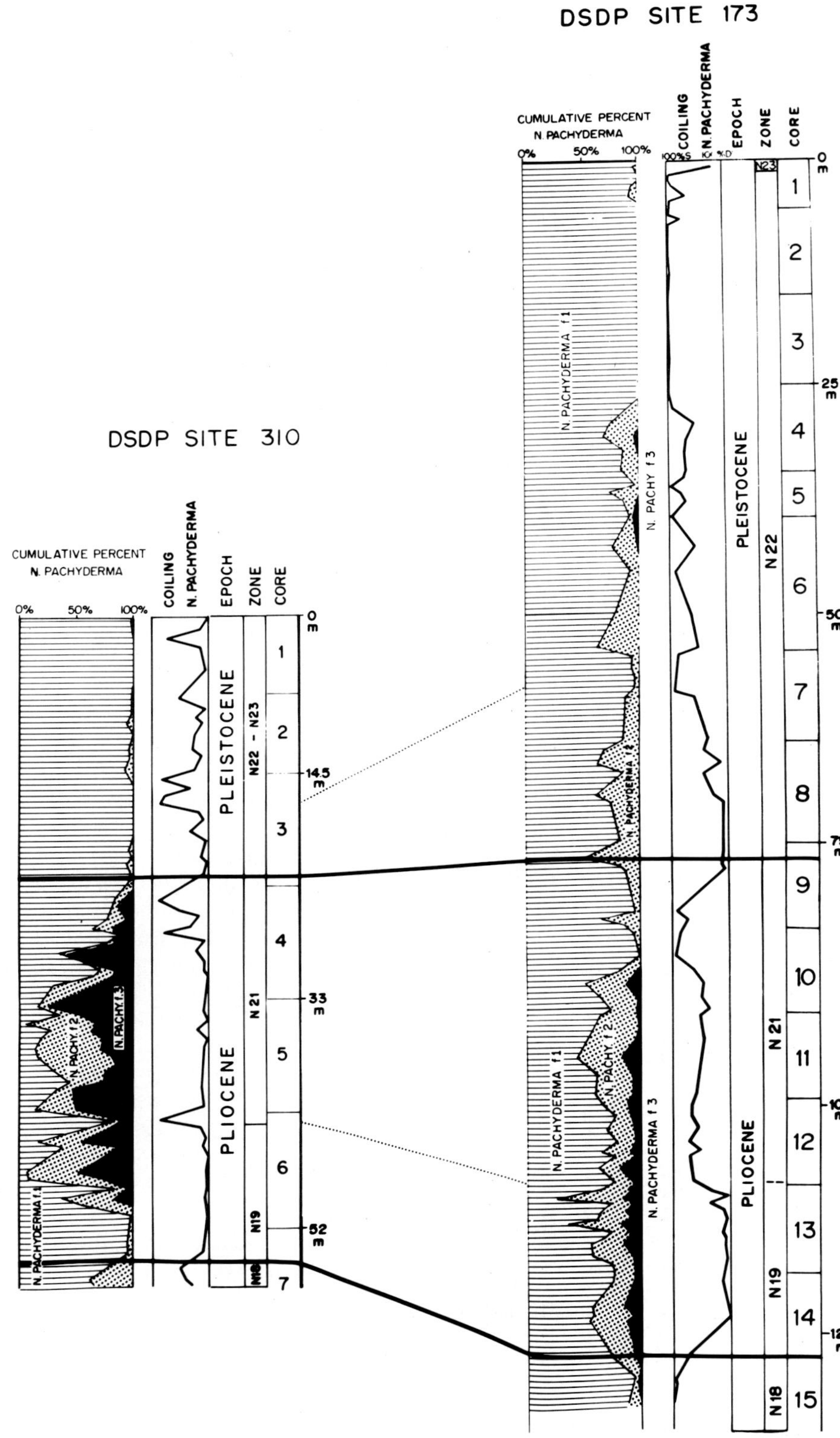

Figure 3. Correlation of DSDP sites 310 and 173 based on cumulative percent of *Neogloboquadrina pachyderma* morphotypes form 1, form 2, and form 3 (Keller, 1978a), coiling ratio curves of this species, and index taxa (from Keller, 1978c). Solid lines connecting the two sections represent the Pliocene-Pleistocene boundary and the planktonic foraminiferal zone boundary N18/N19; dotted lines correlate major climatic events.

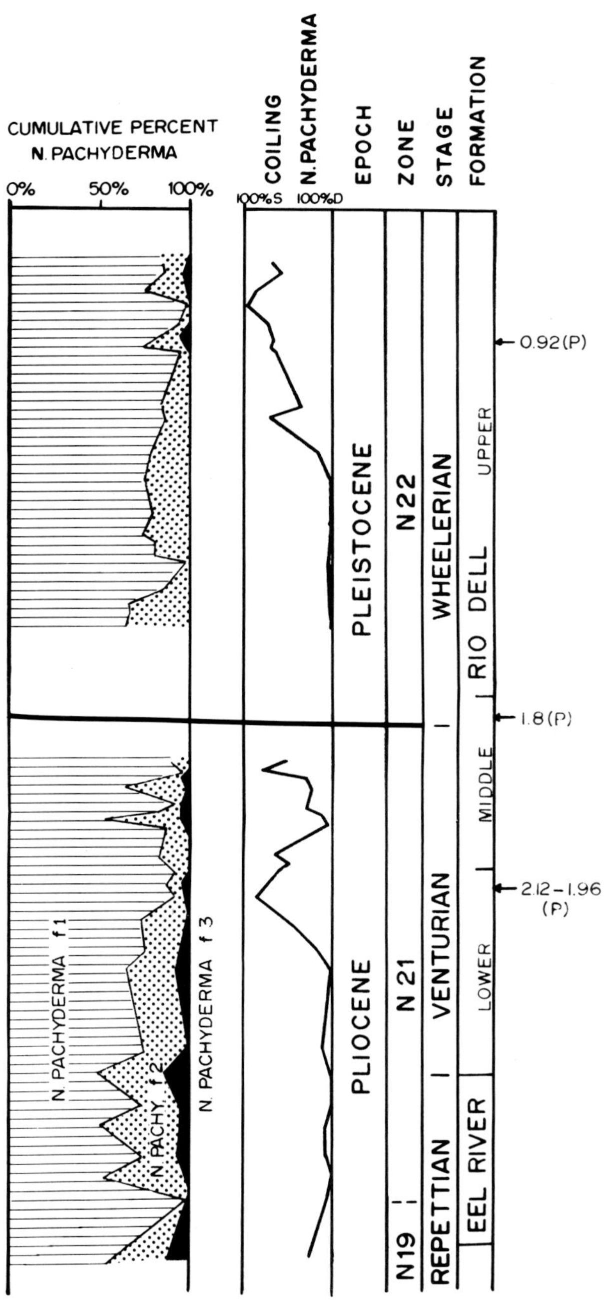

Figure 4. Cumulative percent distribution of *Neogloboquadrina pachyderma* morphotypes form 1, form 2, and form 3 (Keller, 1978a), coiling ration, age, zones, and formations at Centerville Beach, northern California. Paleomagnetic-radiometric dates (P) are from Kodama (1977). Dotted lines mark cold events as correlated with similar events at site 173 (Fig. 3).

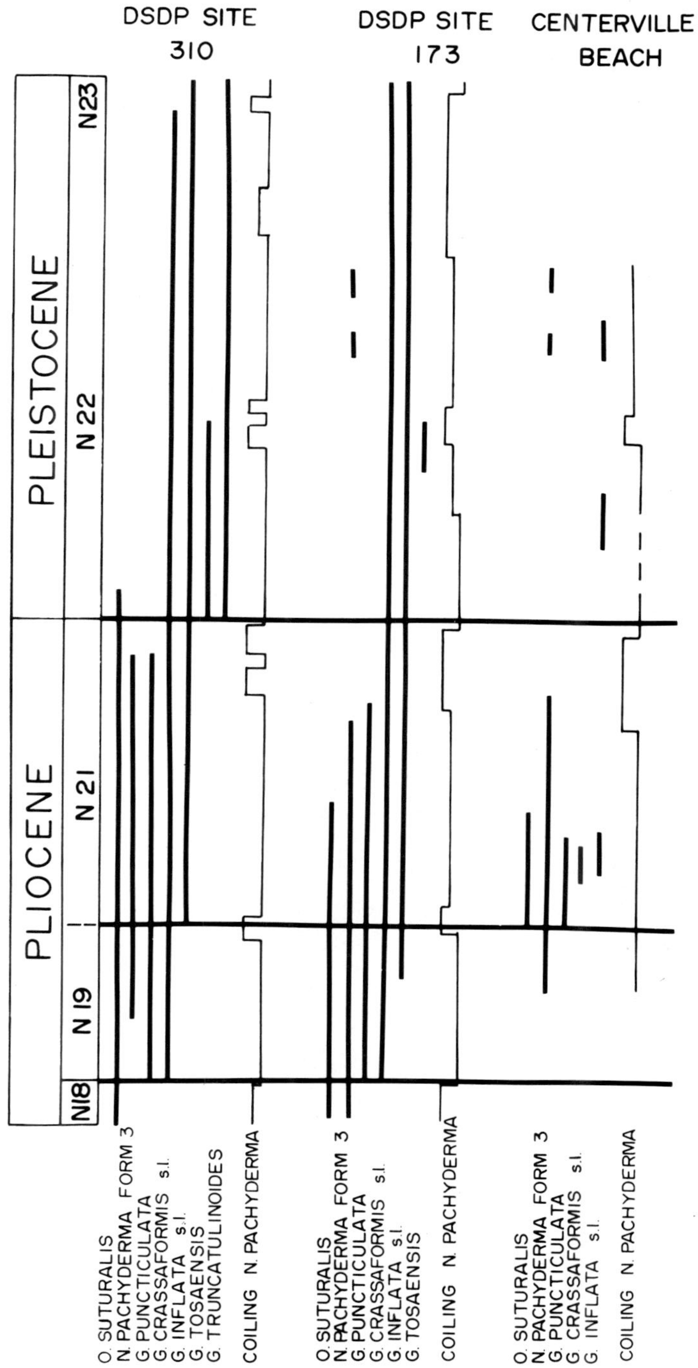

Figure 5. Ranges of selected species and sinistral, dextral coiling intervals of *Neogloboquadrina pachyderma* at DSDP sites 310, 173, and the Centerville Beach section, northern California.

that the standard Neogene zones cannot be directly identified within this sequence due to the absence of defining low-latitude taxa—estimated placement of zonal boundaries and zones per se is accomplished by extrapolation and correlation of those zones between DSDP sites 310 and 173 (Fig. 5).

Faunal correlations based on coiling ratio curves and distribution of *Neogloboquadrina pachyderma* morphotypes (Keller, 1978a) place the Pliocene/Pleistocene boundary within the upper part of the middle member of the Rio Dell Formation (Fig. 4). Recent magnetostratigraphic work by Kodama (1977) supports this correlation; however,

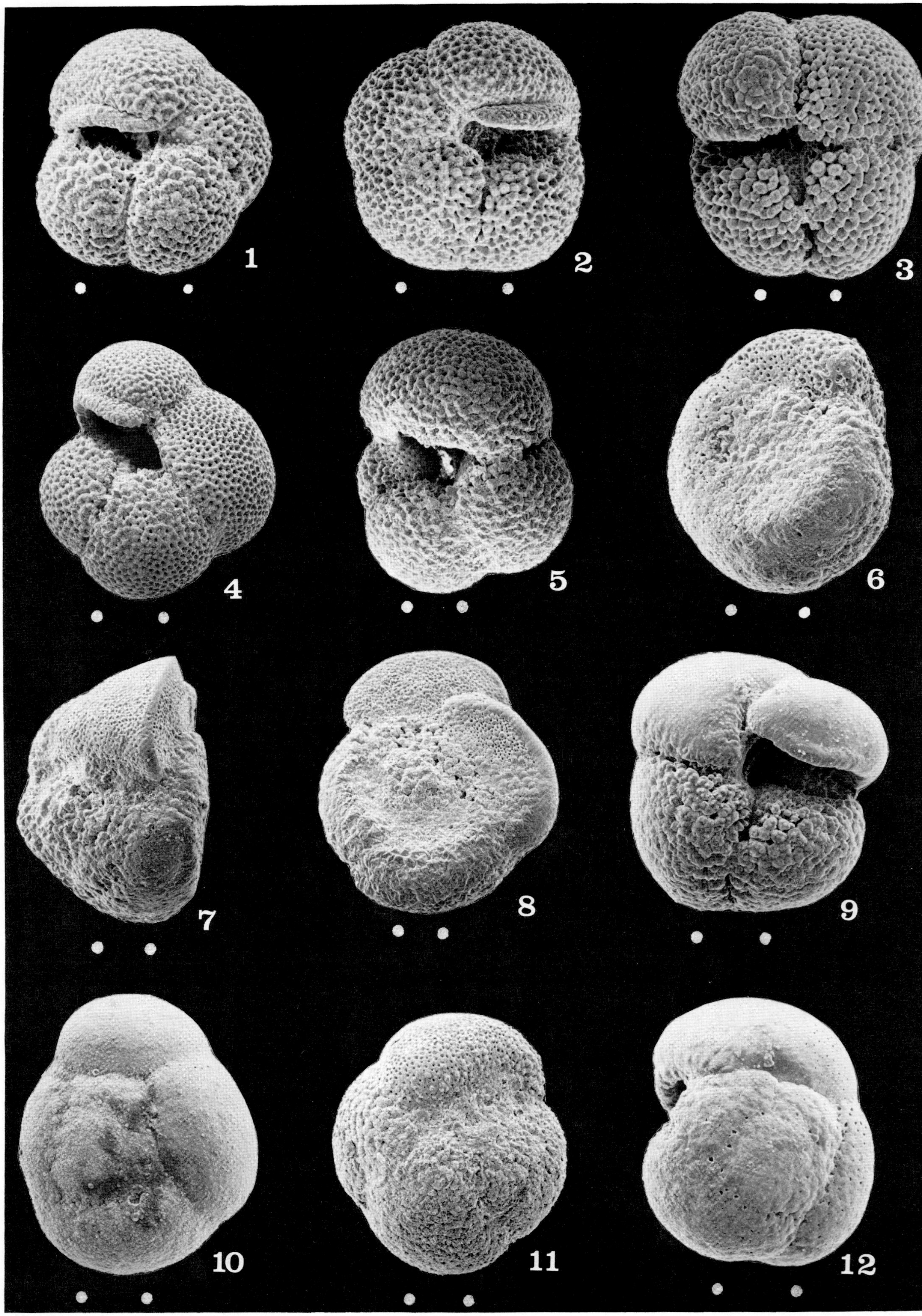
1
2
3
4
5
6
7
8
9
10
11
12

PLATE 1 (facing page)

Distance between two dots represents 100 μm.

Figure

1, 2. ***Neogloboquadrina pachyderma* form 1 (Ehrenberg)**
1. Umbilical view, dextral 173-10-4 (52-54 cm) upper Pliocene
2. Umbilical view, sinistral 173-10-4 (52-54 cm) upper Pliocene

3. ***Neogloboquadrina pachyderma* form 2 (Ehrenberg)**
3. Umbilical view, dextral 173-10-4 (52-54 cm) upper Pliocene

4, 5. ***Neogloboquadrina pachyderma* form 3 (Ehrenberg)**
4. Umbilical view, dextral 173-10-6 (52-54 cm) upper Pliocene
5. Umbilical view, dextral 173-11-1 (52-54 cm) middle Pliocene

6. ***Globorotalia tosaensis* Takayanagi and Saito**
6. Spiral view 173-3-3 (101-103 cm) upper Pleistocene

7, 8. ***Globorotalia crassaformis oceanica* Cushman and Bermudez**
7. Side view 173-7-2 (52-54 cm) lower Pleistocene
8. Spiral view 173-7-2 (52-54 cm) lower Pleistocene

9, 12. ***Globorotalia inflata* (d'Orbigny) modern variety**
9. Umbilical view 173-2-1 (110-112 cm) upper Pleistocene
12. Spiral view 173-2-3 (114-116 cm) upper Pleistocene

10, 11. ***Globorotalia punticulata* (d'Orbigny)**
10. Spiral view 173-13-3 (108-110 cm) lower Pliocene
11. Umbilical view 173-12-5 (15-17 cm) lower Pliocene

magnetic work by Dodd and others (1977) places this boundary somewhat higher in the lowermost part of the upper member of the Rio Dell Formation.

In mid-latitudes of the northeast and central North Pacific DSDP sites 173 and 310, the Pliocene/Pleistocene boundary (N21/N22) is preceded by a cold event which at site 173 lasted approximately from 2.2 to 1.8 m.y., based on correlations with a similar event identified by oxygen-isotope measurements in the eastern equatorial Pacific (Shackleton and Opdyke, 1977; Fig. 3). This cold event is also evident in faunal assemblages at Centerville Beach (Fig. 4) and is characterized by predominantly sinistrally coiling forms of *Neogloboquadrina pachyderma*. Magnetostratigraphic work by Kodama (1977) indicates that a short normal event is coincident with this cold interval, which suggests a date of 2.12 to 1.96 m.y. (50 m above Bear Gulch; Fig. 4). This date is in good agreement with the age estimated for this cold interval at site 173 (Fig. 3; Keller, 1978c). Planktonic foraminifera within sediments overlying the late Pliocene cold event are poorly preserved, thus the Pliocene/Pleistocene boundary cannot be placed precisely. However, a similarly dissolved interval following the late Pliocene cold event is present at site 173 where other microfossil data indicate that this boundary occurs within the lower part of the dissolution-affected interval (Fig. 3). Thus, based on faunal correlation and coiling ratio curves, the Pliocene/Pleistocene boundary occurs in the uppermost part of the middle member of the Rio Dell Formation (vicinity of Guthrie Creek), in agreement with the paleomagnetic data of Kodama (1977) and earlier estimates based exclusively on planktonic foraminifera (Ingle, 1976).

The upper member of the Rio Dell Formation consists of lower Pleistocene deposits. Magnetostratigraphic work within the uppermost part of the Rio Dell Formation indicates an age of 0.92 m.y. just above Fleener Creek (Kodama, 1977; Fig. 4). According to these data, the top of the Rio Dell Formation is no younger than 0.7 m.y. (Dodd and others, 1977; Kodama, 1977).

FAUNAL ASSEMBLAGES OF SITE 173 AND CENTERVILLE BEACH

Late Neogene planktonic foraminiferal faunas at site 173 and in the Centerville Beach sections are dominated by *Neogloboquadrina pachyderma*. This species consistently comprises more than 50% of most planktonic foraminiferal populations and reaches 80% to 90% during cooler intervals identified by predominantly sinistrally coiling forms of this species (Bandy, 1960; Ingle, 1967, 1977a). The rest of the faunal assemblages are characterized by species of *Globigerina* (*G. bulloides, G. quadrilatera, G. umbilicata, G. falconensis*), and *Globigerinita* (*G. glutinata, G. uvula*), with minor occurrences of nonkeeled species of *Globorotalia* (Pl. 1); the latter forms never exceed 10% of the faunal assemblages. *Neogloboquadrina pachyderma* s.l. has been divided into three morphotypes called form 1, form 2, and form 3, which appear to be temperature related (Keller, 1978a). Paleoclimatic-paleoceanographic implications of these morphotypes in the central and northeastern Pacific Ocean were discussed in a recent paper by Keller (1978a).

CORRELATION OF DSDP SITES 310, 173, AND CENTERVILLE BEACH

The biostratigraphy and paleoecology of DSDP site 310 has recently been discussed by Vincent (1975) and Keller (1978b). These same aspects of DSDP site 173 have been presented by Ingle (1973a, 1973b, 1977a, 1977b) and Keller (1978c). Correlation of DSDP sites 173 and 310 has been proposed and documented by Keller (1978c). Figure 3 illustrates the correlation between sites 173 and 310 on the basis of the similarity in the distribution of *Neogloboquadrina pachyderma* and coiling ratio curves. Figure 5 shows a summary diagram representing key species common to all three sections studied (310, 173, Centerville Beach) and dextral-sinistral coiling variations of *N. pachyderma*. In the following discussion, correlation between the three sections is based primarily on recognition of climatic events and associated foraminiferal assemblages, particularly the quantitative distribution of *N. pachyderma* morphotypes (Figs. 3, 4).

Zone N18 and the lower part of zone N19 cannot be identified in the Centerville Beach section because of the absence or rarity of planktonic foraminifera. However, in all three sections, lower Pliocene zone N19 is characterized by almost exclusively dextral-coiling forms of *N. pachyderma* (Figs. 3, 4), which indicate warm temperatures during this interval. All three morphotypes of *N. pachyderma* are present, with the warmer water form 3 averaging 10% and form 2 averaging 40% of *N. pachyderma* s.l. at site 173 and the Centerville Beach section (Fig. 4). Higher averages are indicated for site 310 in the central North Pacific (Fig. 3). Warm climatic conditions during this interval are also supported by recent oxygen-isotope measurements from equatorial Pacific cores which suggest that no large ice sheets accumulated in the Northern Hemisphere between 3.2 and 3.5 m.y. ago (Shackleton and Opdyke, 1977); this allowed the well-documented northward excursion of a warm-temperate planktonic foraminiferal biofacies along the Pacific Coast of North America (Ingle, 1967, 1973b, 1977a).

The N19/N21 boundary at sites 310 and 173 is characterized by the cold event marked by predominantly sinistrally coiling forms of *N. pachyderma* form 1. A similar cool event can be recognized in the faunal assemblages at the Centerville Beach section, which places this zonal boundary within the lower half of the Eel River Formation (Fig. 4). At all three sections, zone N21 is marked by the presence of *Globorotalia puncticulata, G. crassaformis, G. inflata,* and *Orbulina suturalis* (Fig. 5).

A warm interval characterizes the early part of N21 in all three sections with gradual cooling occurring toward the late Pliocene cold event as indicated by increasing percentages of sinistral-coiling *N. pachyderma. Neogloboquadrina pachyderma* forms 2 and 3 constitute as much as 40% of *N. pachyderma* s.l. during the early part of this zone at DSDP site 173 and Centerville Beach; a higher percentage of these morphotypes at site 310 indicates warmer temperatures in the central North Pacific. A sudden decrease of these morphotypes to less than 10% (<20% at site 310) occurs during a late Pliocene cold event (Figs. 3, 4). This cold event is dominated by almost exclusively sinistrally coiling forms of *N. pachyderma* form 1 in all three sections.

The early Pleistocene is marked by almost exclusively dextrally coiling forms of *N. pachyderma* with form 1 dominating at all three sites. This temperate interval is followed by a short intense cold event that can also be recognized in all three sections (Figs. 3, 4) and is marked by almost exclusively sinistrally coiling forms of *N. pachyderma* form 1 with less than 10% of form 2 present. A similar cold event has been detected elsewhere in marginal northeastern Pacific sections (Ingle, 1973a, 1977b) and paleomagnetically dated as occurring at 1.2 m.y. (Kent and others, 1971). Succeeding this cold event is a warming trend as indicated by dextral-sinistral coiling forms of *N. pachyderma* form 3 within the Pleistocene (Fig. 4). Similar occurrences of this morphotype are observed at approximately the same interval at DSDP site 173 as indicated by a radiolarian datum of 0.9 m.y. (Kling, 1973) between the two occurrences. An angular unconformity separates the upper member of the Rio Dell Formation from the overlying nonmarine and shallow marine Hookton Formation at Centerville Beach (Ogle, 1953); this precludes detailed analysis of later Pleistocene marine events at this locality.

ACKNOWLEDGMENTS

This paper has greatly benefited from many suggestions and discussions with J. A. Barron and R. Z. Poore at the U.S. Geological Survey, Menlo Park, and K. Kodama, Department of Geophysics, Stanford University. Samples from DSDP sites were provided by the National Science Foundation. Samples from the Centerville Beach section were collected by Ingle and Y. Takayanagi as part of the United States–Japan Cooperative Science Program supported by National Science Foundation Grant GF195 and the Japan Society for the Promotion of Science. We wish to thank John Armentrout, Ron Echols, Kristin McDougall, William Orr, and Rose Trombley for reviewing the manuscript.

REFERENCES CITED

Bandy, O. L., 1960, The geologic significance of coiling ratios in the foraminifera *Globigerina pachyderma* (Ehrenberg): Journal of Paleontology, v. 34, p. 671–681.

Berggren, W. A., and Van Couvering, J. A., 1974, The late Neogene: Biostratigraphy, geochronology and paleoclimatology of the last 15 million years in marine and continental sequences: Palaeogeography, Palaeoclimatology, Palaeoecology, v. 16 (1/2), p. 1–216.

Blow, W. H., 1969, Late middle Eocene to Recent planktonic foraminiferal biostratigraphy, *in* Bronnimann, P., and Renz, H. H., eds., Proceedings of the International Conference on Planktonic Microfossils, 1st, Geneva 1967: Leiden, E. J. Brill, v. 1, p. 199–421.

Bukry, D., 1973, Coccolith and silicoflagellate stratigraphy, Deep Sea Drilling Project, Leg 18, eastern North Pacific, *in* Initial reports of the Deep Sea Drilling Project, Volume 18: Washington, D.C., U.S., Government Printing Office, p. 817–832.

Dodd, J. R., Mead, J., and Stanton, R. J., Jr., 1977, Paleomagnetic stratigraphy of Pliocene Centerville Beach section, northern California: Earth and Planetary Science Letters, v. 34, p. 381–386.

Foreman, H. P., 1975, Radiolarians from the North Pacific, Deep Sea Drilling Project, Leg 32, *in* Initial reports of the Deep Sea Drilling Project, Volume 32: Washington, D.C., U.S. Government Printing Office, p. 579–676.

Ingle, J. C., Jr., 1967, Foraminiferal biofacies and the Miocene-Pliocene boundary in southern California: American Paleontology Bulletin, v. 52, no. 236, p. 217–394.

——1969, Pliocene planktonic foraminifera from northern California and paleo-oceanographic implications: Geological Society of America Special Paper 121, p. 130–131.

——1973a, Neogene foraminifera from the northeastern Pacific Ocean, Leg 18, Deep Sea Drilling Project, *in* Initial reports of the Deep Sea Drilling Project, Volume 18: Washington, D.C., U.S. Government Printing Office, p. 517-567.

——1973b, Summary comments on Neogene biostratigraphy, physical stratigraphy, and paleo-oceanography in the marginal northeastern Pacific Ocean, *in* Initial Reports of the Deep Sea Drilling Project, Volume 18: p. 693-701.

——1976, Late Neogene paleobathymetry and paleoenvironments of the Humboldt Basin, northern California, *in* Fritsche, A. E., and others, eds., The Neogene Symposium: Society of Economic Paleontologists and Mineralogists, Pacific Section, p. 53-61.

——1977a, Late Neogene marine events and the Pliocene-Pleistocene boundary in the marginal North Pacific: Giornale de Geologia, ser. 2a, v. XLI, Fas I-II, p. 359-374.

——1977b, Summary of late Neogene planktic foraminiferal biofacies, biostratigraphy, and paleoceanography of the marginal North Pacific Ocean: International Congress on Pacific Neogene Stratigraphy, 1st, Tokyo, Japan, May 1976, Proceedings, p. 177-182.

Ingle, J. C., Jr., and Takayanagi, Y., 1968, Lower Pliocene planktonic biofacies in the marginal North Pacific: International Geological Congress, 23rd, Prague, Proceedings, Abstract Volume, p. 277.

Keller, Gerta, 1978a, Morphologic variations of *Neogloboquadrina pachyderma* (Ehrenberg) in sediments of the marginal and central North Pacific Ocean and paleoclimatic interpretation: Journal of Foraminiferal Research, v. 8, no. 7, p. 208-224.

——1978b, Late Neogene biostratigraphy and paleoceanography of DSDP site 310 central North Pacific and Correlation with the Southwest Pacific: Journal of Foraminiferal Research, v. 8, no. 4, p. 332-349.

——1978c, Neogene planktonic foraminiferal biostratigraphy and paleoceanography of the central and northeastern North Pacific: Correlation of DSDP sites 173 and 310: Marine Micropaleontology, v. 3.

Kent, D., Opdyke, N. D., Ewing, M., 1971, Climate change in the North Pacific using ice-rafted detritus as a climatic indicator: Geological Society of America Bulletin, v. 82, p. 2741-2754.

Kling, S. A., 1973, Radiolaria from the eastern North Pacific, Deep Sea Drilling Project, Leg 18, *in* Initial reports of the Deep Sea Drilling Project, Volume 18: Washington, D.C., U.S. Government Printing Office, p. 617-672.

Kodama, K., 1977, The magnetostratigraphy of the Rio Dell Formation, Centerville Beach, California [Ph.D. dissert.]: Stanford, California, Stanford University.

Koizumi, I., 1975, Diatom events in late Cenozoic deep-sea sequences in the North Pacific: Geological Society of Japan Journal, v. 81, no. 9, p. 567-578.

Ogle, B. A., 1953, Geology of the Eel River Valley area, Humboldt County, California: California Division of Mines Bulletin 164, 128 p.

Orr, W. and Zaitzeff, J. B., 1971, A new planktonic species from the California Pliocene: Journal of Foraminiferal Research, v. 1, p. 17-19.

Piper, D., Normark, W., and Ingle, J. C., Jr., 1976, The Rio Dell Formation: a Plio-Pleistocene basin slope deposit in northern California: Sedimentology, v. 23, p. 309-328.

Shackleton, M., and Opdyke, N., 1977, Oxygen isotope and paleomagnetic evidence for early Northern Hemisphere glaciation: Nature, v. 270, p. 216.

Vincent, E., 1975, Neogene planktonic foraminifera from the central North Pacific Leg 32, *in* Initial reports of the Deep Sea Drilling Project, Volume 32: Washington, D.C., U.S. Government Printing Office, p. 765-791.

Wise, S. W., 1973, Calcareous nannofossils from cores recovered during Leg 18, Deep Sea Drilling Project: Biostratigraphy and observations of diagenesis, *in* Initial reports of the Deep Sea Drilling Project, Volume 18: Washington, D.C., U.S. Government Printing Office, p. 569-615.

MANUSCRIPT RECEIVED BY THE SOCIETY MARCH 14, 1980
MANUSCRIPT ACCEPTED MAY 7, 1980

Printed in U.S.A.

Geological Society of America
Special Paper 184
1981

Correlation and ages of Cenozoic chronostratigraphic units in Oregon and Washington

JOHN M. ARMENTROUT
Mobil Exploration and Producing Services, Inc., Post Office Box 900, Dallas, Texas 75221

ABSTRACT

New paleontologic and radiometric data and a synthesis of redefined provincial biostratigraphic frameworks provide the basis for a new geologic time scale for the Cenozoic strata of Oregon and Washington.

The new geologic time scale integrates provincial molluscan, foraminiferal, land-mammal, and megafloral biochronologies of the Pacific Northwest. The integrated biostratigraphic framework incorporates new data on planktonic microfossils from western Oregon and Washington and is correlated with the Cenozoic worldwide geologic time scale. The biostratigraphic correlations are augmented by integration of radiometric age data.

The new provincial geologic time scale revises the 1944 chart on "Correlation of the Marine Cenozoic Formations of Western North America" and provides a basis for time-stratigraphic correlations within the context of the current worldwide geologic time scale.

INTRODUCTION

Correlation and age assignment of Oregon and Washington Cenozoic sequences are based on provincial biochronologies. Correlations were first based on mollusks (Addicott, this volume). Biostratigraphic frameworks were later added for foraminifera, land-mammals, and megafloras (see Rau, this volume; Wood and others, 1941; Wolfe, this volume). Calibration of the provincial biostratigraphic frameworks by reference to the European series-epoch and stage-age units was attempted by Weaver and others (1944). However, recent studies (Addicott, 1976a; Armentrout, 1975, 1978; Berggren and others, 1978; Lander, 1977; Rau, this volume; Repenning 1980; and Wolfe, this volume) have revised definition and placement of boundaries recognized within the provincial biochronologies and have calibrated them by correlation to the worldwide geologic time scale. The correlations proposed by these studies require changes in the provincial usage of the European series-epoch and stage-age units.

This study correlates the revised provincial Cenozoic biostratigraphic frameworks and, using new data, correlates these provincial biostratigraphies to the worldwide geologic time scale. The resulting Cenozoic geologic time scale for Oregon and Washington provides a basis for time-stratigraphic correlations within the context of the current worldwide geologic time scale.

DISCUSSION

The three figures presented in this study show (Fig. 1) the correlation of the provincial Pacific Northwest marine and nonmarine biostratigraphic frameworks with each other and with a million-year time scale; (Fig. 2) the correlation, using planktonic microfossils, of the provincial marine biostratigraphic framework with the worldwide time scale; and (Fig. 3) a synthesis of Figures 1 and 2 presenting a new Cenozoic geologic time scale for Oregon and Washington. Figures 1 and 2 are annotated with either a reference citation or data-source citation documenting each critical boundary or data point(s) used in constructing the figures.

The definition of the provincial biochronologies follows the most recent redefinition or recalibration of the stages and ages. The foraminiferal framework follows Rau (this volume). The molluscan framework follows Addicott (1976a), Armentrout (1975, 1978), and Turner (1938). The foraminiferal and molluscan biostratigraphic frameworks have been correlated within the same stratigraphic sections and are therefore shown with no graphical separation (see Addicott 1976a; Armentrout, 1975). The megafloral framework follows Wolfe (this volume). The land-mammal framework is modified from Wood and others (1941) with all modifications annotated.

The provincial biochronologic frameworks are calibrated to a million-year time scale through the use of radiometric dates on Pacific Northwest rock units. Dates cited are from units that are within or which can be tied physically to fossiliferous sequences. The radiometric dates used in this synthesis include potassium-argon analyses from both whole rocks and mineral separates. All potassium-argon analyses are recalibrated to the new decay and abundance constants recommended by the International Union of Geological Sciences (IUGS) Subcommission on Geochronology (Steiger and Jäger, 1977) with the conversion factors recommended by Dalrymple (1979).

Correlation of the marine and nonmarine biostratigraphic frameworks has been checked by plotting data points of interbedded or superpositionally associated assemblages of two or more biochronologies (annotations 23 to 38). The available data show that within the resolution of each faunal or floral biostratigraphic interpretation, the provincial biostratigraphic frameworks correlate as shown in Figure 1.

The provincial molluscan and foraminiferal biochronologies are correlated with the worldwide geologic time scale by using planktonic foraminifera, calcareous nannofossils, diatoms, and radiometric dates. The data points used and their range of resolution are shown in Figure 2 (annotations 50 to 70).

The worldwide geologic time scale used is based on the time scales proposed by Berggren and Van Couvering (1974) and Hardenbol and Berggren (1978) with modification of boundary definitions as annotated on Figure 2.

The resulting correlation of the provincial biochronologies with the worldwide time scale permits a comparison of the million-year assignments to the European series-epoch and stage-age boundaries (Fig. 3). Provincial usage agrees with worldwide usage except for the time interval between 24 and 40 m.y. B.P. Radiometric data from Oregon and Washington suggest that the Eocene/Oligocene Series boundary is approximately 32 m.y. B.P. rather than 37 m.y. B.P. as proposed by Hardenbol and Berggren (1978) (see annotation 44). Because the correlation chart of Figure 3 is specifically for Oregon and Washington, provincial data have been used to recalibrate the worldwide time scale. The fact that other workers are also proposing younger ages for the Eocene/Oligocene Series boundary (Pomerol, 1978; Odin, 1978; Wolfe, this volume; Harris, 1979; Fullagar and others, 1980) suggests that the Oregon-Washington and European time scales are accurately correlated.

The precise definition and wide use of the Berggren and Van Couvering (1974) and Hardenbol and Berggren (1978) time scales permit easy recalibration of the biochronology here presented. As additional modifications of the geologic time scale are accepted, the Cenozoic geologic time scale for Oregon and Washington can be recalibrated easily.

Correlations using the Cenozoic geologic time scale for Oregon and Washington have been made successfully for faunal and rock sequences from central California to the Alaskan Peninsula (Armentrout, unpub. data). Faunal provincialism increased during the Neogene and makes correlations more difficult, but the use of interdisciplinary data allows reasonable correlations.

The precision of correlations based on the provincial geologic time scale presented in this paper depends on the degree of resolution of the faunal and/or floral biostratigraphy used. Employment of an interdisciplinary approach and careful attention to the superpositional relationship of samples will result in improved resolution.

SUMMARY

The provincial Cenozoic biostratigraphic frameworks used in Oregon and Washington are correlated with each other and with the worldwide geologic time scale. The resulting Cenozoic geologic time scale for Oregon and Washington permits the correlation of time-equivalent faunal and rock sequences. Although the worldwide geologic time scale and the provincial biochronologies are certain to undergo further refinement, the time scale presented reflects the current state of the art.

Figure 1 (facing page). Correlation of Pacific Northwest, North American, biostratigraphic standards. Circled numbers refer to numbered annotations. Numbers in large, open triangles and diamonds refer to annotations concerning biostratigraphic boundaries. Small, solid triangles represent specific radiometric dates: dates from rocks within biostratigraphically controlled sections are plotted within the stage column; dates from rocks correlated into biostratigraphically controlled sections are plotted to the right of the stage column. Data point ties (horizontal lines and arrows) represent the co-occurrence of two or more biochronologically significant groups within the same stratigraphic sequence. Orientation of the arrows is for convenience in matching data point ties. The vertical lines with end brackets represent the age resolution of the taxonomic group evaluated. Patterned areas denote informal units.

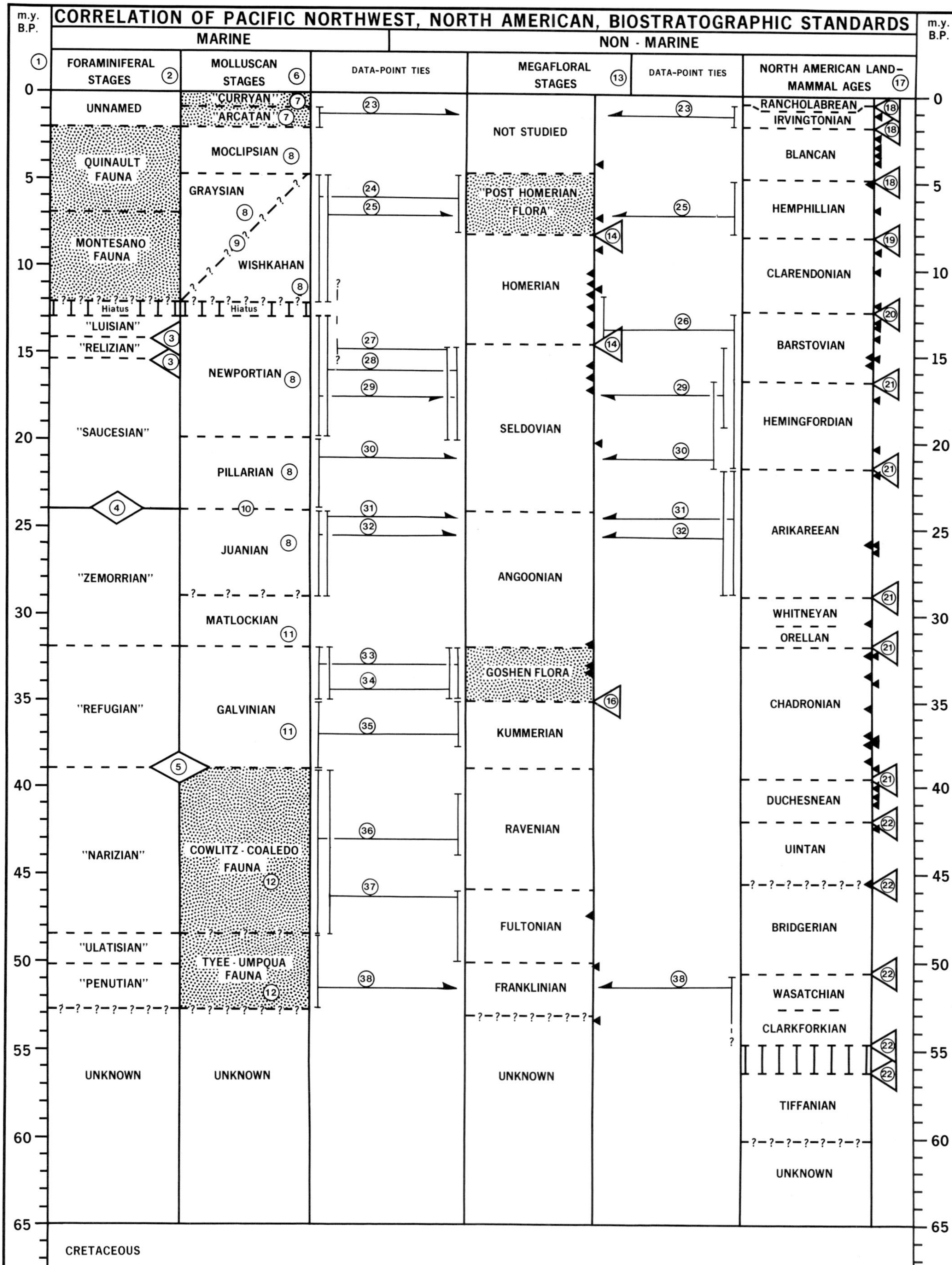
CORRELATION OF PACIFIC NORTHWEST, NORTH AMERICAN, BIOSTRATOGRAPHIC STANDARDS
m.y. B.P.
MARINE
NON - MARINE
FORAMINIFERAL STAGES
MOLLUSCAN STAGES
DATA-POINT TIES
MEGAFLORAL STAGES
DATA-POINT TIES
NORTH AMERICAN LAND-MAMMAL AGES
UNNAMED
QUINAULT FAUNA
MONTESANO FAUNA
Hiatus
"LUISIAN"
"RELIZIAN"
"SAUCESIAN"
"ZEMORRIAN"
"REFUGIAN"
"NARIZIAN"
"ULATISIAN"
"PENUTIAN"
UNKNOWN
"CURRYAN"
"ARCATAN"
MOCLIPSIAN
GRAYSIAN
WISHKAHAN
Hiatus
NEWPORTIAN
PILLARIAN
JUANIAN
MATLOCKIAN
GALVINIAN
COWLITZ - COALEDO FAUNA
TYEE - UMPQUA FAUNA
UNKNOWN
NOT STUDIED
"POST HOMERIAN" FLORA
HOMERIAN
SELDOVIAN
ANGOONIAN
GOSHEN FLORA
KUMMERIAN
RAVENIAN
FULTONIAN
FRANKLINIAN
UNKNOWN
RANCHOLABREAN
IRVINGTONIAN
BLANCAN
HEMPHILLIAN
CLARENDONIAN
BARSTOVIAN
HEMINGFORDIAN
ARIKAREEAN
WHITNEYAN
ORELLAN
CHADRONIAN
DUCHESNEAN
UINTAN
BRIDGERIAN
WASATCHIAN
CLARKFORKIAN
TIFFANIAN
UNKNOWN
CRETACEOUS
0 5 10 15 20 25 30 35 40 45 50 55 60 65

ANNOTATIONS FOR FIGURE 1

1. Million years before present = m.y. B.P.
2. Foraminiferal stages follow Rau's (this volume) usage of the West Coast benthic foraminiferal stages and local assemblages. I have placed the names of the California stages in quotation marks to denote that Rau (this volume) recognizes these stages using provincially defined zones for Oregon and Washington. The positioning of the boundaries is based on radiometric data (annotation 3) and planktonic correlations shown in Figure 2.
3. Radiometric dates (recalculated) for the provincial "Relizian and "Luisian" Stages are from basalts interbedded with foraminifer-bearing marine sediments (Snavely and others, 1973). Similar dates for the Relizian and Luisian Stages in California are reported by Turner (1970) and Howell (1976).
4. Radiometric date (recalculated) for the lower Saucesian Stage is from California (Turner, 1970) (see Hardenbol and Berggren, 1978; Addicott and Poore, 1980).
5. Radiometric dates for the provincial "Narizian"/"Refugian" and "Cowlitz and Coaledo Fauna"/Galvinian Stage boundaries are based on K-Ar dating of basalts interbedded with fossiliferous marine rocks. Snavely and MacLeod (1974) reported a date of 37.5 ± 3.6 m.y. B.P. (recalculated) for the Yachats Basalt, a unit interbedded within marine siltstone and sandstone (Nestucca Formation) which contain late "Narizian" foraminifera. In several areas, the Yachats Basalt is overlain by an unnamed sandstone containing early "Refugian" foraminifera. Armentrout and others (1980) have reported a K-Ar date of 38.5 ± 1.6 m.y. B.P. from a basalt mapped as Crescent(?) Formation just south of Porter, Washington. The basalt is overlain by marine sandy siltstone (Lincoln Creek Formation) containing early Galvinian mollusks and early "Refugian" foraminifera (Armentrout, 1973, 1975; Rau, 1958). Both dates are considered to be close to the time of crystallization, although slight chloritic alteration of the rocks suggests that they might be minimum dates. I place the "Narizian"/"Refugian" and "Cowlitz and Coaledo Fauna"/Galvinian Stage boundaries at approximately 39 m.y. B.P.
6. Molluscan stages follow the work of individual authors noted in annotations 7 to 12. The molluscan stages have been correlated with the foraminiferal stages within stratigraphic sections containing age-diagnostic faunas of both taxonomic groups (Addicott, 1976a; Armentrout 1973, 1975). Therefore, stage-boundary relationships are well established and the two stage frameworks are illustrated without separation.
7. The "Arcatan" and "Curryan" stages follow Roth (1979); I have placed these stages in quotation marks because Roth's work is to date unpublished.
8. The Juanian to Moclipsian Stages follow Addicott (1976a) with modifications (annotations 9 and 10).
9. The Wishkahan and Graysian Stages are modified from Addicott (1976a) on the basis of planktonic microfossil data shown in Figure 2 (annotations 50 to 54).
10. The Juanian/Pillarian and "Zemorrian"/"Saucesian" Stage boundaries are considered coincident. Addicott (1976b) and Addicott and Poore (1980) placed the Juanian/Pillarian Stage boundary within the lowermost part of the "Saucesian" Stage on the basis of one sample containing a "Saucesian" Stage foraminiferal fauna from within Juanian Stage strata 45 m below the Juanian/Pillarian Stage boundary. Armentrout (1973, 1975) found the Juanian/Pillarian boundary (= *Echinophoria apta*/*Liracassis petrosa* Zone boundary) to be coincident with the "Zemorrian"/"Saucesian" Stage boundary. I consider these boundaries to be essentially coeval, within the resolution of the two taxonomic disciplines, and therefore present them as coincident.
11. The Galvinian and Matlockian Stages follow Armentrout (1975, 1978). Armentrout (1975) introduced the Matlockian Molluscan Stage, identified by the faunas of the *Echinophoria rex* and *E. apta* Zones, and considered the Matlockian Molluscan Stage coeval with the entire "Zemorrian" Foraminiferal Stage. Addicott (1976a, 1976b) introduced the Juanian Molluscan Stage as coeval with the upper Matlockian *E. apta* Zone. To eliminate this dual nomenclature within the molluscan biostratigraphic framework, Armentrout (1978) restricted the Matlockian Stage to the *E. rex* Zone; this left the Juanian Stage as the only stage name for strata of the *E. apta* Zone.
12. The "Tyee/Umpqua" and "Cowlitz-Coaledo Faunas" were described by Turner (1938), who studied the molluscan biostratigraphy of the lower Tertiary rocks of western Oregon. Although Turner did not define zones or stages, his work provides the basis for recognizing two time-stratigraphically significant faunas based on the mollusks of the Tyee and Umpqua Formations and the Coaledo and Cowlitz Formations. I have compared Turner's (1938) work with new data which support Turner's original biostratigraphic subdivisions.
13. The Megafloral Stages follow Wolfe (this volume). The "Post-Homerian Flora" assemblages are age-equivalent to the Clamgulchian Stage of Wolfe (this volume). Wolfe (1979, personal commun.) considered the Clamgulchian Stage to be based on a flora restricted to Alaska. All older megafloral stages can be recognized throughout western North America. All radiometric data points (solid triangles) are recalculated from Wolfe (this volume).
14. Radiometric dates from Turner and others (1980) support Wolfe's (this volume) placement of the Homerian/Clamgulchian and Seldovian/Homerian Megafloral Stage boundaries.
15. The "Goshen Flora"/Angoonian Stage boundary is placed at 32.5 m.y. B.P. by Wolfe (this volume) based on K-Ar dating of the early Angoonian Willamette flora (31.8 m.y. B.P.) and Bridge Creek floras (31.9 and 32.3 m.y. B.P.), and the upper "Goshen"-LaPorte flora (33.1 and 33.4 m.y. B.P.). Wolfe (this volume) correlates the "Goshen Flora"/Angoonian Megafloral Stage boundary with the "Refugian"/"Zemorrian" Foraminiferal Stage boundary. I consider these boundaries coincident within the resolution of the sample distribution, and place the boundary at 32.0 m.y. B.P.
16. The Kummerian Stage/"Goshen Flora" boundary is placed at 35.0 m.y. B.P. by Wolfe (this volume); this conclusion is supported by an additional K-Ar date of 35.4 m.y. B.P. on lavas 65 m stratigraphically above Wolfe's Iron Mountain flora (= upper Clarno flora of Kummerian age) (C. B. Hanson, 1979, personal commun.; age determination recalculated from 34.5 ± 0.9 m.y. B.P. provided by Curtis and Drake, University of California, Berkeley) of the upper part of the Kummerian Megafloral Stage (Wolfe, this volume).
17. North American Land-mammal Ages follow Wood and others (1941) with the boundary positions modified. The numbers in the large triangles refer to the authors of the modifications (annotations 18 to 22). The radiometric data points (recalculated: solid triangles) are from a variety of sources summarized in Repenning (1979) for 0 to 7 m.y. B.P., Lander (1977) for 6 to 42 m.y. B.P., Woodburne and Robinson (1977) for 13 to 45 m.y. B.P., and Berggren and others (1977) for 42 to 57 m.y. B.P.
18. Boundary placement follows Repenning (1979).
19. Boundary placement follows Lander (1977 and 1979, personal commun.) and Repenning and Tedford (1977).
20. Boundary placement follows Lander (1977 and 1979, personal commun.) and Repenning (1979, personal commun.).
21. Boundary placement follows Woodburne and Robinson (1977).
22. Boundary placement follows Berggren and others (1977).
23. Mollusks assigned to the "Arcatan" molluscan stage (Roth, 1979) occur in the same rocks as vertebrates assigned to the

Irvingtonian Land-mammal Age (Repenning, 1979, personal commun.): Port Orford Formation, section south of Cape Blanco, Oregon.

24. Graysian Molluscan Stage type section (Addicott, 1976a) contains a pollen flora assigned to the "Post-Homerian Flora" megafloral stage (Wolfe, 1979, personal commun.): Montesano Formation, Wishkah River section, Washington.
25. Mollusks assigned to the Wishkahan Molluscan Stage (Addicott, 1976a) occur in the same rocks as vertebrates assigned to the early part of the Hemphillian Land-mammal Age (Repenning, 1979, personal commun.): Empire Formation type section, Coos Bay, Oregon.
26. Strata containing the Trout Creek flora (Graham, 1965), assigned to the lower part of the Homerian Megafloral Stage (Wolfe, 1979, personal commun.), are interbedded with strata containing a vertebrate fauna (localities M32 and M33, Walker and Repenning, 1965) assigned to the "Mascall Fauna" of Hemingfordian (Lander, 1977 and 1979, personal commun.) or Barstovian Land-mammal Age (Repenning, 1979, personal commun.): Miocene tuffaceous strata, Adel quadrangle, Harney County, Oregon.
27. A megaflora (LaMotte, 1936), assigned to the upper part of the Seldovian Megafloral Stage (Wolfe, 1979, personal commun.) occurs stratigraphically below rocks containing mollusks assigned to the Wishkahan Molluscan Stage (Addicott, 1976a): lower part of Montesano Formation, Wishkah River section, Washington.
28. Mollusks assigned to the Newportian Molluscan Stage (Addicott, 1976a, 1980) occur in strata above and below a tuffaceous sandstone containing a megaflora assigned to the upper(?) part of the Seldovian Megafloral Stage (Wolfe, 1979, personal commun.): sandstone of Floras Lake, section south of Cape Blanco, Oregon.
29. Mollusks assigned to the Newportian Molluscan Stage (Addicott, 1976a) occur in strata which also contain foraminifera assigned to the upper part of the "Saucesian" Foraminiferal Stage and "Relizian"(?) Foraminiferal Stage (Rau, 1979, personal commun.), and land-mammals (Munthe and Coombs, 1979) from a "Mascall-like" fauna assigned to the upper part of the Hemingfordian and lower part of the Barstovian Land-mammal Age (Lander, 1977 and 1979, personal commun.; Repenning, 1979, personal commun.): Iron Mountain beds, Astoria Formation, north of Yaquina Head, Newport, Oregon.
30. Mollusks assigned to the Pillarian Molluscan Stage (Addicott, 1976a) occur in rocks interbedded with strata containing foraminifera assigned to the "Saucesian" Foraminiferal Stage (Rau, 1979, personal commun.) and vertebrates assigned to the Hemingfordian(?) Land-mammal Age (Repenning, 1979, personal commun.): Nye Formation, Newport, Oregon.
31. Mollusks assigned to the Juanian Molluscan Stage (Addicott, 1976a) occur in rocks interbedded with strata containing marine mammals assigned to the Arikareean Land-mammal Age (Repenning, 1979, personal commun.): Yaquina Formation, Yaquina Bay section, Newport, Oregon.
32. Mollusks assigned to the Juanian Molluscan Stage (Addicott, 1976a, 1976b) occur in strata containing marine mammals assigned to the Arikareean Land-mammal Age (Repenning, 1979, personal commun.): Sooke Formation, type area, southern Vancouver Island, British Columbia and upper part of the Twin River Formation, Twin River area, Olympic Peninsula, Washington.
33. Mollusks assigned to the upper part of the Galvinian Molluscan Stage occur in strata correlated with rocks containing the Bilyeu Creek megaflora (Klucking 1962) assigned to the "Goshen Flora" megafloral stage (Wolfe, 1979, personal commun.): marine beds (Eugene(?) Formation) east of Salem, Oregon.
34. Mollusks (Hickman, 1969) assigned to the upper part of the Galvinian Stage (Armentrout, 1973, 1975) occur in rocks interpreted to be interbedded with strata containing the Goshen flora (Chaney and Sanborn, 1933) assigned to the "Goshen Flora" megafloral stage (Wolfe, 1979, personal commun.): Eugene(?) Formation, Goshen, Oregon.
35. Foraminifera assigned to the "Refugian" Foraminiferal Stage (K. McDougall, 1979, personal commun.) occur in strata containing faunas of the lower part of the Galvinian Molluscan Stage (Armentrout, 1973, 1975) and a megaflora assigned to the upper part of the Kummerian Megafloral Stage (Wolfe, 1979, personal commun.): Keasey Formation, Haydite quarry, Columbia County, Oregon.
36. Foraminifera assigned to the "Narizian" Foraminiferal Stage (Rau, in Snavely and others, 1958; U.S.G.S. loc. 9106) occur in strata interbedded with rocks containing a megaflora assigned to the middle part of the Ravenian Megafloral Stage (Wolfe, 1979, personal commun.): McIntosh Formation, Centralia-Chehalis area, Washington.
37. Mollusks (F. S. MacNeil in Vine, 1962) assigned to the "Cowlitz-Coaledo Fauna" molluscan assemblage occur in strata interbedded with rocks containing a megaflora assigned to the Fultonian Megafloral Stage (Wolfe, 1979, personal commun.): Tiger Mountain Formation, King County, Washington.
38. Strata correlative with the "Ulatisian" Foraminiferal Stage (Thoms, 1975) and the "Tyee-Umpqua Fauna" molluscan assemblage (Turner, 1938), contain a single Tapiroidea or Rhinocerotoidea humerus closely comparable in morphology and size to material from genera of Bridgerian or Uintan Land-mammal Age (C. B. Hanson and D. Taylor, 1979, personal communs.; C. B. Hanson, 1980, written commun.): Roseburg Formation, Umpqua Group, Oakland, Oregon.

ANNOTATIONS FOR FIGURE 2

39. Million years before present = m.y. B.P.
40. Chronostratigraphic units: Division of the Neogene System from Hardenbol (unpub., after Ryan and others, 1974) taken from Vail and others (1977); Paleogene System after Hardenbol and Berggren (1978). Calibration of geochronologic units between 25 and 40 m.y. B.P. (late Eocene and Oligocene time) differs from that of Hardenbol and Berggren (1978). The difference results from the use of 32 m.y. B.P. for the Eocene/Oligocene Series boundary rather than the more widely used 37 m.y. B.P. (see annotation 44). The differences are shown in Figure 3.
41. The Pliocene/Pleistocene Series Boundary Stratotype, at LaCastella, Italy, is correlated by multiple overlapping criteria to a level equivalent to, or slightly higher than, the top of the Olduvai polarity subzone (Haq and Berggren, 1977; Van Couvering, 1978) dated at 1.67 m.y. B.P. (Mankinen and Dalrymple, 1979). See also discussion of Damon and others (1976), Haq and Berggren (1978), and Jenkins 1978).
42. The Miocene/Pliocene Series Boundary is placed at the base of the Zanclean Stage (Cita, 1975) (correlative with the base of the Tabianian Stage; Hardenbol, unpub., after Ryan and others, 1974) within planktonic foraminiferal Zone N18, in upper calcareous nannoplankton Zone NN12, and in the lowermost, reversed event of the Gilbert Polarity Zone (Salvador and others, 1979) which is dated at approximately 5.3 m.y. B.P. (Van Couvering and others, 1976).
43. The Oligocene/Miocene Series Boundary: By international agreement, this boundary is defined at the base of the Aquitanian stratotype (Van Couvering, 1978), which is within the middle of planktonic foraminiferal Zone N4, not at its base (Addicott and Poore, 1979). Nannofossil studies indicate the middle of Zone N4 is approximately coeval with the NP25/NN1 zone boundary (Van Couvering and Berggren, 1977). Hardenbol and Berggren (1978) calibrated the NP25/NN1 Zone boundary at approximately 24 m.y. B.P.
44. The Eocene/Oligocene Series Boundary is placed at the Priabonian/Rupelian Stage boundary which falls within planktonic foraminiferal Zone P17, and within the lower part of calcareous nannoplankton Zone NP21. Hardenbol and Berggren (1978) consider this boundary to be at 37 m.y. B.P. Pomerol (1978), Odin (1978), Harris (1979), Wolfe (this volume), and Fullagar and others (1980) present radiometric age determinations from carefully selected glauconites and igneous rocks which suggest an age of 32 m.y. B.P. for the boundary. Radiometric data from Oregon and Washington for the late Eocene to Oligocene interval (Fig. 1, annotations 5, 15, and 16) suggest an age of 32 m.y. B.P. for the boundary, which is compatible with the proposed recalibration of the Eocene to Oligocene stages by Pomerol, Odin, Harris, Wolfe, Fullagar, and others. I have used Oregon and Washington data for calibrating the "Narizian"/"Refugian" Foraminiferal Stage boundary and "Cowlitz and Coaledo Fauna"/Galvinian Molluscan Stage boundary, and the "Refugian"/"Zemorrian" Foraminiferal Stage and Galvinian/Matlockian Molluscan Stage boundaries (Fig. 1, annotations 5, 15, and 16). Planktonic microfossil data are used to correlate the provincial foraminiferal and molluscan stage boundaries with the European series-epoch and stage-age boundaries (Fig. 2, annotations 58 through 68). The Bartonian/Priabonian and Rupelian/Chattian Stage boundaries are calibrated with the million-year scale according to the limits imposed by radiometric data from Oregon and Washington (Figs. 2 and 3). The differences between this calibration and that proposed by Hardenbol and Berggren (1978) are shown in Figure 3.
45. The Paleocene/Eocene Series Boundary is placed at the Thanetian/Ypresian Stage boundary which falls within planktonic foraminiferal Zone P6 and is approximately coeval with the calcareous nannoplankton Zone NP9/NP10 boundary. The Paleocene/Eocene Series boundary is dated at 53.5 m.y. B.P. by Hardenbol and Berggren (1978).
46. The Cretaceous/Paleocene Series Boundary is placed at the base of the Danian Stage, which is considered to be coeval with the base of planktonic foraminiferal Zone P1 and calcareous nannoplankton Zone NP1. The boundary is dated at 65 m.y. B.P. by Hardenbol and Berggren (1977).
47. Worldwide Biostratigraphic Units: Correlation of planktonic foraminiferal zones (PFZ) and calcareous nannoplankton zones (CNZ) with chronostratigraphic units follows Hardenbol and Berggren (1977) and Hardenbol (unpub., after Ryan and others, 1974) (see Vail and others, 1977).
48. North Pacific Diatom Zones are those defined by Koizumi (1973) and modified by and correlated with the planktonic foraminiferal zones by Barron (1981) (see Barron, this volume, Fig. 7).
49. Oregon-Washington foraminiferal and molluscan stages are calibrated with the million-year time scale in Figure 1.
50. Diatoms assigned to the *Nitzschia reinholdii* Zone (Barron and Armentrout, 1980; Barron, this volume) occur in strata containing mollusks assigned to the Wishkahan Molluscan Stage (Addicott, 1976a): upper part of Empire Formation in the Coos Head section, Coos Bay, Oregon.
51. Diatoms assigned to the *Thalassiosira antiqua* Zone (Barron and Armentrout, 1980; Barron, this volume) occur in strata containing mollusks assigned to the Wishkahan Molluscan Stage (Addicott, 1976a, 1980): middle part of Empire Formation, Coos Bay, Oregon, and upper part of Empire Formation, section south of Cape Blanco, Oregon.
52. Diatoms assigned to the *Denticulopsis hustedii* Zone (Barron, this volume) occur within the type section of the Graysian Molluscan Stage (Addicott, 1976a): Montesano Formation, Wishkah River section, Washington.
53. Diatoms assigned to Subzone d of the *Denticulopsis hustedii-D. lauta* Zone occur within the type section of the Graysian Molluscan Stage (Addicott, 1976a): Montesano Formation, Wishkah River section, Washington.
54. Diatoms (Barron and Armentrout, 1980; Barron, this volume) of the Wishkahan Molluscan Stage (early late Miocene; Addicott, 1976a) are younger than diatoms of the latest Miocene Graysian Molluscan Stage (latest Miocene; Addicott, 1976a). This relationship suggests that the Wishkahan and Graysian Molluscan Stages are coeval, as shown in Figures 1, 2, and 3 of this paper.
55. Diatoms assigned to Subzone b of the *Denticulopsis lauta* Zone and Subzone a of the *Denticulopsis hustedtii-D. lauta* Zone (Barron, 1979, personal commun,; 1980, written commun.) occur between basalt flows correlated with the Frenchman Springs flow (14.9 m.y. B.P.) and Pomona Member (12.3 m.y. B.P.) of the Columbia River Basalt Group (A. R. Niem and T. H. Murphy, 1980, written commun.; radiometric dates recalibrated after Dalrymple, 1979): sandstone and siltstone at Cliffton, Astoria area, Oregon.
56. Diatoms assigned to Subzone b of the *Denticulopsis lauta* Zone occur in strata containing foraminifera assigned to the *Baggina washingtonensis* Zone (Rau, 1967) which correlates with the upper "Saucesian" and lower "Relizian" Foraminiferal Stages (Rau, this volume): Astoria(?) Formation, Middle Fork of the Wishkah River section, Washington.

Figure 2 (facing page). Correlation of Pacific Northwest marine biostratigraphic units with worldwide biostratigraphic and chronostratigraphic units. See Figure 1 caption for explanation. PFZ = Planktonic Foraminiferal Zones; CNZ = Calcareous Nannoplankton Zones. Patterned areas denote informal units.

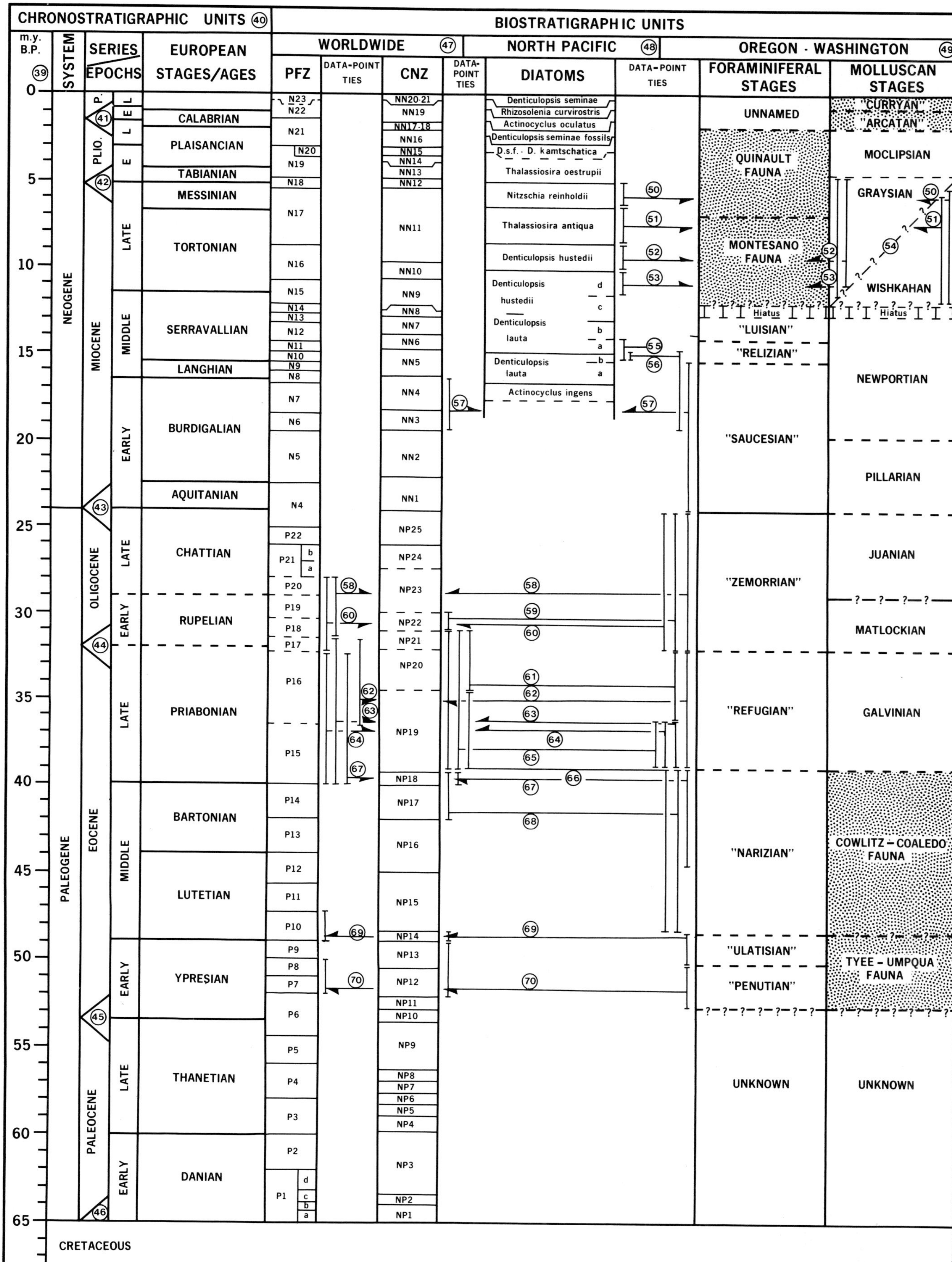

CHRONOSTRATIGRAPHIC UNITS 40
BIOSTRATIGRAPHIC UNITS
m.y. B.P. 39
SYSTEM
SERIES EPOCHS
EUROPEAN STAGES/AGES
WORLDWIDE 47
PFZ
DATA-POINT TIES
CNZ
DATA-POINT TIES
NORTH PACIFIC 48
DIATOMS
DATA-POINT TIES
OREGON - WASHINGTON 49
FORAMINIFERAL STAGES
MOLLUSCAN STAGES
0
5
10
15
20
25
30
35
40
45
50
55
60
65
NEOGENE
PALEOGENE
CRETACEOUS
PLIO.
P.
MIOCENE
OLIGOCENE
EOCENE
PALEOCENE
EARLY
MIDDLE
LATE
L
E
CALABRIAN
PLAISANCIAN
TABIANIAN
MESSINIAN
TORTONIAN
SERRAVALLIAN
LANGHIAN
BURDIGALIAN
AQUITANIAN
CHATTIAN
RUPELIAN
PRIABONIAN
BARTONIAN
LUTETIAN
YPRESIAN
THANETIAN
DANIAN
41
42
43
44
45
46
N23
N22
N21
N20
N19
N18
N17
N16
N15
N14
N13
N12
N11
N10
N9
N8
N7
N6
N5
N4
P22
P21 b a
P20
P19
P18
P17
P16
P15
P14
P13
P12
P11
P10
P9
P8
P7
P6
P5
P4
P3
P2
P1 d c b a
NN20-21
NN19
NN17-18
NN16
NN15
NN14
NN13
NN12
NN11
NN10
NN9
NN8
NN7
NN6
NN5
NN4
NN3
NN2
NN1
NP25
NP24
NP23
NP22
NP21
NP20
NP19
NP18
NP17
NP16
NP15
NP14
NP13
NP12
NP11
NP10
NP9
NP8
NP7
NP6
NP5
NP4
NP3
NP2
NP1
Denticulopsis seminae
Rhizosolenia curvirostris
Actinocyclus oculatus
Denticulopsis seminae fossils
D.s.f. - D. kamtschatica
Thalassiosira oestrupii
Nitzschia reinholdii
Thalassiosira antiqua
Denticulopsis hustedii
Denticulopsis hustedii d c
Denticulopsis lauta b a
Denticulopsis lauta b a
Actinocyclus ingens
50
51
52
53
54
55
56
57
58
59
60
61
62
63
64
65
66
67
68
69
70
UNNAMED
QUINAULT FAUNA
MONTESANO FAUNA
Hiatus
"LUISIAN"
"RELIZIAN"
"SAUCESIAN"
"ZEMORRIAN"
"REFUGIAN"
"NARIZIAN"
"ULATISIAN"
"PENUTIAN"
UNKNOWN
"CURRYAN"
"ARCATAN"
MOCLIPSIAN
GRAYSIAN
WISHKAHAN
Hiatus
NEWPORTIAN
PILLARIAN
JUANIAN
MATLOCKIAN
GALVINIAN
COWLITZ - COALEDO FAUNA
TYEE - UMPQUA FAUNA
UNKNOWN

57. Nannofloras assigned to the interval of Zones NN3 and NN4 (Armentrout and Worsley, 1980) occur in strata containing foraminifera assigned to the *Epistominella parva* Zone (Rau, 1958). The *E. parva* Zone is correlated to the "Saucesian" Foraminiferal Stage (Rau, this volume): Astoria(?) Formation, Willapa River section, southwestern Washington.
58. Planktonic foraminifera assigned to the interval of Zones P18 to P20 (Poore, 1979, personal commun.) occur in strata containing foraminifera assigned to the "Zemorrian" Foraminiferal Stage (Rau, 1958 and this volume): Lincoln Creek Formation, Willapa River section, Washington.
59. Nannofloras assigned to Zone NP22 (Armentrout and Worsley, 1980) occur in strata containing foraminifera assigned to the "Zemorrian" Foraminiferal Stage (K. McDougall, 1979, personal commun.): Lincoln Creek Formation, Willapa River section, Washington.
60. Planktonic foraminifera assigned to the interval of Zones P17 to P20 (Poore, 1979, personal commun.) occur in strata containing foraminifera assigned to the "Zemorrian" Foraminiferal Stage (K. McDougall, 1979, personal commun.): Lincoln Creek Formation, Willapa River section, Washington.
61. Nannofloras assigned to the interval of Zones NP20 and NP21 (Armentrout and Worsley, 1980) occur in strata containing foraminifera assigned to the *Eponides kleinpelli* Zone (Rau, 1958) which is correlated with the upper part of the "Refugian" Foraminiferal Stage (Rau, this volume): Lincoln Creek Formation, Willapa River section, Washington.
62. Planktonic foraminifera assigned to the interval of Zones P16 and P17 and nannofloras assigned to the interval Zones NP19 to NP21 (Warren and Newell, this volume) occur in strata containing foraminifera assigned to the "Refugian" Foraminiferal Stage (Tipton, 1975): upper part of Bastendorff Formation, Coos Bay, Oregon.
63. Planktonic foraminifera assigned to the interval of Zones P15 to P17 (Poore, 1979, personal commun.) occur in strata containing foraminifera assigned to the lower part of the "Refugian" Foraminiferal Stage (K. McDougall, 1979, personal commun.): lower part of Keasey Formation, Wolf Creek and Rock Creek sections, Columbia County, Oregon.
64. Planktonic foraminifera assigned to the interval Zones P15 to P16 (Poore, 1979, personal commun.) occur in strata containing foraminifera assigned to the lower part of the "Rufugian" Foraminiferal Stage (Rau, 1958; K. McDougall, 1979, personal commun.): middle part of Lincoln Creek Formation, Willapa River section, Washington.
65. Nannofloras assigned to the interval of Zones NP19 to NP21 (Armentrout and Worsley, 1980) occur in strata containing faunas assigned to the lower part of the "Refugian" Foraminiferal Stage (Rau, 1964) and faunas assigned to the lower part of the Galvinian Molluscan Stage (Armentrout, 1973, 1975): lower Lincoln Creek Formation, Canyon River section, Satsop River area, Washington.
66. Nannofloras assigned to Zone NP19 (Armentrout and Worsley, 1980) occur in strata containing foraminifera assigned to the "Narizian" Foraminiferal Stage (K. McDougall, 1979, personal commun.). The sample was collected 1 m below Lincoln Creek Formation strata containing foraminifera of the "Refugian" Foraminiferal Stage (Rau, 1958) and mollusks of the lower part of the Galvinian Molluscan Stage (Armentrout, 1973, 1975): upper unnamed formation of late Eocene age (McIntosh Formation of Rau, 1958), Willapa River section, Washington.
67. Nannofloras assigned to Zone NP18 and planktonic foraminifera assigned to the interval of Zones P15 to P16 (Warren and Newell, this volume) occur in strata containing foraminifera assigned to the upper part of the "Narizian" Foraminiferal Stage (Tipton, 1975): middle part of Bastendorff Formation, Coos Bay, Oregon.
68. Nannofloras assigned to the interval of Zones NP17 to NP18 (Armentrout and Worsley, 1980) occur in strata containing foraminifera assigned to the "Narizian" Foraminiferal Stage (K. McDougall, 1979, personal commun.). This occurrence is immediately below the nannoflora of annotation 65: upper part of the unnamed formation of late Eocene age (McIntosh Formation of Rau, 1958), Willapa River section, Washington.
69. Planktonic foraminifera assigned to Zone P10 and calcareous nannofossils assigned to Zone NP14 (Miles, this volume) occur in strata containing foraminifera assigned to the "Ulatisian" Foraminiferal Stage (Bird, 1967): Lorane Shale of the Elkton Formation, Elkton area, Oregon.
70. Planktonic foraminifera assigned to the interval of Zones P7 and P8 and calcareous nannofossils assigned to the interval of Zones NP12 and NP13 (Miles, this volume) occur in rocks containing foraminifera assigned to the "Penutian" Foraminiferal Stage (Thoms, 1975): Lookingglass Formation, Umpqua Group, Roseburg area, Oregon.

Figure 3 (facing page). Correlation of Cenozoic chronostratigraphic units for Oregon and Washington, based on Figures 1 and 2. Patterned areas denote informal units.

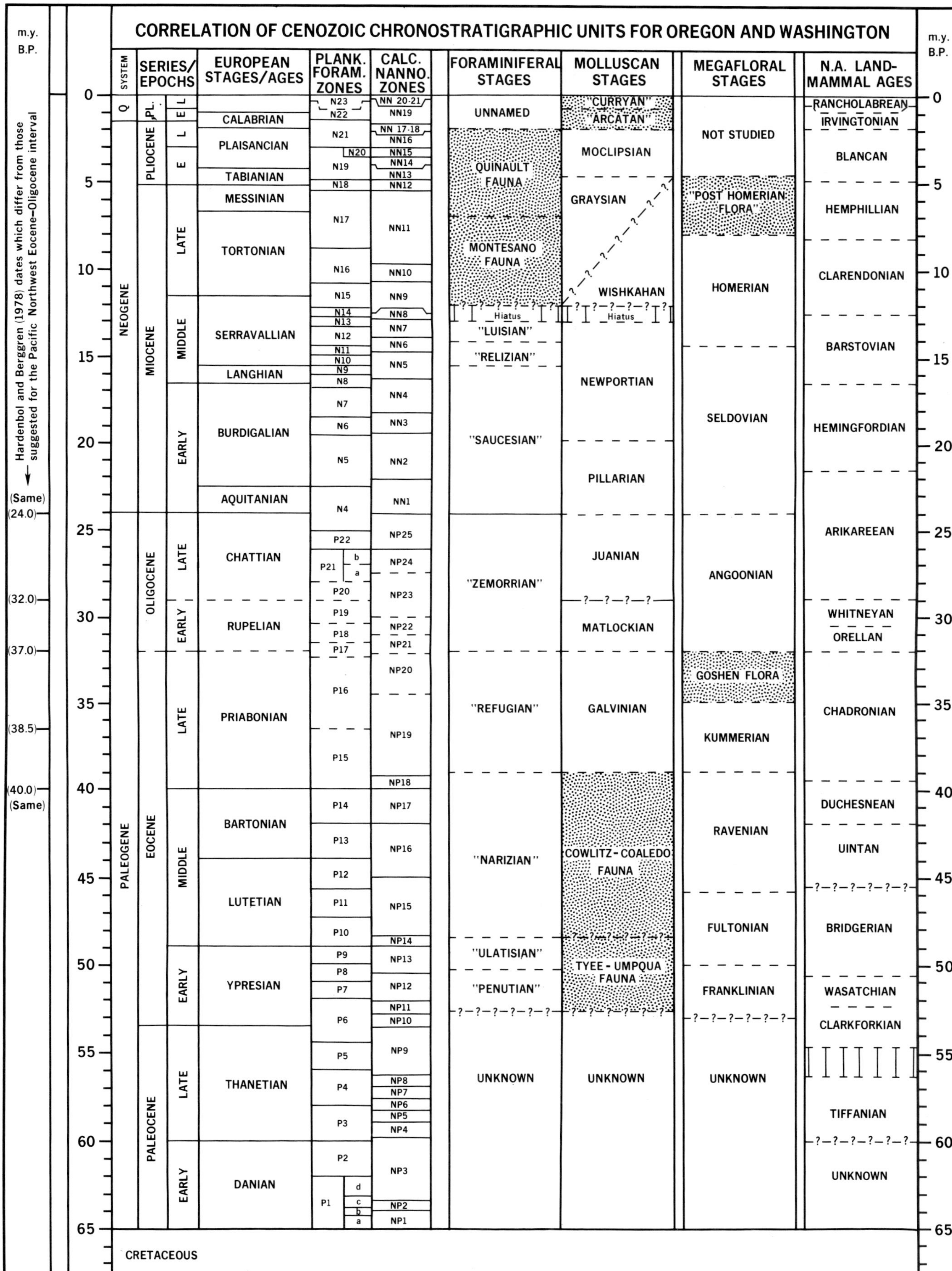
CORRELATION OF CENOZOIC CHRONOSTRATIGRAPHIC UNITS FOR OREGON AND WASHINGTON
m.y. B.P.
Hardenbol and Berggren (1978) dates which differ from those suggested for the Pacific Northwest Eocene–Oligocene interval
(Same)
(24.0)
(32.0)
(37.0)
(38.5)
(40.0)
(Same)
0
5
10
15
20
25
30
35
40
45
50
55
60
65
SYSTEM
SERIES/EPOCHS
EUROPEAN STAGES/AGES
PLANK. FORAM. ZONES
CALC. NANNO. ZONES
FORAMINIFERAL STAGES
MOLLUSCAN STAGES
MEGAFLORAL STAGES
N.A. LAND-MAMMAL AGES
Q
NEOGENE
PALEOGENE
PL.
PLIOCENE
MIOCENE
OLIGOCENE
EOCENE
PALEOCENE
L
E
LATE
MIDDLE
EARLY
CALABRIAN
PLAISANCIAN
TABIANIAN
MESSINIAN
TORTONIAN
SERRAVALLIAN
LANGHIAN
BURDIGALIAN
AQUITANIAN
CHATTIAN
RUPELIAN
PRIABONIAN
BARTONIAN
LUTETIAN
YPRESIAN
THANETIAN
DANIAN
N23
N22
N21
N20
N19
N18
N17
N16
N15
N14
N13
N12
N11
N10
N9
N8
N7
N6
N5
N4
P22
P21
a
b
P20
P19
P18
P17
P16
P15
P14
P13
P12
P11
P10
P9
P8
P7
P6
P5
P4
P3
P2
P1
a
b
c
d
NN 20-21
NN19
NN 17-18
NN16
NN15
NN14
NN13
NN12
NN11
NN10
NN9
NN8
NN7
NN6
NN5
NN4
NN3
NN2
NN1
NP25
NP24
NP23
NP22
NP21
NP20
NP19
NP18
NP17
NP16
NP15
NP14
NP13
NP12
NP11
NP10
NP9
NP8
NP7
NP6
NP5
NP4
NP3
NP2
NP1
UNNAMED
QUINAULT FAUNA
MONTESANO FAUNA
Hiatus
"LUISIAN"
"RELIZIAN"
"SAUCESIAN"
"ZEMORRIAN"
"REFUGIAN"
"NARIZIAN"
"ULATISIAN"
"PENUTIAN"
UNKNOWN
"CURRYAN"
"ARCATAN"
MOCLIPSIAN
GRAYSIAN
WISHKAHAN
Hiatus
NEWPORTIAN
PILLARIAN
JUANIAN
MATLOCKIAN
GALVINIAN
COWLITZ-COALEDO FAUNA
TYEE-UMPQUA FAUNA
UNKNOWN
NOT STUDIED
"POST HOMERIAN FLORA"
HOMERIAN
SELDOVIAN
ANGOONIAN
GOSHEN FLORA
KUMMERIAN
RAVENIAN
FULTONIAN
FRANKLINIAN
UNKNOWN
RANCHOLABREAN
IRVINGTONIAN
BLANCAN
HEMPHILLIAN
CLARENDONIAN
BARSTOVIAN
HEMINGFORDIAN
ARIKAREEAN
WHITNEYAN
ORELLAN
CHADRONIAN
DUCHESNEAN
UINTAN
BRIDGERIAN
WASATCHIAN
CLARKFORKIAN
TIFFANIAN
UNKNOWN
CRETACEOUS

ACKNOWLEDGMENTS

I am indebted to those researchers who provided unpublished data and especially to the following reviewers whose criticism has greatly improved this paper: W. W. Rau, State of Washington; W. O. Addicott, K. McDougall, J. A. Barron, R. Z. Poore, C. A. Repenning, E. J. Moore, and J. A. Wolfe, U.S. Geological Survey; C. B. Hanson and E. B. Lander, University of California-Berkeley; A. R. Niem and T. H. Murphy, Oregon State University: G. A. Miles, Exxon Company; and R. J. Echols, R. Evans, and L. B. Gibson, Mobil Exploration and Producing Services, Inc.

This paper was prepared during a National Research Council Post-Doctoral Associateship at the United States Geological Survey, Menlo Park.

REFERENCES CITED

Addicott, W. O., 1976a, Neogene molluscan stages of Oregon and Washington, *in* Fritsche, A. E., and others, eds., Neogene Symposium: Society of Economic Paleontologists and Mineralogists, Pacific Section, San Francisco, California, April 1976, p. 95-115.

——1976b, New molluscan assemblages from the upper Twin River Formation, western Washington: Significance in Neogene biostratigraphy: U.S. Geological Survey Journal of Research, v. 4, p. 437-447.

——1980, Miocene stratigraphy and fossils, Cape Blanco, Oregon: Oregon Geology, v. 42, p. 87-97.

——1981, Significance of pectinids in Tertiary biochronology of the Pacific Northwest: Geological Society of America Special Paper 184 (this volume).

Addicott, W. O., and Poore, R. Z., 1979, the Paleogene/Neogene boundary in eastern North Pacific marine sequences: Pacific Science Association, XIV Pacific Science Congress, Khabarovsk, USSR, August 1979, Abstracts of the Papers, v. 2, p. 5-6.

Armentrout, J. M., 1973, Molluscan paleontology and biostratigraphy of the Lincoln Creek Formation, late Eocene-Oligocene, southwestern Washington [Ph.D. dissert.]: Seattle, University of Washington, 478 p.

——1975, Molluscan biostratigraphy of the Lincoln Creek Formation, southwest Washington, *in* Weaver, D. E., ed., Future energy horizons of the Pacific Coast; Paleogene symposium and selected technical papers: American Association of Petroleum Geologists, Society of Economic Paleontologists and Mineralogists, and Society of Economic Geologists, Pacific Sections, Annual Meeting, Long Beach, California, April 1975, p. 14-28.

——1978, Progress report on Cenozoic correlations, West Coast North America: Geological Society of America Abstracts with Programs, v. 10, p. 360.

Armentrout, J. M., and Worsley, T. R., 1980, Tertiary calcareous nannofossils from southwestern Washington: Stratigraphic Newsletters (in press).

Armentrout, J. M., and others, 1980, Geologic field trip guide for the Cenozoic stratigraphy and late Eocene paleoecology of southwestern Washington: Oregon Department of Geology and Mineral Industries Bulletin 101, p. 79-119.

Barron, J. A., 1981, Late Cenozoic diatom biostratigraphy and paleoceanography of the middle-latitude eastern North Pacific, DSDP Leg 63, *in* Initial reports of the Deep Sea Drilling Project, Volume 63: Washington, D.C., U.S. Government Printing Office (in press).

——1981, Marine diatom biostratigraphy of the Montesano Formation near Aberdeen, Washington: Geological Society of America Special Paper 184 (this volume).

Barron, J. A., and Armentrout, J. M., 1980, Late Miocene diatom florules of the Empire Formation, Coos Bay, Oregon: Geological Society of America Abstracts with Programs, v. 12, p. 95-96.

Berggren, W. A., and Van Couvering, J. A., 1974, Biostratigraphy, geochronology and paleoclimatology of the last 15 million years in marine and continental sequences: Palaeogeography, Palaeoclimatology, Palaeoecology, v. 16, p. 1-216.

Berggren, W. A., and others, 1977, Revised Paleogene polarity time scale: Journal of Geology, v. 86, p. 67-81.

Bird, K. J., 1967, Biostratigraphy of the Tyee Formation (Eocene), southwestern Oregon [Ph.D. dissert.]: University of Wisconsin, 209 p.

Chaney, R. W., and Sanborn, E. I., 1933, the Goshen flora of west central Oregon: Carnegie Institition of Washington Publication 439, 103 p.

Cita, M. B., 1975, The Miocene/Pliocene boundary: History and definition, *in* Saito, T., and Burckle, L. H., eds., Late Neogene epoch boundaries: New York, Micropaleontology Press, p. 1-30.

Dalrymple, G. B., 1979, Critical tables for conversion of K-Ar ages from old to new constants: Geology, v. 7, p. 558-560.

Damon, P. E., Izett, G. A., and Naeser, C. W., 1976, Pliocene and Pleistocene geochronology: Geology, v. 4, p. 591-593.

Fullagar, P. D., Harris, W. B., and Winters, J., 1980, Rb-Sr glauconite ages, Claibornian and Jacksonian Strata (Eocene, southeastern Atlantic Coastal Plain: Geological Society of America Abstracts with Programs, v. 8 (in press).

Graham, A., 1965, The Sucker Creek and Trout Creek Miocene floras of southeastern Oregon: Kent, Ohio, Kent State University Bulletin, v. 53, 147 p.

Haq, B. U., and Berggren, W. A., 1977, Corrected age of the Pliocene/Pleistocene boundary: Nature, v. 269, p. 483-488.

——1978, Corrected age of the Pliocene/Pleistocene boundary; reply to Jenkins: Nature, v. 272, p. 287-288.

Hardenbol, J., and Berggren, W. A., 1978, A new Paleogene numerical time scale, *in* Cohee, G. V., and others, eds., Contributions to the geologic time scale: American Association of Petroleum Geologists Studies in Geology, no. 6, p. 213-234.

Harris, W. B., 1979, Rb-Sr glauconite ages and revisions of the Eocene time-scale, southeastern Atlantic Coastal Plain: Geological Society of America Abstracts with Programs, v. 11, p. 439.

Hickman, C.J.S., 1969, The Oligocene marine molluscan fauna of the Eugene Formation, Oregon: University of Oregon Museum of Natural History Bulletin 16, 112 p.

Howell, D. G., 1976, A review of the estimates for the radiometric ages for the Relizian Stage of the Pacific Coast, *in* Fritsche, A. E., and others, eds., Neogene symposium: Society of

Economic Paleontologists and Mineralogists, Pacific Section, San Francisco, California, April 1976, p. 13-32.

Jenkins, D. G., 1978, Corrected age of the Pliocene/Pleistocene boundary: Nature, v. 272, p. 287.

Klucking, E. P., 1962, An Oligocene flora from the western Cascades [M.S. thesis]: Berkeley, University of California, 241 p.

Kozumi, I., 1973, The late Cenozoic diatoms of sites 183-193, Leg 19, *in* Initial reports of the Deep Sea Drilling Project, Volume 19: Washington, D.C., U.S. Government Printing Office, v. 19, p. 805-855.

LaMotte, R. S., 1936, The upper Cedarville flora of northwestern Nevada and adjacent California: Carnegie Institution of Washington Publication 455, p. 57-142.

Lander, E. B., 1977, A review of the Oreodonta (Mammalia, Artiodactyla), Parts I, II, and III [Ph.D. dissert.]: Berkeley, University of California, 474 p.

Mankinen, E. A., and Dalrymple, G. B., 1979, Revised geomagnetic polarity time scale for the interval 0-5 m.y. B.P.: Journal of Geophysical Research, v. 84, p. 615-626.

Miles, G. A., 1981, Planktonic foraminifers of the lower Tertiary Roseburg, Lookingglass, and Flournoy Formations (Umpqua Group), southwest Oregon: Geological Society of America Special Paper 184 (this volume).

Munthe, J., and Coombs, M. C., 1979, Miocene dome-skulled chalicotheres (Mammalia, Perissodactyla) from the western United States: A preliminary discussion of a bizarre structure: Journal of Paleontology, v. 53, p. 77-91.

Odin, G. S., 1978, Isotopic dates for a Paleogene time scale, *in* Cohee, G. V., and others, eds., Contributions to the geologic time scale: American Association of Petroleum Geologists Studies in Geology, no. 6, p. 247-257.

Pomerol, C., 1978, Critical review of isotopic dates in relation to Paleogene stratotypes, *in* Cohee, G. V., and others, eds., Contributions to the geologic time scale: American Association of Petroleum Geologists Studies in Geology, no. 6, p. 235-245.

Rau, W. W., 1958, Stratigraphy and foraminiferal zonation in some of the Tertiary rocks of southwestern Washington: U.S. Geological Survey Oil and Gas Investigations Chart OC-57, 2 sheets.

——1964, Foraminifera from the northern Olympic Peninsula, Washington: U.S. Geological Survey Professional Paper 374-G, p. G1-G33.

——1967, Geology of the Wynoochee Valley quadrangle, Grays Harbor County, Washington: Washington Division of Mines and Geology Bulletin 56, 51 p.

——1981, Pacific Northwest Tertiary benthic foraminiferal biostratigraphic framework—An overview: Geological Society of America Special Paper 184 (this volume).

Repenning, C. A., 1980, Faunal exchanges between Siberia and North America: Canadian Journal of Anthropology, v. 1, no. 1 (in press).

Repenning, C. A., and Tedford, R. H., 1977, Otarioid seals of the Neogene: U.S. Geological Survey Professional Paper 992, 93 p.

Roth, B., 1979, Late Cenozoic marine invertebrates from northwest California and southwest Oregon [Ph.D. dissert.]: Berkeley, University of California, 803 p.

Ryan, W.B.F., and others, 1974, A paleomagnetic assignment of Neogene stage boundaries and the development of isochronous datum planes between the Mediterranean, the Pacific and Indian Oceans in order to investigate the response of the world ocean to the Mediterranean "salinity crisis": Rivista Italiana di Paleontologia e Stratigrafia, v. 80, p. 631-688.

Salvador, A., and others, 1979, Magnetostratigraphic polarity units—A supplementary chapter of the ISSC International Stratigraphic Guide: Geology, v. 7, p. 578-583.

Snavely, P. D., Jr., and others, 1958, Geology and coal resources of the Centralia-Chehalis district, Washington: U.S. Geological Survey Bulletin 1053, 159 p.

Snavely, P. D., Jr., and MacLeod, N. S., 1974, Yachats Basalt—An upper Eocene differentiated volcanic sequence in the Oregon Coast Range: U.S. Geological Survey Journal of Research, v. 2, p. 395-403.

Snavely, P. D., Jr., MacLeod, N. S., and Wagner, H. C., 1973, Miocene tholeiitic basalts of coastal Oregon and Washington and their relations to coeval basalts of the Columbia Plateau: Geological Society of America Bulletin, v. 84, p. 387-424.

Steiger, R. H., and Jäger, E., 1977, Subcommission on geochronology: Convention on the use of decay constants in geo- and cosmochronology: Earth and Planetary Science Letters, v. 36, p. 359-363.

Thoms, R. E., 1975, Biostratigraphy of the Umpqua Group, southwestern Oregon, *in* Weaver, D. E., ed., Future energy horizons of the Pacific Coast; Paleogene symposium and selected technical papers: American Association of Petroleum Geologists, Society of Economic Paleontologists and Mineralogists, and Society of Economic Geologists, Pacific Sections, Annual Meeting, Long Beach, California, April 1975, p. 513-562.

Tipton, Ann, 1975, Foraminiferal biostratigraphy of the late Eocene to early Oligocene type Bastendorff Formation, near Coos Bay, Oregon, *in* Weaver, D. E., ed., Future energy horizons of the Pacific Coast; Paleogene symposium and selected technical papers: American Association of Petroleum Geologists, Society of Economic Paleontologists and Mineralogists, and Society of Economic Geologists, Pacific Sections, Annual Meeting, Long Beach, California, April 1975, p. 563-585.

Turner, D. L., 1970, Potassium-argon dating of Pacific Coast Miocene foraminiferal stages, *in* Bandy, O. L., ed., Radiometric dating and paleontological zonation: Geological Society of America Special Paper 124, p. 91-129.

Turner, D. L., and others, 1980, Radiometric dating of ash partings in Alaskan coal beds and upper Tertiary paleobotanical stages: Geology, v. 8, p. 92-96.

Turner, F. E., 1938, Stratigraphy and Mollusca of the Eocene of western Oregon: Geological Society of America Special Paper 10, 130 p.

Vail, P. R., Mitchum, R. M., Jr., and Thompson, S., III, 1977, Seismic stratigraphy and global changes of sea level, Part 4: Global cycles of relative changes of sea level, *in* Payton, C. E., ed., Seismic stratigraphy-applications to hydrocarbon exploration: American Association of Petroleum Geologists Memoir 26, p. 83-97.

Van Couvering, J. A., 1978, Status of late Cenozoic boundaries: Geology, v. 6, p. 169.

Van Couvering, J. A., and Berggren, W. A., 1977, Biostratigraphic basis of the Neogene time scale, *in* Kauffman, E. G., and Hazel, J. E., eds., Concepts and methods of biostratigraphy: Stroudsburg, Pennsylvania, Dowden, Hutchinson and Ross, p. 283-306.

Van Couvering, J. A., and others, 1976, The terminal Miocene event: Marine Micropaleontology, v. 1, p. 263-286.

Vine, J. D., 1962, Stratigraphy of Eocene rocks in part of King County, Washington: Washington Division of Mines and Geology, Report of Investigations no. 21, 20 p.

Walker, G. W., and Repenning, C. A., 1965, Reconnaissance geologic map of the Adel quadrangle, Lake, Harney, and Malheur Counties, Oregon: U.S. Geological Survey Miscellaneous Geologic Investigations Map I-446.

Warren, A. D., and Newell, J. H., 1981, Calcareous plankton biostratigraphy of the type Bastendorff Formation, southwest Oregon: Geological Society of America Special Paper 184 (this volume).

Weaver, C. E., and others, 1944, Correlation of the marine Cenozoic formations of western North America: Geological Society of America Bulletin, v. 55, p. 569-598.

Wolfe, J. A., 1981, A chronologic framework for Cenozoic megafossil floras of northwestern North America and its relation to marine geochronology: Geological Society of America Special Paper 184 (this volume).

Wood, H. E., II, and others, 1941, Nomenclature and correlation of the North American continental Tertiary: Geological Society of America Bulletin, v. 52, p. 1-48.

Woodburne, M. O., and Robinson, P. T., 1977, a new late Hemingfordian mammal fauna from the John Day Formation, Oregon, and its stratigraphic implications: Journal of Paleontology, v. 51, p. 750-757.

MANUSCRIPT RECEIVED BY THE SOCIETY APRIL 3, 1980
MANUSCRIPT ACCEPTED MAY 7, 1980

Printed in U.S.A.

Late Paleogene and Neogene molluscan faunas of the Gulf of Alaska region: Correlation, paleoclimatology and paleozoogeography[1]

RICHARD C. ALLISON
Division of Geosciences
University of Alaska
Fairbanks, Alaska 99701

ABSTRACT

Although warmer water Cenozoic molluscan faunas of the Gulf of Alaska differ compositionally from those of the western United States, correlations are possible. Cooler water faunas are difficult to correlate because they contain many endemic and "pre-circumboreal"-circumboreal elements. The strong latitudinal temperature gradient of the Pliocene and Pleistocene prevents direct correlation with faunas of the conterminous United States, but some correlations with the Bering Sea transgressions are possible. Because of differing thermal histories and water depths, western Gulf faunas differ from northeastern Gulf faunas. Mollusk-bearing units in the western Gulf are the Stepovak (late Galvinian, Matlockian?), "Narrow Cape" of Sitkinak Island (Juanian), Narrow Cape Newportian), Unga Conglomerate (late Newportian? to early Wishkahan?), upper Bear Lake (Wishkahan), Tachilni (Graysian?), and Tugidak (Beringian-Anvilian). Mollusk-bearing units in the northeast Gulf are the Poul Creek (late Galvinian to Pillarian), mainland Yakataga (Newportian to Beringian-Anvilian), and Middleton Island Yakataga (Anvilian). Newportian and Wishkahan (Miocene) temperatures of the western Gulf indicate warm temperate water comparable to that of the western United States. Newportian and younger faunas of the northeast Gulf indicate anomalous cooler water, probably caused by local glaciation. Water temperatures colder than those of the present probably did not occur until the Beringian (late Pliocene-Pleistocene). "Pre-circumboreal" mollusks first appear in the Juanian (late Oligocene) of the Gulf, but the first opening of Bering Strait apparently occurred in Wishkahan (late Miocene) time; re-opening in Beringian time caused significant changes in Gulf faunal composition.

---

[1]Allison, Richard C., 1978, Late Oligocene through Pleistocene molluscan faunas in the Gulf of Alaska region: Veliger, v. 21, no. 2, p. 171-188, 2 figs., 2 tables.

Upper Eocene foraminiferal biostratigraphy and paleoecology of northwestern Oregon

KRISTIN McDOUGALL
U.S. Geological Survey
Menlo Park
California 94025

ABSTRACT

Upper Eocene benthic foraminiferal zones in Washington and California can be correlated because of the transitional nature of the faunas of northwestern Oregon. The mixing of the northern and southern faunas is most apparent there in the Keasey and Cowlitz Formations. Planktonic foraminifers and calcareous nannofossils, although scarce, allow these local zonations to be placed in an international framework.

Correlation of sections in northwestern Oregon provides a paleoecological framework from which interpretations of the Keasey-Cowlitz sequence and other upper Eocene sequences can be made. The Keasey-Cowlitz sequence suggests a shallowing trend with depths ranging from bathyal to neritic. Faunas indicative of middle to lower bathyal conditions are composed of species of *Alabamina, Bulimina, Oridorsalis,* and *Uvigerina,* whereas species of *Guttulina, Hoeglundina, Nodosaria, Pullenia* (compressed), and *Uvigerina* (costated) represent the upper bathyal to outer neritic, and species of *Cyclogyra, Elphidium, Nonionella,* and *Quinqueloculina* represent the middle and inner neritic. Planktonic species are present during minor transgressions of the sea.

Late Eocene and Oligocene benthic foraminiferal correlations between California and the Pacific Northwest

ANN TIPTON
Resource Associates of Alaska, Inc.
5926 McIntyre Street
Golden, Colorado 80401

ABSTRACT

An essential task in biochronology is defining the geographic limits of stages and zones. During the late Eocene and Oligocene, the benthic foraminiferal faunas of the Pacific Coast of North America defined a single zoogeographic province within which the Narizian, Refugian, and Zemorrian Stages are clearly recognizable as belts of strata bearing the diagnostic congregations of foraminifera on which those stages were originally based. At the zonal level of refinement, however, the upper Eocene-Oligocene foraminiferal sequences of northern Oregon and Washington are different in detail from those of California. The differences consist of, first, the absence or extreme rarity of a taxon in one region that is zonally significant in the other and, second, differences in stratigraphic range of zonally diagnostic taxa known from both regions. The faunas of California and southern Oregon define a southern zoogeographic subprovince within which the zonations of Mallory, Tipton, and Kleinpell are valid for the Narizian, Refugian, and Zemorrian Stages, respectively. Northernmost Oregon and Washington are part of a northern subprovince where instead the zonations of Rau are applicable. The base of the Zemorrian Stage as defined by Rau in the Pacific Northwest is essentially correlative with the base of the type Zemorrian in California, but the base of Rau's lower Refugian *Sigmomorphina schencki* Zone may be older by one subzone or less than the base of Tipton's *Valvulineria tumeyensis* Zone, lower Refugian of California.

Tertiary diatoms from southwestern Washington

LLOYD H. BURCKLE
Lamont-Doherty Geological Observatory of Columbia University
Palisades, New York 10964

ABSTRACT

Two hundred and fifty-five samples from the Eocene and Oligocene of southwestern Washington were examined for diatoms. Of these, 55 contained enough material for floral analysis. The Eocene and lower Oligocene are characterized by the absence of diatoms or by the sporadic occurrence of partially dissolved forms such as *Pyxilla, Xanthiopyxis,* and *Stephanopyxis.* Many diatoms occur in the upper Oligocene (middle Zemorrian) in both the Canyon River and Satsop River section. Most forms are benthic or members of the neritic plankton with one uppermost Oligocene (lowermost Miocene?) sample containing abundant fresh-water forms. The presence of diatoms in the late Oligocene probably reflects increased upwelling during that time, which resulted in increased surface productivity and an increase in diatom abundance in bottom sediments.

Vertical distribution of diatoms through these sections suggests the following scenario for the Eocene through Oligocene of this area: the late Eocene and early Oligocene are characterized by a warm, shallow-water environment. This is followed in the late Oligocene by a cool-temperate environment featuring increased upwelling and, possibly, increased runoff from the continents. The terminal Oligocene reflects a return to warmer-water conditions.

Pacific Coast Cenozoic middle- and high-latitude Echinoidea

J. WYATT DURHAM
Department of Paleontology
University of California
Berkeley, California 94720

ABSTRACT

The diversity (43 genera, some with numerous species) of West North American Tertiary and Pleistocene echinoids from San Diego to the Bering Sea is greater than that of those now living (14 genera, 18 species) on the continental shelf in the same area. The fossils include: *Eucidaris, Goniocidaris, Hesperocidaris, Polydiadema(?), Arbacia, Salenia, Lytechinus, Glyptocidaris, Echinometra, Strongylocentrotus, Areosoma,* an unidentified echinothurioid, *Amblypygus, Echinocyamus, Eoscutella, Kewia, Vaquerosella, Remondella, Tenuirachnius, Astrodapsis, Echinarachnius, Scutellaster, Merriamaster, Dendraster, Scutaster, Encope, Mellita, Echinolampas, Cassidulus, Eupatagus, Hypsopatagus, Paleopneustes, Archaeopneustes, Megapetalus, Spatangus, Lovenia, Linthia, Brissus, Brisaster, Schizaster, Periaster, Agassizia (Anisaster),* and an Aeropsid. The living *Centrostephanus, Allocentrotus, Brissopsis,* and *Nacospatangus* have not yet been recognized as fossils but can be expected. *Amblypygus, Echinolampas, Glyptocidaris, Goniocidaris,* and *Salenia* are biogeographically important. *Glyptocidaris* is the only Phymosomatid genus with a living representative (Japan). Crushed specimens of *Brisaster* in the upper Eocene Keasey Formation represent the earliest record of this extant genus. In the middle and upper Tertiary, the Scutellinoid "sand dollar" stock diversified rapidly (9 genera) and then declined. Only *Dendraster* and *Echinarachnius* survive. *Astrodapsis* and *Dendraster,* and to a lesser extent *Kewia* and *Scutellaster,* are represented by numerous species. *Strongylocentrotus* and *Echinarachnius* migrated from their North Pacific homeland to the North Atlantic in the later Cenozoic. *Echinocyamus pusillus* (Pleistocene, St. Paul Island) appears to be an immigrant from the North Atlantic.

Molluscan paleoecology and biostratigraphy of Cenozoic deep-water facies in the Pacific Northwest

CAROLE S. HICKMAN
Department of Paleontology
University of California
Berkeley, California 94720

ABSTRACT

A major difficulty of the provincial marine molluscan biostratigraphic column for the Cenozoic of the Pacific Northwest is its admixture of deep- and shallow-water zones that are not suitable for correlation except within the bathymetric facies from which they were defined. The desirability of a separate biostratigraphic zonation of the deep-water (>200 m) facies is emphasized by the abundance of lower neritic and bathyal mollusk assemblages in thick, predominantly massive, fine-grained tuffaceous sedimentary sequences along the margin of the northeastern Pacific. Previous bathymetric interpretations of these sequences have been controversial because they have been based on methods inadequate to differentiate deep-water faunas from shallow cold-water faunas. A new method of bathymetric interpretation is based on comparison of the taxonomic structure (percent species composition of major mollusk groups) of modern and fossil faunas of different depth zones. Changes in taxonomic structure are related to changes in trophic structure with depth and are independent of temperature. Major herbivorous and suspension feeding taxa disappear below the level of light penetration and primary productivity, while deposit feeding and carnivorous taxa increase in prominence. Tropic structure, at both higher and lower taxonomic levels, is remarkably constant in deep-water faunas of the Pacific Northwest and is characterized by parallel assemblages in which species are replaced by close congeners in successively younger intervals of zonal magnitude. Gastropods of the family Turridae are particularly useful in detailed biostratigraphy. *Ptchosyrinx* Thiele, 1925, *Parasyrinx* Finlay, 1924, *Procerapex* Hickman, 1976, and *Acamptogenotia* Rovereto, 1899, exemplify uses of the group in biostratigraphy.

Index